Environmental Sampling and Analysis for Technicians

MARIA CSUROS

Environmental Sampling and Analysis for Technicians

LEWIS PUBLISHERS
Boca Raton Ann Arbor London Tokyo

Library of Congress Cataloging-in-Publication Data

Csuros, Maria.
Environmental sampling & analysis for technicians / Maria Csuros.
p. cm.
Includes bibliographical references and index.
ISBN 0-87371-835-6
1. Pollution—Measurement. 2. Pollution—Measurement—Laboratory manuals. I. Title. II. Title: Environmental sampling and analysis for technicians.
TD193.C78 1994
628′.028′7—dc20

94-865
CIP

International Standard Book Number 0-87371-835-6
Library of Congress Card Number 94-865
Printed in the United States of America 1 2 3 4 5 6 7 8 9 0
Printed on acid-free paper

Preface

A clean and healthy environment has always been a concern to most of us. Critical decisions in regard to the protection of our surroundings are based on data collected and derived from laboratory measurements of environmental pollutants.

During the past several years dramatic changes have taken place in environmental regulations, including enhanced laboratory performance and sample collection methods. Well-trained and knowledgeable technical staff is imperative for successful pollution control and environmental management; therefore, we have seen an increase in related educational programs, special training, and refresher courses.

Even though there are many detailed, discipline-specific texts, EPA and DER methodology releases, they give information only for selected areas of the necessary knowledge, and their language makes them unusable for teaching purposes. I have always had to rely on reprints, copies, and other source materials to supplement and support my lectures and laboratory classes. The lack of an extensive textbook in this field prompted me to compile all the information available, combined with my 30 years of laboratory experience, in a book form.

Although the primary audience of this publication is environmental laboratory technician students, I believe that it will be useful as a supplementary information source for more general courses in environmental studies and for a variety of job related training programs. As a practical handbook, it will also assist laboratory technicians in their everyday chores.

All in all, this book will provide a valuable advantage to environmental education and special training programs.

The text contains all required rules in environmental sample collection, related field activities, and proper sample custody. Special emphasis was placed on the philosophy and practical approach of quality assurance and quality control applied in analytical measurements, as established by the Florida Department of Environmental Regulations Quality Assurance Section ("DER Requirements for Quality Assurance" DER-QA-001/91). An overview of the occurrence, source, and fate of toxic pollutants and their control by regulations and standards will provide a helping hand to the readers in understanding analytical reports.

The Author

Maria Csuros is an environmental chemist with many years of varied experience. She received her Master of Science degree from Jozsef Attila University, Szeged, Hungary, in chemistry. She received a grant from the Environmental and Public Health Institution, Budapest, Hungary, for her postgraduate work specializing in environmental chemistry. Most of her professional life revolved around environmental testing laboratories and teaching.

She first served in Hungary at the Environmental and Public Health Laboratory as laboratory supervisor of the Water Department, with her main area of interest in the prevention and elimination of methaemoglobinaemia caused by high nitrate content of private well waters.

She spent six years in Benghasi, Libya, as part of an international team studying the health effects of brackish water quality drinking water.

After immigrating to the United States, she worked as a chemist and laboratory supervisor in private industry, and for the past four years she has dedicated her knowledge and time to Environmental Education. During these years, she has designed and developed a strong Environmental Science program, which is focused on Environmental Sampling and Analysis. Presently she is the coordinator of this program. She also teaches chemistry and environmental science courses.

She also appears at environmental educational seminars for local, state, and federal agencies as well as numerous regional companies.

She lives in Pensacola, Florida, with her husband, Csaba (a professor of microbiology and anatomy and physiology), and their two German shepherds.

She has two sons and four grandsons.

Acknowledgments

It has been a pleasure to thank those who have contributed in their own special way to the completion of this project. I begin with fond words of thanks to my husband, who has given me constant support, stimulating ideas and helpful suggestions in various ways. My warm and very special thanks to my sons, Geza and Zoltan, for their love, encouragement and technical assistance in the preparation of tables and figures.

I gratefully acknowledge the initiation, encouragement and support of Jon Lewis, Lewis Publishers. I appreciate his personal interest in this project, his cheerful spirit and helping hand. I also want to thank Susan Fox, Project Editor, for her careful work and patience, and Shayna Murry, Designer, for the design work.

To all of you, I thank you!

Dedication

To the memory of my mother

Table of Contents

Part 1

Collecting, Preserving, and Handling Environmental Samples

Chapter 1
General Considerations in Sampling

It is an old axiom, that the result of any test procedure can be no better than the sample on which it is performed.

The objective of sampling is to collect a portion of material that represents the actual sample composition. The quality of data depends upon six major activities:

1. formulating the particular objectives for a sampling program
2. collecting representative samples
3. proper sample handling and preservation
4. adhering to adequate chain-of-custody and sample identification
5. participating quality assurance and quality control (QA/QC) in the field
6. properly analyzing the sample

These areas are equally important for insuring that environmental data are of the highest validity and quality.

1.1 Objectives of a Sampling Program

Project Scope and Purpose

The purpose of a project is to be defined, clearly stated in the project plan, and hence in the related sampling program. Well-designed and well-implemented sampling programs are vital to pollution control.

The following six phases are essential to smooth project management:

- planning
- permitting
- compliance
- enforcement
- design
- research and development

Type of Sampling Program

The type of sampling program depends on the program objective, and accordingly, programs may be designed and characterized as follows:

- reconnaissance survey
- point source characterization
- intensive survey
- fixed station monitoring
- network monitoring
- groundwater monitoring
- special surveys

Sampling programs should contain all of the following objectives

- Site identification; include a site map that identifies the sampling locations. The location of the sampling site is critical to obtain representative samples
- Sample source (groundwater, drinking water, surface water, wastewater, sediment, soil, etc.)
- Number and matrix of the samples
- Duration of the survey
- Frequency of sampling (monthly, quarterly, etc.)
- Type of samples (grab or composite samples)
- Method of sample collection (manual, automatic)
- Needed analytical parameters with method numbers and references
- Field measurements
- Field quality control (QC) requirements
- Sample collector(s)

Type of Samples

Samples can be divided into four types:

Grab samples — Grab sample is an individual sample, collected at a particular time and place. This type of sample represents conditions at the time it was collected. Therefore, a grab sample should not be used as a basis for a decision about pollution abatement. However, some sources are quite stable in composition, and may be represented well by single grab samples.

Composite samples — Composite sample refers to a mixture of grab samples collected at the same point at different times. A series of smaller samples are collected in a single container and blended for analysis. The mixing process averages the variations in sample composition and minimizes analytical effort and expense. When a time factor is being taken into consideration, grab samples should be collected in suitable sampling intervals, chosen according to the expected changes. When composition depends on location, collect grab samples from appropriate spots. Composite samples reflect the average characteristics during the sampling period and in most cases, a 24 hour period is standard. The volume of the samples should be constant (for example, 200 mls each time) in constant time intervals (for example, every hour), and mixed well at the end of the composite period.

Samples for Volatile Organic Compounds (VOCs), Oil and Grease, Total Recoverable Petroleum Hydrocarbons (TRPHs), and Microbiology testing should never be composited and should only be a grab sample!

Duplicates — Duplicate samples are collected for checking the preciseness of the sampling process.

Split samples — Split samples are taken for checking analytical performance. The sample is taken in one container, mixed thoroughly, and halved into another properly cleaned container. Preserve both samples as needed. Both halves are now samples that represent the same sampling point and are called split samples.

Nature of Sample Collection

Manual Sample Collection

In the case of the lack of automatic samplers, or in case of samples collected for immediate field tests, collect samples directly into the actual container. If the sample cannot be sampled directly in its container, an intermediate vessel should be used. It must be as clean as the sample container and must be made from the required material for that particular parameter. (Selection of the material of the sampling device according to the collected parameters is discussed later). The sample is collected by lowering one properly cleaned device on a rope, pole, or chain into the sample medium. In some cases, it is best to use a pump, either power or hand operated, to withdraw the sample. For most parameters, rinsing the sampling device three times is sufficient, except if the bottles are prepreserved, and are for analyses such as VOCs, Oil and Grease, TRPH, and Microbiology.

Sample Collection with Automatic Samplers

A wide variety of automatic samplers are commercially available. When sampling a large number of locations, the use of automatic samplers are more practical, help to reduce human errors, and are able to keep the samples cool to 4°C during the sampling period. The disadvantage of automatic sampling is the cost of the equipment.

The Material of the Sampling Device Should be Selected According to the Requested Parameters

For *inorganic parameters*, that do not need preservation, use plastic, glass, teflon, stainless steel, aluminum, or brass. For *nutrients*, use plastic, glass, teflon, stainless steel, aluminum, or brass. For *trace metals*, use plastic, stainless steel, or teflon. For *extractable organics*, use glass, aluminum, brass, stainless steel, or teflon. For *volatile organics*, use glass, stainless steel, or teflon. For *microbiological samples*, use a presterilized sample container.

Maintenance and Calibration of Sampling Equipment

To insure proper operation of the automatic samplers, the correct maintenance and calibration (as described in the manufacturer's guide) must be followed. A maintenance log should be used to record all of the activities, such as batteries and desiccant checks, any repair, etc. with the name and signature of the person who performed these activities. After returning from the field, the checked and cleaned sampling equipment should be properly stored.

1.2. General Sampling Rules

Sample collectors must understand and apply the following general rules in sampling:

1. Samples must be collected from the least to the most contaminated sampling locations within the site.

2. Disposable latex gloves should be worn when sampling and new, unused gloves must be used for each separate sampling point. For sampling hazardous materials, *rubber gloves* are recommended.
3. For compositing or mixing samples, use a bowl and spatula to thoroughly mix the sample. For trace organics and metal analysis, the material of these tools should be stainless steel, glass, or teflon. Samples should be mixed thoroughly and sectioned and the quantity of each subsample should be recorded.

The preferred order in sample collection is the following:

1. VOCs
2. Extractable organics, including Oil and Grease and TRPH
3. Total Metals
4. Dissolved Metals
5. Microbiological samples
6. Inorganic nonmetals

For aqueous matrices, sampling equipments and containers are rinsed with sample before the actual sample is taken! Exceptions are samples for VOCs, Oil and Grease, TRPH, Microbiological testing, and any samples collected in prepreserved containers.

Step-by-step, easy to follow, Standard Operation Procedures (SOP) for sampling should be available. Text may include all activities used to collect samples from the arrival on site through delivery to the laboratory. The title, revision date, sections, subsections, and page number(s) of the reference material, used in the preparation of the sampling SOP, should be incorporated.

1.3. Preparation of Sampling Equipments

Written, regulatory outlined step-by-step cleaning procedures, called "DECON" (for decontamination) should be performed. Equipments should be cleaned before sampling, and at the field between samples. At the end of the field trip, sample collection equipments must be labeled as "rinsed, ready to in-house cleaning". After sufficient cleaning in the laboratory, they should be labeled as "in-house cleaned, ready for field" with date and signature of the cleaner. Both house and field cleaning should be documented properly. Detergents specified for cleaning are **ALCONOX** (or equivalent) with <5% phosphate, or **LIQUINOX** (or equivalent) that is phosphate- and ammonia-free.

The solvent to be used in routine cleaning should be pesticide grade isopropanol. Analyte-free water is to be used as rinsing water, and for blanks preparation. The purity and reliability of the analyte-free water is shown by the results of the blank.

Outline for House Cleaning of Sampling Equipments

- Wash with hot soapy tap water and scrub with a brush.
- Rinse thoroughly with hot tap water.
- Rinse with 10 to 15% nitric acid (HNO_3). If nutrients are of interest, after the HNO_3 rinse, 10 to 15% hydrochloric acid (HCl) rinse is required or the HNO_3 rinse may be replaced with HCl rinse. Acid rinse should never be applied to stainless steel or any metallic equipment!
- Rinse thoroughly with deionized water.
- Rinse thoroughly with pesticide-grade isopropanol.
- Rinse thoroughly with analyte free water.
- Air-dry completely.
- Wrap in aluminum foil for storage and transportation.

Outline for Field Cleaning of Sampling Equipments

- Use the same procedure as in-house cleaning, with the exception of hot water.
- First wipe or scrub the equipment to remove particles with the appropriate soap solution, rinse with tap water, followed by deionized water, and finally air-dry.
- For heavily contaminated equipment, use Acetone or Acetone-Hexane-Acetone rinse before regular decon.
- The rinse with analyte-free water is recommended, but optional.
- When only inorganic parameters are of interest, equipment may be rinsed with analyte-free water and with the sample water.
- If proper cleaning of the equipment is impossible, it should be properly disposed until effective cleaning is possible.

Cleaning of Purging Equipments (Submersible Pumps and Non-Teflon Hoses)

Wipe or scrub to remove particles with appropriate soap solution, rinse with tap water, rinse with deionized water, and air dry as long as possible before purging next well. Care should be taken to completely clean the exterior of the pump, and the exterior and interior surfaces of tubing.

Decontamination of Teflon Tubing

Always clean in the laboratory and never in the field.

- Soak tubing in hot soapy water and use a brush to remove any particulate if necessary.
- Rinse tubing exterior and ends liberally with tap water.
- Rinse tubing surfaces and ends with 10 to 15% HNO_3.
- Rinse with tap water.
- Rinse with pesticide grade methanol or isopropanol.
- Rinse with analyte-free water.
- Place tubing on clean aluminum foil.
- With teflon inserts, connect all of the hose used on site. Using the field-use peristaltic pump, assemble the system used in the field, but use a larger size bottle that has the same cap size as the collection bottles. (A large size bottle such as the type containing solvents or acids is suitable.)
- Pump copious amounts of hot, soapy water through the connected tubing. Follow this with tap water.
- With the pump running, draw at least 1 liter of 1+1 HNO_3 through the tubing. Close valve, and stop the pump. Let the acid remain in the tubing for 15 to 20 minutes. Pump an additional 1 to 2 liters of acid through the tubing, followed by 1 to 2 liters of tap water.
- Pump 1 liter of pesticide grade methanol or isopropanol through the tubing. Let solvent remain in the tubing 15 to 20 minutes. Pump an additional 1 to 2 liters of solvent through the system.
- Finally, rinse with 2 to 3 liters of analyte-free water.
- Leave the teflon inserts between the pre-cut lengths and cap or connect the remaining end.
- After the interior has been sufficiently cleaned, the exterior needs a final rinse with analyte-free water.
- Wrap in aluminum foil and store in a clean, dry area. Label with the date of cleaning.

Documentation of this cleaning should be in a bound notebook.

1.4. Preparation of Sample Containers

The material that the sample container is composed of should be chosen so it will not react with the sample. It should be resistant to leakage and breakage and should have the proper volume necessary for the analyte(s) of interest. Plastic containers are the best for sampling inorganic parameters. The containers must have tight screw-type lids. Glass and teflon containers with teflon lined caps are suitable for organic analytes. However, there are some disadvantages. Glass is breakable and teflon is quite expensive. For purgeable organics, use 40 ml borosilicate glass vials with screw cap and teflon backed silicon septum. Sterile plastic cups, individually wrapped, or sterile whirl-pack plastic bags are commercially available for microbiological samples.

Sample containers may be cleaned by the sampling organization or purchased from commercial vendors as precleaned containers. All records for these containers (lot numbers, certification statements, date of receipt, etc.) and their uses must be documented.

Cleaning Procedures for Sample Containers

These regulated cleaning procedures must be strictly followed to eliminate sample contamination by the sample container.

Physical Properties and Mineral Analysis

Bottle type: Plastic or glass, minimum of half gallon capacity.

Soap: LIQUINOX or equivalent.

- Wash bottles and caps with hot soapy water, and rinse liberally with tap water until suds are no longer present.
- Rinse bottles and caps with laboratory pure water at least 3 to 5 times. Drain and store tightly capped until used.

Nutrients, Demands, and Radiological Analysis

Bottle type: Plastic or glass.

Soap: LIQUINOX or equivalent (phosphate- and ammonia-free).

- Wash bottles and caps in hot soapy water and rinse liberally with tap water until soap suds are no longer present.
- Rinse bottles and caps with 1+1 HCl, then follow by rinsing 3 to 5 times with laboratory pure water.
- Drain and store bottles tightly capped until use.

Metals

Bottle type: Plastic bottle with lid.

Soap: Should be metal free ACATIONOX or equivalent.

- Wash bottles and caps in hot soapy water, and rinse liberally with tap water, until soap suds are gone.
- Rinse bottles and caps with 1+1 HCl, followed by tap water rinse.
- Rinse bottles and caps with 1+1 HNO_3. Rinse three times with liberal amount of laboratory pure water.
- Drain and cap tightly until use.

Extractable Organics

Bottle type: 1 liter narrow necked glass bottle with Teflon lined caps. **Plastic bottles and plastic or rubber lined caps are not acceptable!**

Soap: ALCONOX or equivalent.

Do not use liquid or powdered detergent that has been stored in a plastic container!

- Wash bottles and caps in hot soapy water. **Do not use brushes with rubber or plastic parts! The use of plastic gloves while washing or rinsing organic bottles is not recommended since they are a good source of contamination.**
- Rinse bottles five times with tap water until all soap is gone.
- Rinse each bottle with 10 ml of pesticide grade acetone, cap tightly and shake approximately 10 seconds. Care should be taken not to allow the interior plastic portion of the cap to come in contact with the acetone.
- The final aqueous rinse should be with organic-free water. There should be no acetone smell in the bottle. This means, rinse about five times.
- Drain bottles and cap until use.

Volatile Organic Compounds (VOCs)
Bottle type: 40 ml glass vial with teflon lined septum.
Soap: ALCONOX or equivalent.
Do not use liquid or powdered detergent that has been stored in a plastic container!

- Wash vials, caps, and septums in hot soapy water, using the same precautions as described under extractable organics.
- Rinse liberally with tap water and laboratory pure water.
- Finally, rinse with pesticide grade methanol.
- Dry vials, caps, and septums in oven at 105°C for more than 60 minutes.
- Cool in inverted position and cap immediately after bottles are cool enough to handle.

1.5. Sample Preservation

Preservation is necessary for all samples according to Federal Register 40 CFR, Part 136 (Table 1-1). Table 1-1 contains the type of sample containers, preservations, maximum holding time, and amount of sample for each parameter.

Sample preservation may be accomplished by ready, prepreserved bottles, obtained from the laboratory. Additional preservatives should be available in the field, if the measured pH of the preserved sample indicates the need for more preservative. Samples may be preserved in the field after sample collection. If the sample is preserved in the field, the following protocols should be followed:

- Preservatives should be added with pipet or premeasured droppers to each sample container.
- Preservative should be reagent grade, or higher grade chemical.
- Fresh preservatives must be obtained prior to each sampling trip.
- After the addition of preservatives, the preserved sample should mix thoroughly and a pH check should be performed. Narrow range pH paper is used on an aliquot of preserved sample poured out into a disposable container. If the pH value indicates, add more preservative until the pH is satisfactory. The preservative must originate from the same source as the original preservative. The pH check, and the quantity of the additional preservative should be documented.
- The same amount of additional preservative has to be added to all corresponding blank samples!

Table 1-1. Recommendation for Collecting and Preservation of Samples

Parameters	Volume ml	Container	Preservative	Holding time
Physical properties				
Color	50	P,G	Cool, 4°C	48 hours
Conductance	100	P,G	Cool, 4°C	28 days
Hardness	100	P,G	HNO_3 to pH < 2	6 months
Odor	200	G only	Cool, 4°C	24 hours
pH	25	P,G	None required	Analyze immediately
Residue filtrable	100	P,G	Cool, 4°C	48 hours
Residue non-filtrable	100	P,G	Cool, 4°C	7 days
Residue total	100	P,G	Cool, 4°C	7 days
Residue volatile	100	P,G	Cool, 4°C	7 days
Settleable matter	1000	P,G	Cool, 4°C	48 hours
Temperature	1000	P,G	None required	Analyze immediately
Turbidity	100	P,G	Cool, 4°C	48 hours
Metals				
Dissolved	200	P,G	Filter on site HNO_3 to pH < 2	6 months
Suspended	200	P,G	Filter on site HNO_3 to pH < 2	6 months
Total	100	P,G	HNO_3 to pH < 2	6 months
Chromium +6	200	P,G	Cool, 4°C	24 hours
Mercury dissolved			HNO_3 to pH < 2	28 days
Total	100	P,G	HNO_3 to pH < 2	28 days
Inorganic non metallic				
Acidity	100	P,G	Cool, 4°C	14 days
Alkalinity	100	P,G	Cool, 4°C	14 days
Bromide	100	P,G	None required	28 days
Chloride	100	P,G	None required	28 days
Chlorine	1000	P,G	None required	Analyze immediately
Cyanides	500	P,G	Cool, 4°C NaOH to pH > 12	14 days
Fluoride	300	P,G	None required	28 days
Iodide	100	P,G	Cool, 4°C	24 hours
Nitrogen ammonia	400	P,G	Cool, 4°C H_2SO_4 to pH < 2	28 days
Kjeldahl	500	P,G	Cool, 4°C H_2SO_4 to pH < 2	28 days
Nitrate plus nitrite	100	P,G	Cool, 4°C H_2SO_4 to pH < 2	28 days
Nitrate	100	P,G	Cool, 4°C	48 hours
Nitrite	50	P,G	Cool, 4°C	48 hours
Dissolved oxygen probe	300	G bottle and top	None required	Analyze immediately
Winkler	300	G bottle and top	Fix on site, store in dark	8 hours
Phosphorus ortho-P dissolved	50	P,G	Filter on site Cool, 4°C	48 hours
Hydrolyzable	50	P,G	Cool, 4°C H_2SO_4 to pH < 2	28 days
Total	50	P,G	Cool, 4°C H_2SO_4 to pH < 2	28 days
Total dissolved	50	P,G	Filter on site Cool 4°C H_2SO_4 to pH < 2	24 hours
Silica	50	P only	Cool, 4°C	28 days
Sulfide	500	P,G	Cool, 4°C 2 ml zinc acetate + 2N NaOH to pH > 9	7 days
Sulfite	100	P,G	None required	Analyze immediately
Sulfate	100	P,G	Cool, 4°C	28 days

Table 1-1. Recommendation for Collecting and Preservation of Samples (continued)

Parameters	Volume ml	Container	Preservative	Holding time
Organics				
BOD	1000	P,G	Cool, 4°C	48 hours
COD	50	P,G	Cool, 4°C	28 days
Oil and grease	1000	G only	Cool, 4°C H_2SO_4 to pH < 2	28 days
Organic carbon	50	P,G, G brown	Cool, 4°C H_2SO_4 to pH < 2	28 days
Phenolics	500	G only	Cool, 4°C H_2SO_4 to pH < 2	28 days
Surfactants	500	P,G	Cool, 4°C	48 hours
Purgeable halocarbons	40	G, teflon lined septum	Cool, 4°C 0.008% $Na_2S_2O_3$*	14 days
Purgeable aromatics	40	G, teflon lined septum	Cool, 4°C 0.008% $Na_2S_2O_3$* HCl to pH < 2	14 days
Acrolein and Acrylonitrile	40	G, teflon lined septum	Cool, 4°C 0.008% $Na_2S_2O_3$* pH 4 to 5	14 days
Phenols	1000	G, teflon lined cups	Cool, 4°C 0.008% $Na_2S_2O_3$*	7 days until extraction, 40 days after extraction
Phthalate esters	1000	G, teflon lined cups	Cool, 4°C 0.008% $Na_2S_2O_3$*	7 days until extraction, 40 days after extraction
Nitrosamines	1000	G, teflon lined cups	Cool, 4°C 0.008% $Na_2S_2O_3$* store in dark	7 days until extraction, 40 days after extraction
PCBs	1000	G, teflon lined cups	Cool, 4°C	7 days until extraction, 40 days after extraction
Nitroaromatics and isophorone	1000	G, teflon lined cups	Cool, 4°C store in dark	7 days until extraction, 40 days after extraction
Polynuclear aromatic hydrocarbons	1000	G, teflon lined cups	Cool, 4°C store in dark	7 days until extraction, 40 days after extraction
TCDD (Dioxin)	1000	G, teflon lined cups	Cool, 4°C 0.008% $Na_2S_2O_3$*	7 days until extraction, 40 days after extraction
Chorinated hydrocarbons	1000	G, teflon lined cups	Cool, 4°C	7 days until extraction, 40 days after extraction
Pesticides	1000	G, teflon lined cups	Cool, 4°C pH 5 to 9	7 days until extraction, 40 days after extraction
Soil, sediment, sludge				
Organic extractable	8 oz.	Widemouth G teflon lined cup	Cool, 4°C	ASAP
Organic volatile	8 oz.	Widemouth G teflon lined cup	Cool, 4°C	ASAP
Metal	1 pint	P	Cool, 4°C	6 months
Fish samples		Wrap in Al foil**	Freeze	ASAP
Chemical wastes	8 oz.	Widemouth G, ** teflon lined cap	None	ASAP
Bacteriology Total and fecal coli Fecal streptococcus	100	P,G sterile	Cool, 4°C 0.008% Na_2SO_4	6 hours
Radiological Alpha, Beta, Radium	1000	P,G	HNO_3 pH < 2	6 months

** Required if residual chlorine is present.*

*** Plastic containers may be used if only metals are required.*

ASAP = As soon as possible; P = Polyethylene container; G = Glass container. Sample preservation = Sample preservation should be performed immediately upon sample collection. For composite samples, each aliquot should be preserved at the time of collection. When use of an automated sampler makes it impossible to preserve each aliquot, then chemical samples may be preserved by maintaining 4°C until compositing and sample splitting is completed. Holding time = Samples should be analyzed as soon as possible after collection. Those listed should be the maximum time that samples may be held before analysis and still be considered valid. Dissolved parameters = Samples should be filtered immediately on site before preservation. Reference 40 CFR Part 136

- Acid preservation should be done in a well-ventilated area to avoid build-up of acid fumes and toxic gases released from the samples. Any unusual reaction should be noted in the field documentation.
- Avoid spattering or spilling acids. Wipe up any spill immediately and flush the area with a generous amount of water.
- All chemicals transported to the field should be properly stored in the laboratory. Acids should be stored in acid-storage cabinets and solvent should be stored in solvent-storage cabinets. Chemicals should be separated according to their chemical character. All chemicals and reagents must be reagent or higher grade.
- All chemicals should be transported to the field stored in properly cleaned plastic or Teflon containers, to avoid breakage, and should be segregated from sample containers to avoid accidental contamination.

Special Preservation Techniques

Volatile Organic Compounds (VOCs)

If residual chlorine is present, sodium thiosulfate ($Na_2S_2O_3$) should be added to the sample vial first. The vial is then filled to at least half volume with the sample, acid is added, and finally the vial is filled as described in Section 1.6. Do not mix the two preservatives together in an intermediate vessel.

Chlorophyll "a"

The sample should be filtered (in the laboratory) within 24 hours after collection. Magnesium carbonate ($MgCO_3$) should be added to the filter while the last quantity of the sample passes through, and analyze. The filter may be freezed for later analysis.

Cyanide

If residual chlorine is present, add 0.6 g of ascorbic acid ($C_6H_8O_6$). If sulfide is present, samples must be pretreated in the field or must be taken to the laboratory unpreserved at 4°C for analysis within 24 hours. Sulfide must be checked in the field by lead acetate paper. If sulfide is present, indicated by the developing black color of the paper, add cadmium nitrate until yellow precipitation of cadmium sulfide (CdS) appears. The sample should then be filtered and preserved with sodium hydroxide (NaOH) until pH 12.

If possible, one member of the sampling team should take all notes, label bottles, write documentation, and preserve samples, while the other members of the team take the samples.

1.6. Special Sampling Procedures

Total Metals

Take the sample, and preserve with 3 ml of 1+1 HNO_3 or 1.5 ml of concentrated HNO_3. Samples with high buffer capacity, and high alkaline samples may need more acids, as the pH measurement indicates. In this case, the additional acids should also be added to the blank and documented as mentioned previously in the general preservation techniques. **The holding time for preserved samples is 6 months.**

Dissolved Metals

Samples must be filtered through a 0.45 u membrane filter prior to preservation. After the sample is filtered, acidify the filtrate, as for total metals. Filtering the sample is possible in the field, or after being transported to the laboratory. In this case, after the sample collection, transport the sample to the laboratory as soon as possible.

Suspended Metals

Unpreserved samples must be filtered through a O.45 u membrane filter, and the filter is retained for further analysis. The sample may be filtered in the field, or in the laboratory, but in this case, after collection, the sample should be taken to the laboratory as soon as possible, and filtered. Filter paper should be acid washed and dried!

VOCs

Because of the high sensitivity of the instrumentation used in trace organic chemical analysis and the low concentration of organic compounds being analyzed, special attention must be taken when collecting samples for trace organic analyses. Collection of VOCs needs special care, due to the volatile character of the analytes. Collect samples in 40 ml glass vials with Teflon lined septums. The procedure for filling and sealing containers during sampling is as follows:

- Slowly fill container to overflowing.
- Carefully set the container on a level surface.
- Place the septum Teflon side down on the convex sample meniscus.
- Seal the sample with the screw cap.
- To insure that the sample has been properly sealed, invert the sample and lightly tap the lid on a solid surface. The absence of entrapped air bubbles indicates a proper seal. If air bubbles are present, discard the sample, and collect the sample again in the same manner as stated above.
- The sample must be hermetically sealed until it is analyzed.
- Maintain samples at 4°C during transportation and in storage prior to analysis.
- If the sample is taken from a water tap, turn on the water and permit the system to flush. When the temperature of the water has been stabilized, adjust the flow to about 500 ml/min and collect samples as outlined above (detailed sampling procedure for VOCs from tap water is in Chapter 3).

The holding time for preserved samples is 14 days. If bubbles are present, sample should be discarded, a new sample should be collected and checked for bubbles! Samples for VOCs are never composited! Samples must be iced or refrigerated at 4°C from the time they are collected until analysis.

Purgeable Halocarbons (EPA Method-601)

When collecting samples for purgeable halocarbons, the following preservation techniques should be followed: if residual chlorine is present, add $Na_2S_2O_3$ to the sample vial first then fill and seal the bottle as described previously.

Purgeable Aromatics (EPA Method-602)

When collecting samples for purgeable aromatics, the following preservation techniques should be followed: if residual chlorine is present, add $Na_2S_2O_3$ to the sample vial first. Then fill the vial to at least half volume with the sample, adjust the pH to about 2 by adding 1+1 HCl, and finally fill the vial and seal as described previously. It is not recommended to mix the two preservatives (and sample) together in an intermediate vessel.

Acrolein and Acrylonitrile (EPA Method-603)

When collecting samples for acrolein and acrylonitrile, the following preservation technique should be followed: if residual chlorine is present, add $Na_2S_2O_3$ to the sample vial first. Then fill vial with the sample to at least half volume, add 1+1 HCl until pH is adjusted to 4 to 5, and finally fill and seal the vial the usual way. It is not recommended to mix the two preservatives (and sample) together in an intermediate container. Samples for Acrolein analysis receiving no pH adjustment must be analyzed within 3 days of sampling.

Extractable Organics

Collect grab samples in glass containers. Conventional sampling practices should be followed except that the sample bottle must not be prewashed with the sample before collection. Composite samples should be collected as outlined in the general sampling procedures. Automatic sampling equipment must be free of tygon or other potential sources of organic contamination. Collect samples in a 1 liter narrow-necked glass bottle with Teflon lined caps. Plastic bottles or plastic or rubber lined caps are not acceptable. **All samples must be extracted within 7 days, and completely analyzed within 40 days of extraction.**

Phenols (EPA Method-604)

Samples must be iced or refrigerated at 4°C from the time of collection until extraction. At the sampling location, fill the sampling container and add 80 mg of $Na_2S_2O_3$ per 1 liter of sample for chlorinated sample sources.

Benzidines (EPA Method-605)

Samples must be iced or refrigerated at 4°C from the time of collection until extraction. Benzidine and Dichlorobenzidine are easily oxidized by materials such as free chlorine. For chlorinated wastes, immediately after collection, add 80 mg of $Na_2S_2O_3$ per 1 liter of sample. If 1,2-diphenylhydrazine is likely to be present, adjust the pH of the sample to pH 4.0 ± 0.2 to prevent the rearrangement of benzidine.

All samples must be extracted within seven days. Extracts may be held up to seven days before analysis if stored under an inert (oxidant free) atmosphere. The extracts must be protected from light.

Phthalate esters (EPA Method-606)

Samples must be iced or refrigerated at 4°C from the time of collection until extraction. **All samples must be extracted within 7 days and completely analyzed within 40 days of extraction.**

Nitrosamine (EPA Method-607)

Samples must be iced or refrigerated at 4°C from the time of collection until extraction. If residual chlorine is present, add 80 mg of $Na_2S_2O_3$ per 1 liter sample. If diphenylnitrosamine is to be determined, adjust the pH of the sample to pH 7.00 to 10.00, using NaOH or sulfuric acid (H_2SO_4). Record the added milliliter of acids or bases. **All samples must be extracted within 7 days and completely analyzed within 40 days of extraction. Extracts must be stored in the dark!**

Organochlorine Pesticides and Polychlorinated Biphenyls (PCBs) (EPA Method-608)

Samples must be iced or refrigerated at 4°C after collection until extraction. If the samples are not extracted within 72 hours of collection, the pH of the sample should be adjusted to a range of pH 5.00 to 9.00 with NaOH or H_2SO_4. If Aldrin is to be determined, and if residual chlorine is present, add 80 mg of $Na_2S_2O_3$ per 1 liter of sample. **Samples must be extracted within 7 days of collection and completely analyzed within 40 days of extraction.**

Nitroaromatics and Isophorone (EPA Method-609)
Samples must be iced or refrigerated at 4°C after sample collection until extraction. **Samples must be extracted within 7 days of collection and completely analyzed within 40 days of extraction.**

Polynuclear Aromatic Hydrocarbons (PAHs) (EPA Method-610)
Samples must be iced or refrigerated at 4°C after sample collection until extraction. Samples must be collected in dark bottles. PAHs are known to be light sensitive; therefore, samples, extracts, and standards should be stored in amber or foil wrapped bottles to minimize photolytic decomposition. Fill the sample bottle and, if residual chlorine is present, add 80 mg $Na_2S_2O_3$ per 1 liter sample. **Samples must be extracted within 7 days of collection, and completely analyzed within 40 days of extraction.**

Haloethers (EPA Method-611)
Samples must be iced or refrigerated at 4°C after sample collection until extraction. If residual chlorine is present, add 80 mg of $Na_2S_2O_3$ per 1 liter sample. **Samples must be extracted within 7 days of collection and completely analyzed within 40 days after extraction.**

Chlorinated Hydrocarbons (EPA Method-612)
Samples must be iced or refrigerated at 4°C after sample collection until extraction. **Samples must be extracted within 7 days of collection and completely analyzed within 40 days of extraction.**

2,3,7,8-Tetrachlorodibenzo-p-Dioxin (EPA Method-613)
Samples must be iced or refrigerated at 4°C after sample collection until extraction. If residual chlorine is present, add 80 mg of $Na_2S_2O_3$ per 1 liter sample. Protect the samples from light from the time of collection until analysis. **Samples must be extracted within 7 days of collection and completely analyzed within 40 days of extraction.**

Purgeables (GC/MS EPA Method-624)
Samples must be iced or refrigerated at 4°C from the time of collection until analysis. Samples should be collected in 40 ml VOC bottles. If the sample contains residual-free chlorine, add $Na_2S_2O_3$ to the empty sample container (10 mg/40 ml) just prior to sampling. Fill bottle with sample as previously stated under the heading "Volatile Organic Compounds (VOCs)."

Some aromatic compounds, notably benzene, toluene, and ethylbenzene are susceptible to rapid biological degradation under certain environmental conditions. Refrigeration alone may not be adequate to preserve these compounds in wastewaters for more than 7 days. For this reason, a separate sample should be collected and acidified to pH 2 by the addition of 1+1 HCl when these aromatics are to be determined. The technique for the addition of $Na_2S_2O_3$ and HCl to the sample vial is the same as described under the sampling procedures for VOCs. **Samples should be analyzed within 14 days of collection.**

Base/Neutrals, Acids, Pesticides (GC/MS EPA Method 625)
Samples must be iced or refrigerated at 4°C after sample collection, until extraction, and must be protected from light. If the sample contains residual chlorine, add 80 mg of $Na_2S_2O_3$ per 1 liter sample. **Samples must be extracted within 7 days of collection and completely analyzed within 40 days of extraction.**

Nutrients, Demands, Nonmetallic Inorganics, Physical Properties, and Radiological Analysis

Follow the instructions given in Table 1-1 for containers and preservatives. Follow the general sample collection instructions and fill the bottles to within 1/2 inch of the cap. Transport immediately to the laboratory.

Microbiological Examination

Collect samples in sterile containers, and leave ample air space in the bottle (at least 2.5 cm) for adequate mixing prior to examination. Keep sample bottles closed until the moment they are to be filled. Remove stopper or cap, taking care to avoid soiling. Do not handle stopper, cap, or the neck of the bottle during sampling. Protect them from contamination. Fill the bottle without rinsing with the sample, replace stopper or cap immediately. If the water sample is taken from a tap, open tap fully and let water run for 2 to 3 minutes or for a sufficient time to permit clearing the service line.

Take samples from *surface waters* or *reservoirs* by holding the bottle near its base and submerging it below the surface. Turn bottle until neck points slightly upward and the mouth is directed toward the current.

When sampling *chlorinated sources* add sufficient $Na_2S_2O_3$ to the sample bottle before sterilization. To a 120 ml container you will need 0.1 ml of 10% $Na_2S_2O_3$ solution. It will neutralize a sample containing up to 15 mg/l residual chlorine. Sterile caps or sterile "Coli bags" with $Na_2S_2O_3$ tablets are commercially available.

Samples for microbiological testing must not be composited!!! Samples should be in the laboratory within 6 hours of sampling. Holding time is only 6 hours.

Oils and Greases, Total Recoverable Petroleum Hydrocarbons (TRPH)

Never use composite samples! Never rinse sample container before actual sampling! Never use intermediate vessel for sample collection! Never take samples from water surfaces, only from a well-mixed area!

Collect samples in a 1 liter glass container, and acidified with H_2SO_4 to pH less than 2. Store in refrigerator. **Holding time is 7 days until extraction, and 40 days for the extracts until final analysis.**

1.7. Field Sample Custody, Field Records

All sampling events should be documented and equipped with a *Chain-of-Custody* form. It ensures that the samples are collected, transferred, stored, analyzed, and destroyed only by authorized personnel. Each custodian or sampler must sign, record, and date the transfer. It includes the name of the sampling project, collector's signature, sampling location, sampling site, sampling point, date, time, type of sample, number of containers, and analysis required. The chain-of-custody is used to track samples from collection through analysis. It is kept on record for correction actions when needed.

Field records are taken for all data generated during sample collection. These field records are kept on a chain-of-custody form (Table 1-2), sample label (Figure 1-1), field notebook (Table 1-3), sample field log (Table 1-4), preservative preparation log (Figure 1-2), QC and field spike solution preparation log (Figure 1-3).

A sample label will be affixed to all sample containers. This is an important part of sample identification. It should be waterproof, and all of the information should be written in waterproof ink. A field notebook should also be specially designed for field work, with waterproof paper and a hard cover. Entries into all field records are to be written in waterproof ink, and any errors in all documents should be deleted by a single cross-line through the incorrect information with the date and initial of the person making the corrections.

Field Records

- Name of sample collector and all personnel participating in the sample collection.
- Date and time of sampling.
- Field conditions: weather, exact description of sample site, or any important information necessary for the correct representation of the sample location.
- Specific description of the sample location: address, sample site, exact sampling points, etc.
- Sample type: grab or composite sample. If it is a composite sample, record the appropriate time intervals, volume of the subsamples, and the duration of the composition.
- The requested analytical parameters, the type and number of containers, and preservation techniques. (For example: total metals, 1/2 gallon plastic container, preserved with 3 ml 1+1 HNO_3 per 1 liter sample).
- Preservative preparation, and information about the used chemicals (Figure 1-2).
- How pH checked on the preserved sample, and the value of the measured pH. If additional chemicals are used for correct pH, how many milliliters of extra preservatives were added, how was the blank prepared by the addition of this extra preservative?
- The sequential order in which samples are taken. Each sample has its "sequence number".
- Beside the sequence number, each sample should be given a field identification number (FID) with the analysis requested.
- If duplicate samples are taken, properly identify with FD1 and FD2.
- If split samples are taken, correctly describe and identify by FS1 and FS2.
- Information about the preparation and the true value of the field Quality Control (QC) samples, used to check the accuracy of the field tests (Figure 1-3) (if applicable).
- Spiked samples are marked as FSp1 and FSp2, if duplicate spiked samples are collected. Documentation for the preparation of spiked samples is on Figure 1-3 (if applicable).
- Field measurement data for temperature, pH, conductivity, dissolved oxygen (DO), residual chlorine.
- List of the purging and sampling equipment used.
- Field decontamination performed.
- Documentation for monitoring wells:
  - well casing composition and diameter of well casing
  - water table depth and well depth
  - calculation used for volume purged
  - total volume of water purged
  - date and time well was purged
  - measurements to monitor stabilization of wells
  - at least 3 volumes must be purged. If field measurements are taken, purging shall continue until the measurements are stable. If no measurements are taken, at least 5 well volumes must be purged before the sample collection can begin.

Table 1-2. Chain of Custody

Field I.D. ______________________ Site Name: ______________________

Date Sample Received: ______________________ Address ______________________

Sampier(s) ______________________ Laboratory: ______________________

Sample Identity	Date Sampled	Sample Container Description								Total	Remarks

Total Number of Containers ______

Relinquished By: ______________ Organization: ______________ Received By: ______________ Organization: ______________

Date: ______________ Time: ______________ Date: ______________ Time: ______________

Relinquished By: ______________ Organization: ______________ Received By: ______________ Organization: ______________

Date: ______________ Time: ______________ Date: ______________ Time: ______________

Delivery Method: ______________________________ (attach shipping bills, if any)

Use extra sheets if necessary

Field Sequence No.____

Field Sample No. ___________Date_____________ Time _____________

Sample Location ____

Sample Source ____

Preservative used ____

Analyses required ____

____

Collected by____

Remarks____

____

____

Final pH checked ____

Additional preservative used (if applicable)____________________

____

FIGURE 1-1. Sample label.

- Additional documentation for surface water:
  - depth samples were taken
- Additional documentation for wastewater effluent:
  - beginning and ending times for composite sampling, if applicable
- Additional documentation for soil and sediment:
  - depth samples were taken
- Additional documentation for drum sampling:
  - type of drum and description of contents
  - if stratified, what layer(s) was sampled
- How samples are transported to the laboratory; packing, cooling, separated, carrier, etc.
- Sample transmittal form (usually the chain-of-custody form) must include the following information:
  - site name and address
  - date and time of sample collection
  - name of sampler, responsible for the sample transmittal
  - identification of samples such as field ID number, number of samples, date and time sample collected, intended analysis, preservation, and any comments about the sample or sample container.

Failing to fill out these records properly could result in data invalidation!

1.8. Postsampling Activities

Sample Dispatch

Sample transportation or packing for shipment is the responsibility of field and sample custody personnel. Transferee must sign and record the date and time of

Table 1-3. Typical Field Notebook Page

Date ________________ Time ____________

Sampler's name __________________________ Signature ______________

Other field people ____

Sample location ____

Sample type grab ________
composite _________ Compositing time __________ hrs
Time interval __________ min
Subsample volume __________ mls

Sq No	*FID*	*Preserv. container*	*Analysis required*	*pH*	*T °C*	*Cond umhos /cm*	*DO ppm*	*Cl_2 ppm*	*Com ment*

Sq.No. Sample Sequence Number
FID Field Identification Number
Cond Conductivity
DO Dissolved Oxygen
Cl_2 Chlorine, residual
ppm parts per million, mg/L

Field conditions:

pH check :

Additional preservative used :

Other observations:

transportation on the chain-of-custody form and must keep in mind the following rules when transporting samples to the laboratory:

- Samples should be separated by sampling locations and by analysis type
- VOC samples should be packed in individual plastic bags and stored in a separate cooler that is marked "VOCs only"
- Only wet ice is used for cooling samples for 4°C
- Samples must be delivered to the laboratory as soon as possible

Waste Disposal in the Field

Wastes generated during sampling are separated into properly labeled waste containers and returned to the laboratory for appropriate waste management. Laboratory and field generated wastes are disposed of by certified waste management companies

Table 1-4. Sample Log Sheet

Purpose of Analysis: ______________________ Sample Field ID: ______________

Type of Sample: __________ Sampler: ______________ Date / Time: __________

Sample Site Number	Sample Source Description	Bottle Type	Bottle No	Preservative	Analysis Required							

Remarks:

* **Field Measurements**

contracted by the laboratory. Calibration standards used in field measurements are taken to the laboratory and flushed into the community water treatment system.

1.9. Field Quality Control

The quality of data resulting from sampling activities depends on the following major activities:

- Collecting representative samples
- Use of appropriate equipment
- Proper sample handling and preservation
- Proper chain-of-custody and sample identification procedure
- Proper Quality Assurance (QA) and Quality Control (QC) in the field

Preservative ______________________________

Preparation procedure______________________________

Date prepared ______________________________

Date of Expiration ______________________________

Analyte preserved ______________________________

Informations related to the chemical used:

Name, formula, and grade of the chemical

____

____

Source of the chemical (name of manufacturer)

____

Lot # of the chemical ______________________________
Date chemical received ______________________________
Date container was opened ______________________________
Expiration date ______________________________
Storage of the chemical

____

Check of preservative______________________________

____

______________________________ Preparer

______________________________ Supervisor

FIGURE 1-2. Preservative preparation log.

The first four activities have been previously discussed. This section introduces the required field QA/QC activities to assure that the process is in control, and the highest quality in field work and sample collection is maintained.

QC is a set of procedures that provides precise and accurate analytical results, and QA ensures that these results are adequate for their intended purposes. The goal of QA/QC in sample collection is to prevent the use of improper methodology or sampling techniques, insufficient sample preservation, inadequate identification, and transportation, and will prove the validity of the data from field measurements.

The Field QA/QC Program Included and Documented in the Following Areas

- Sample collection methodology, called Field Standard Operation Procedure (FSOP), with special sample handling procedures. Each method must be accompanied by method numbers, method reference, method detection limits, and accepted limits for precision and accuracy. These methods should be approved by EPA and DER

Analyte spiked ______________________________

Field No. of sample spiked ______________________________

Sample volume spiked ______________________________

Value of spike added ______________________________

Concentration of Stock Spike solution ______________________________

Volume of Spike Stock solution added ______________________________

Source of Spike Stock solution:

Commercial source

Manufacturer : ____________________
Lot # : ____________________
Date received: ____________________
Date expired : ____________________

Laboratory prepared

Date of preparation : ____________________
Expiration date : ____________________

Date Spike Sample prepared ______________________________

Signature of field personnel ______________________________

FIGURE 1-3. Field sample spike preparation log.

- Field QC requirements
- Procedures to record and process data
- Procedures to review and reduce data, based on QC results. Processes to validate and prepare field measurement data for reporting purposes
- Procedures to calibrate and to maintain field instruments and equipment
- Qualification and training of sampling personnel so they are able to handle the following areas:
  - determine the best representative sample site
  - use proper sampling techniques, by choosing grab or composite sampling, select the appropriate sampling equipment, use proper sample preservation, and sample identification
  - use appropriate data recording, and reporting forms
  - calibration and maintenance of field instruments and equipment
  - use QC samples such as duplicate, split, and spiked samples

There is no substitute for field experience.

After the training program, the fresh sample collector must be involved in sampling activities under the direction and supervision of a more experienced person for at least one month, prior to taking full responsibility. Special EPA and DER training workshops are available for training sampling personnel.

The Effectiveness of Sample Collection and Related Field Activities are Supported by Field QC Checks

Analytical work is more than one single process, and may be better characterized as a system, consisting of different independent processes. All components of a system are critically dependent on each other. Sampling operations must also be supported by a well-designed and reliable quality assurance program, including QC checks. The main goal of QA in sample collection is to monitor contamination during the process.

Criteria of Field QC Checks

Equipment Blanks

Equipment blanks are used to detect any contamination from sampling equipment. At least one equipment blank should be collected for every 20 samples per parameter group and per each matrix. Each type of equipment used in sampling must be accompanied with an equipment blank. This blank is prepared in the field before sampling begins, by using the precleaned equipment and filling the appropriate container with analyte-free water. Preservation and documentation of these blanks should be the same as for the collected samples. If equipment is cleaned on site, then additional equipment blanks should be collected for each equipment group.

Field Blanks

Field blanks are collected at the end of the sampling event. Fill an appropriate sample container with analyte-free water and then preserve and document in the same manner as the collected samples.

Trip Blanks

The reason for trip blanks is to verify contaminations that may occur during sample collection and transportation (improperly cleaned sample containers, contaminated reagents, airborne contamination during transportation, etc.). Trip blanks are usually prepared when VOCs are collected. At least one trip blank should be prepared for each VOC method, and analyzed for each cooler used for storing and transporting VOC samples. Trip blanks are blanks of analyte-free water that are prepared by the laboratory. These are transported to the field with the empty VOC sample containers, remaining unopened during the sampling event, and transported back to the laboratory with the collected samples. These blanks need to be properly labeled and documented.

Duplicates

Duplicates are samples collected at the same time from the same source (field duplicates) or aliquot of the same sample that are prepared and analyzed at the same time (laboratory duplicates). During each independent sampling event, at least one sample or 10% of the samples, whichever is greater, must be collected for duplicate analysis. This requirement applies to each parameter group and matrices sampled.

Field Spiked Samples

Field spiked samples are environmental samples that have specific concentrations of various parameters of interest added. Spiked samples are used to measure the performance of the complete analytical system including interference from sample matrix. Field preparation and transportation to the laboratory should be similar to the samples, and marked as FSp. If spike duplicates are collected, their identifications are FSp1 and FSp2. The sample that will be spiked may be selected by specific requirement, by a previous evaluation of the sample site, or an on-site inspection.

Split Samples

Split samples are a replica of the same sample. The split samples are given to two independent laboratories for analysis.

QC Checks in Field Measurements

Duplicate samples are analyzed to calculate the precision of the measurement. QC check standards with a known value are analyzed with a sample set of similar matrix, and used to determine the accuracy of the analysis. Duplicates and QC check standards must be analyzed at a continuing frequency equivalent to 5% of the samples in the analytical set (i.e., one every 20 samples) or analyzed at the beginning of each run to verify the calibration.

The *precision* of the field measurements is based on duplicate analyses. Precision values are expressed as Relative Percent Deviation (RPD), and are calculated by dividing the difference between the two measurement values with the average of the two measurement values and multiplying by 100.

$$RPD = \{(A - B)/[(A+B)/2]\} \times 100 \text{ or shortly } [(A - B)/(A+B)] \times 200$$

The *accuracy* of the field measurements is based on the recovery the true value of the QC check standard, expressed as % Recovery, and calculated as follows:

$$\% R = (\text{Measured Value} \times 100)/\text{True value}$$

The precision and accuracy of each field test has to be proven by these calculated values with each measurement, and documented on the working papers. The calculated RPD and % R values should be inside of their calculated control limits.

Control Limits for Precision

To establish the control limits for each field parameter, 20 RPD data must first be collected. Calculate the average $(\overline{x})$ and the standard deviation (s) of these values, and establish the **Warning Limit** by adding two standard deviations to the mean value, and the **Control Limit** by adding three standard deviations to the mean of these 20 data.

$$\text{Warning Limit} : 0 - (\overline{x} + 2s)$$

$$\text{Control Limit} : 0 - (\overline{x} + 3s)$$

Any RPD value falling above warning limits signals that the analytical system is nearing an out of control situation. Serious consideration and careful conduction of corrective action may need to be performed to find and eliminate the problem. Data falling outside the control limit indicate that the system is out of control. For example:

- RPD data collected for pH measurement
- The calculated mean value, $\overline{x}$ = 2.8% (RPD)
- The calculated standard deviation, s = 2.1% (RPD)

$$\text{Warning limit} : 2.8 + 2\ (2.1) = 0 - 7.0\%\ (RPD)$$

$$\text{Control limit} : 2.8 + 3\ (2.1) = 0 - 9.1\%\ (RPD)$$

Control Limits for Accuracy

Collect 20 accuracy data (% recovery) for a particular measurement. Calculate the mean ($\bar{x}$) and the standard deviation (s) for these recovery values, and determine the warning and control limits. Adding 2 standard deviations to the mean gives the upper warning limit and subtracting 2 standard deviations from the means gives the lower warning limit. Adding three standard deviations to the mean and subtracting three standard deviations from the mean gives the upper and lower control limits, respectively.

Upper Warning limit, UWL : $\bar{x} + 2s$

Lower Warning Limit, LWL : $\bar{x} - 2s$

Upper Control Limit, UCL : $\bar{x} + 3s$

Lower Control Limit, LCL : $\bar{x} - 3s$

Data points falling outside warning limits indicate that the system is approaching an out of control situation and may require a corrective action. Any data falling outside control limits signifies an out of control system. Analysis must be stopped and corrective actions taken immediately before further analysis continues. For example:

- Data collected for % Recovery of pH measurement
- Calculate the mean ($\bar{x}$) = 99.8%
- Calculate the standard deviation (s) = 2.3%
- Determine the QC limits:

Upper Control Limit : 99.8 + 3(2.3) = 106.7%

Upper Warning Limit : 99.8 + 2(2.3) = 104.4%

Lower Warning Limit : 99.8 - 2(2.3) = 92.9%

Lower Control Limit : 99.8 - 3(2.3) = 95.2%

Warning limit : 92.9 - 104.4%

Control Limit : 95.2 - 106.7%

To monitor precision and accuracy data of each field measurement, values are plotted daily on QC Charts. Preparation and interpretation of QC Charts are in the QA/QC section.

1.10. Summary of Field Activities and Preparation for Sample Collection

Field Activities

Correct and appropriate sample collection and field work insure the quality of analytical data. The following is a summary of the necessary activities and documentation for technicians involved in sample collections. All of the following documentation must be available for each sample collection event.

- List of sample collection methods, with method numbers and references
- Step-by-step sampling procedures per matrix, FSOP for:
  - Groundwater well monitoring
  - Potable waters

  - Surface waters
  - Wastewater effluents
  - Sediments, soils, sludges
  - Hazardous wastes
- A list of equipment (material, name, model, and manufacturer)
  - Sampling equipment
  - Equipment for field measurements
  - Equipment cleaning procedures
  - Equipment preventive maintenance log
  - Equipment maintenance log
  - Equipment performance check log
  - Equipment calibration log
  - Source and preparation of calibration standards log
- A list of sample containers
  - Cleaning procedures for sample containers
- Sample preservation, holding time, and proper container table
- Sample custody, identification
  - Sample label
  - Chain-of-custody form
  - Field sample log
  - Individual sample documentation in field notebook (date, time, sample sites, sampling points, sampler(s) name, site conditions, weather, monitor well conditions, construction, depth, purge volume, measurement of stabilization, time of purge, flow rate, field identification number [I.D.] for each sample, parameters required, identification of carrier(s) etc.)
  - Surface water — depth at which samples were collected
  - Sediments, soils — depth at which samples were collected
  - Drum sampling — type of drum, contents, which layer used
- Documentation for preservative preparation form
- List the methodology for field measurements with method number and references. Step-by-step procedure for each field test should be included.
- Field calibration procedures
  - Source of calibration standards
  - Initial calibration standards
  - Continuing calibration
- QC and spike stock solution preparation log
- Field QC samples and field spike sample preparation log
- Precision and Accuracy Control Charts for each field test (optional)
- Documentation for the quality check of deionized water or analyte-free water
- Written statement about the packing and transfer of collected samples to the laboratory

Preparation for Sample Collection

- Understand the sampling plan; all information should be written and discussed with field personnel
- Prepare, clean, and calibrate sampling equipment so it is ready to use
- Check, calibrate, and prepare equipment for field tests
- Prepare sample containers
- Preservative preparation and disposal in safe containers
- Labels and markers

- Laboratory notebook
- pH paper and small disposable cups for pH check of preserved samples
- Prepare all blanks
- Check all calibration standards and their expiration date for freshness. If necessary prepare new ones.
- Check QC samples for availability, check date for freshness. If necessary, prepare new ones.
- Check spike stock solution(s) and check date for freshness. If necessary prepare new ones. Discuss the concentration of the spikes, and select if the sample should be spiked. Calculate the volume of the added spike stock solution for each spiked parameter.
- Take pipets, with the suitable volume, and pipet bulbs
- Empty bottles for splits, duplicates, etc.
- Collect and check the cleanness of glassware for field tests
- Thermometer should be in a protective carrier to avoid breakage
- Ice boxes for sample transport
- Soap for washing sampling equipment (if necessary)
- Soap for washing hands, paper towels, soft tissues

Project Plans

Each Project Plan should contain the following documentation:

- Site identification and history, site map
- Previously collected data (if applicable)
- Scope and purpose of the project — state the overall purpose of the project, and the schedule and scope of the work. Also, note key activities with the beginning and ending date.
- Project organization — each project plan should clearly state the following information related to the particular project:
  - Frequency of sampling
  - Total number of samples
  - Sampling points
  - Analytical parameters with method numbers and related references
  - Method Detection Limits (MDLs)
  - QC objectives
  - Date reports were submitted
- The certification number of the Comprehensive Quality Assurance Plan (CompQAP) and other references for operation of the organization are important parts of the project plans. Education and training of the personnel participating in the field activities of the project is also needed.

Chapter 2
Field Tests

Calibration refers to the process by which the response of an analytical system is related to standards with known composition. The most important requirement of calibration is to use appropriate and accurate standards. Instruments used for field tests should be calibrated prior to the field trip, and in the field before taking the measurement. The use of each instrument is specific and both the calibration and usage should follow the manufacturer's instructions.

2.1. Calibration of Field Instruments

Calibration of field instruments should be performed on a regular basis, and all records should be kept in a calibration log. The record must indicate the name and model of the meter, the method used for calibration, the date and time, number of standards with concentration, the response of the instrument, and the verification of the calibration by a known value Quality Control (QC) sample. This control sample should originate from another source as the calibration standards. Field calibration of each field instrument must be performed at the first sample site and must be verified during the following measurements. Field calibration and checks are recorded on the field working papers, specially designed for each measured parameter (Tables 2-1 to 2-4) or in the field notebook.

2.2. Maintenance of Field Instruments

Each maintenance or repair of a field instrument must be documented in the maintenance logbook. Prior to the transportation to the field, the field technician must verify that the equipment is in good condition, working properly, and that the batteries are charged. Both the routine maintenance checks, all problems, and their solutions have to be noted in the instrument maintenance log book.

2.3. Minimum Quality Control (QC) Activities in Field Measurements

The new Quality Assurance rules (DEP SOP June 92) no longer require the generation of field precision and accuracy data. According to the new rule, more frequent continuing calibrations have to be performed. Once the instrument is calibrated, calibration has to be checked every 4 hours and at the end of the sampling. If the continuing calibration fails, a complete initial calibration should be performed before continuing the work. The following information, related to calibration of field equipments, should be incorporated into the field operation procedures:

Table 2-1. Working Paper for pH Measurement

Method No.: 150.1.
Reference : EPA 600/4-79-020
Model of pH meter :
Date :
Analyst:

Sample I.D. No.	Sample description	Measured pH	Sample temperature °C	Remarks
	Buffer pH 7.00			
	Buffer pH 4.00			
	Buffer pH 10.00			
	Q.C. sample			
	Duplicate			

QC sample true value: ________

Precision (RPD) : ________ Accepted limit ________

Accuracy (% R) : ________ Accepted limit ________

RPD (Relative Percent Deviation) = (A-B / A+B) x 200
A and B = duplicates

% R (% Recovery) = (Measured value x 100) / True value

- Listing of all the equipment
- Calibration and maintenance intervals
- Listing the required calibration standards and their source
- Step-by-step calibration procedure for all equipment
- Records on calibration activities

The preparation of the calibration standards should be kept in record. Calibration standards are either commercially available or prepared in house, according to the Standard Operation Procedure (SOP) of the organization. Each calibration stock or standard, *commercially available* or *laboratory (or house) prepared*, must be accompanied with a log form, containing the name, concentration, test used for, source, date received, date of expiration, date of opening, date of logging, the signature of the record keeping person and the field supervisor as on Figure 2-1. Standards should be labeled correctly with all of the necessary information: name, concentration, date received or prepared, date opened, and expiration date.

Table 2-2. Working Paper for Conductivity Measurement

Method No.: 120.1. Date : ____________
Reference : EPA 600/4-79-020 Analyst: ____________
Model of the Meter : ____________

Samp I.D. No.	Sample description	Sample Temp. ^0C	Reading mhos/cm	Reading umhos/cm	Result at 25^0C	Remark
	Reference standard 0.01 M KCl					
	Q.C. Sample					
	Duplicate					

Q.C. Sample true value : ____________

Precision (RPD) : ____________ Accepted limit ____________
Accuracy (% R) : ____________ Accepted limit ____________

RPD (Relative Percent Deviation) = (A-B / A+B) x 200
A and B = duplicates
% R (% Recovery) = (Measured value x 100) / true value

Conductivity value for 0.01 M KCl should be 1413 umhos/cm at 25^0C

Temperature correction :

If the temperature of the sample is below 25^0C, add 2% of the reading per degree.
If the temperature of the sample is above 25^0C, subtract 2% of the reading per degree.

2.4. The Most Common Field Tests, Instrumentations, and Calibration Procedures

Temperature Measurement

Temperature measurements may be taken with any good mercury-filled thermometer. As a minimum, the thermometer should have a scale marked with every 0.1°C. The Celsius grade (°C) thermometers are universally used in science related fields. Since the Fahrenheit (°F) scale is predominant in the United States, it is necessary to pay special attention to the unit at which the temperature reading was reported.

Table 2-3. Working Paper for Dissolved Oxygen (DO) Measurement

Method No.: 360.1. (Membrane electrode) Ref.: EPA-600/4-79-020
Method No.: 4500-OC (Iodometric method) Ref.: SM 17th Ed.1989

Model of meter ______________ Date ________ Analyst ________
Normality of
Sodium thiosulfate __________ Date ________ Analyst ________

Samp I.D No.	Sample description	Meter reading DO mg/L	Sample ml	Titrant ml	DO mg/L	Remark
	Duplicate					

Precision (RPD) : ______________ Accepted limit : ______________

RPD (relative Percent deviation) = (A-B / A+B) x 200
A and B = duplicates

Calculation for the iodometric method =

mg/L DO = (ml titrant x N x 8,000) / ml sample

N = Normality of titrant (Sodium thiosulfate, $Na_2S_2O_3$)
8,000 = Equivalent weight of oxygen x 1,000

The reported units are easily converted and may get the temperature in the favorable unit. Conversion of Celsius to Fahrenheit, and vice versa, is easy with any scientific calculator, or just use the following formulas:

$$°C = 5/9\ (°F - 32)$$

$$°F = 9/5\ °C + 32$$

The conversion formula from °F to °C results because the zero point on the Celsius scale is 32° below that on the Fahrenheit scale, and the size of the Celsius degree is 5/9 (reduced from 100/180) of the Fahrenheit degree. On the Fahrenheit scale the freezing point of water is assigned a temperature of 32°F. The boiling point is 212°F. Celsius scale defines 0°C as the freezing point of water and 100°C as the boiling point

Table 2-4. Working Paper for Chlorine (Residual) Determination

Method No.: 360.1. EPA-600/4-79-020 (Iodometric method)
Description of KIT used in the field ______________
Normality of Sodium thiosulfate ($Na_2S_2O_3$) ______________

Field measurement : Date ______________ Analyst ______________
Iodometric method : Date ______________ Analyst ______________

Samp I.D. No.	Sample description	Reading Cl_2 mg/L	Sample ml	Titrant ml	Cl_2 mg/L	Remark
	Duplicate					

Precision (RPD) : ______________ Accepted limit ______________

RPD (Relative Percent Deviation) = (A-B / A+B) x 200
A and B = Duplicates

Calculation for iodometric method =

$$\text{mg/L } Cl_2 = [(A+/-B) \times N \times 35{,}450] / \text{ml sample}$$

A = ml titrant for sample
B = ml titrant for blank
N = Normality of titrant (Sodium thiosulfate, $Na_2S_2O_3$)
35,450 = 35.450 (Equivalent weight of chlorine) x 1,000

of water. This means that 100°C degrees are equal to 180°F degrees, so the Celsius degree is nearly twice as large as a degree on the Fahrenheit scale.

Transportation of thermometers to the field should be in metal cases to prevent breakage. Periodically check the thermometers against a precision thermometer, certified by the National Bureau of Standards. The frequency and documentation is described in Part II. (QA/QC programs).

pH Measurement

Measurement of pH is one of the most important and frequently used tests in water chemistry. Practically every phase of the water supply and wastewater treatment, e.g.,

COMMERCIALLY AVAILABLE STOCKS AND STANDARDS

Name and concentration ______________________
Test for used ______________________
Source (manufacturer) ______________________
Lot number ______________________
Date received ______________________
Date opened ______________________
Expiration date ______________________
Certification number ______________________

LABORATORY PREPARED STOCKS AND STANDARDS

Name and concentration ______________________
Test for used ______________________
Date of preparation ______________________
Date of received by field personnel ______________________
Date opened ______________________
Expiration date ______________________

REMARK :

______________________ Date ______________________
Sample collector

______________________ Date ______________________
Field supervisor

FIGURE 2-1. Documentation log form of calibration stock and standard solutions used for field tests.

acid-base neutralization, coagulation, disinfection, and corrosion control, is pH dependent. At a given temperature, the intensity of the acidic and basic character of a solution is indicated by pH.

pH is the negative logarithm of the hydrogen ion concentration,

$$pH = - \log [H^+]$$

The term "concentration of hydrogen ion" is written $[H^+]$. The brackets mean "mole concentration" and H^+ means "hydrogen ion". The concentration of the hydrogen ion is expressed as moles per liter. Water dissociates by a very slight extent into H^+ and OH^- ions. It has been experimentally determined that in pure water $[H^+] = 1.0 \times 10^{-7}M$, and $[OH^-] = 1.0 \times 10^{-7}$, and $[H^+] = [OH^-]$. The negative logarithm of 1.0×10^{-7} is 7.00; therefore, the pH of the pure water is 7.00. Such a solution is neutral, and has no excess of either $[H^+]$ or $[OH^-]$. An excess of $[H^+]$ is indicated by a pH below 7.00 and a solution is said to be acidic. pH values above 7.00 indicate an excess of $[OH^-]$ or an alkaline condition.

pH Papers

Measurement of the pH is possible by using pH indicator paper available with a color scale. The color change of the indicator paper after it has been immersed into the sample is compared with the scale, giving the pH value. This measurement is just an approximation of the pH value.

Comparators

A more accurate reading is possible by using a comparator, accompanied with a pH disc. It is simple and very common in field testing. In this measurement a reagent, called a pH indicator, is added to the sample, and its color is compared to the comparator disc. The pH value is read directly from the disc. The indicator is commercially available and specially prepared to each pH disc.

pH Meter

A pH meter functions by measuring the electric potential between two electrodes that are immersed into the solution of interest. The basic principle is to determine the activity of the hydrogen ions by potentiometric measurement, using a glass electrode and a reference electrode. The most popular, called combination electrodes, incorporate the glass and the reference electrode into a single probe. The *glass electrode* is sensitive to the hydrogen ions, and changes its electric potential with the change of the hydrogen ion concentration. The *reference electrode* has a constant electric potential. The difference in potential of these electrodes, measured in millivolts (MV), is a linear function of the pH of the solution. The scale of the pH meters is designed so that the voltage can be read directly in terms of pH.

A *glass electrode* usually consists of a silver and silver chloride electrode in contact with dilute aqueous hydrochloric acid (HCl), surrounded by a glass bulb that acts as a conducting membrane. The hydrogen ion concentrations of each side of the glass membrane develop a potential difference. The hydrogen ion concentration of the HCl solution inside the electrode is constant; therefore, the potential of the glass electrode depends on the hydrogen ion concentration outside of the glass membrane. The *reference*, also called *"calomel"* electrode, contains elemental mercury (Hg), calomel (Hg_2Cl_2) paste, and Hg metal. This paste is contacted with an aqueous solution of potassium chloride (KCl) solution. KCl serves as a salt bridge between the electrode and the measured solution. The glass electrode is relatively free from *interferences* caused by color, turbidity, colloidal matter oxidants, reductant, or high salt content of the sample. *High sodium content samples* give higher pH reading interference. It can be reduced or eliminated by using "low sodium error electrode". *Coating of oily material or particulate matter can impair electrode response.* This coating usually can be removed by gentle wiping or detergent washing followed by distilled water rinsing. Additional treatment with 1+9 HCl may be applied to remove any remaining film. *Temperature effects* may be eliminated by using pH meters equipped with a temperature compensator.

Calibration of pH Meter and Measurement of pH

The pH measurement method used depends on both the instrument and the electrodes used. Both meter and electrodes should be in good condition to ensure reliable and accurate results. Each meter, along with its electrodes, must be calibrated by pH buffers 7.00 and 4.00 or 10.00 depending on the expected pH range of the sample.

The calibration steps are as follows:

- Meter is in the "on" position. Allow a few minutes for the instrument to warm up. Electrodes should be in the storage solution. For storage of the electrodes, a pH 7.00 buffer solution, or a commercially available pH electrode storage solution may be used. Keeping electrodes in distilled water is not recommended.
- Set the temperature compensate knob to room temperature.
- Remove electrodes from storage solution and rinse with deionized (DI) water, dry by gently blotting with a soft tissue.
- Immerse electrode into pH 7.00 calibration buffer solution, stir at a constant rate, by using magnetic stirrer, to provide homogeneity and suspension of solids. The rate of stirring should minimize the air transfer rate at the air-water interface of the sample. Be sure that the sample volume is sufficient to cover the sensing elements of the electrodes. Read the pH value. If the reading deviates from the actual value (7.00), adjust with the "calibration" knob.
- Remove electrode from pH 7.00 buffer solution, rinse with DI water, wipe, and place into pH 4.00 buffer. Read. If the reading deviates from 4.00, adjust to the exact reading by turning the "slope" knob.
- Remove electrode from pH 4.00 buffer. Rinse with DI water, wipe and immerse into pH 10.00 buffer. Read. pH value should be 10.00 or very close. Do not adjust meter at this point! The instrument is now calibrated.
- After calibration, but before the sample measurement, check the correct performance of the meter by reading a QC standard with a known value of pH. If the reading is correct, the instrument is calibrated, and will give accurate and reliable results on sample reading. QC standards are prepared by the laboratory prior to field tests, or are commercially available.
- If the sample reading shows a high alkalinity character, the meter must be recalibrated by using pH 7.00 and pH 10.00 buffer as calibration standards. After recalibration, read the sample again.

Because of the wide variety of pH meters, detailed operation procedures are given for each instrument. The analyst must be familiar with the operation of the system, and with the instrument functions. If field measurements are being made, the electrode(s) may be immersed in the sample stream to an adequate depth and moved in a manner to insure sufficient sample movement across the electrode sensing element as indicated by drift-free (0.1 pH) readings. Calibration, measurement, and QC data are recorded on pH working paper (Table 2-1).

Measurement of Conductivity

Conductivity is the numerical expression of the ability of an aqueous solution to carry an electric current. The ability depends on the presence of the ions, their total concentrations, mobility, valence, and the temperature of the measurement system. Electrolytic conductivity increases with temperature at a rate of approximately 1.9% deviation per degree Celsius. Significant errors can result from inaccurate temperature measurement.

Solutions of most inorganic acids, bases, and salts are good conductors. Conversely, molecules of organic compounds that do not dissociate in an aqueous solution conduct the current very poorly. The term conductivity is preferred and reported in the unit of micromhos/cm (umhos/cm) or millimhos/cm (mmhos/cm). The International System of Units (SI system) use the unit of milliSiemens/m (mS/m).

1 mS/m = 10 umhos/cm; therefore, 1 uS/cm is equal to 1 umhos/cm.

Conductivity measurements are used to:

- Establish the degree of mineralization
- Estimate Total Filtrable Residue, or Total Dissolved Solids (TDS) in a sample by multiplying the conductivity value with an empirical factor that may vary from 0.55 to 0.9, depending on the soluble components in the water and on the temperature of the measurement. Relatively high factors may be required for saline waters and boiler heaters, whereas lower factors may apply where considerable hydroxide or free acid is present.
- The sum of the analyzed cations and anions, expressed in milliequivalent per liter unit multiplied by 100 must give the approximate value of the measured conductivity in umhos/cm.
- The sum of the cations or sum of the anions expressed in milliequivalent per liter is equal by multiplying conductivity (umhos/cm) by 0.01.

Choose an instrument capable of measuring conductivity with an error not exceeding 1% or 1 umhos/cm. The conductivity bridge range must be from 1 to 1000 umhos/cm. The conductivity cell constant is 1.0. The platinum electrode type is recommended for laboratory or field measurements, and the nonplatinum electrode types are used for continuing monitoring or field studies. The performance of the conductometer and the electrode is checked by measuring the conductivity of a 0.01 M potassium chloride (KCl) reference solution, which at 25°C has a conductivity of 1413 umhos/cm. The KCl standard solution should be stored in glass stoppered borosilicate glass containers, and will stay fresh for 6 months. If a faulty reading is indicated, preparation of a new standard is recommended. It is best to apply two KCl standards, bracketing the expected conductivity value of the samples. The reading should be within 2% of the expected value. Apply continuing calibration standards with the previous criteria.

- Electrode is stored in DI water
- After instrument is in the "on" position, wait for a few minutes to warm up.
- Rinse electrode with DI water, wipe gently, and immerse into the 0.01 M KCl solution. Measure the temperature of the solution and read the conductivity.
- After a satisfying reading, rinse electrode with DI water, wipe to dry, and immerse into a QC standard solution. Measure the temperature of the sample and read the conductivity.
- If the reading indicates that the meter is working properly, then reading of the samples may start. If the instrument indicates a malfunction, then stop the measurement, document the problem, and conduct corrective action, with detailed and appropriate record.

Applying Temperature Correction

If the temperature of the sample is below 25°C, add 2% of the actual reading per each degree difference. If the temperature of the sample is above 25°C, subtract 2% of the actual reading per each degree difference. Also the following formula may be used to correct conductivity values for the reported temperature.

Conductivity at 25°C is equal to measured conductivity/1 + 0.0191(t – 25).

Resistivity

The reciprocal value of the conductivity is the resistivity.

For example:

Conductivity	**Resistivity**
694 umhos/cm	1/694 umhos.cm (0.0014)
0.694 mmhos/cm	1/0.694 mmhos.cm (1.44)
0.000694 mho/cm	1/0.000694 ohm.cm (1440)

Working paper for conductivity measurement is on Table 2-2.

Measurement of Dissolved Oxygen (DO)

Dissolved Oxygen (DO) levels in natural and wastewaters depend on the physical, chemical, and biochemical activities in the water body. The analysis for DO is a key test in water pollution and waste treatment process control.

Measurement by Membrane Electrode

Membrane electrodes are suited for analysis in the field. Their portability and ease of operation and maintenance make them particularly convenient for field application. Oxygen sensitive membrane electrodes are composed of two solid metal electrodes in contact with a supporting electrolyte separated from the test solution by an oxygen selective membrane. A variety of membrane electrodes are commercially available. In all these instruments, the "diffusion current" is linearly proportional to the concentration of molecular oxygen. DO probes are temperature sensitive, and temperature compensation is normally provided by the manufacturer. Membrane probes have a temperature coefficient of 4 to 6% per degree Celsius, depending on the membrane employed. Each instrument is accompanied with the manufacturer's instructions and must be followed for acceptable results. Be careful when changing the membrane to avoid contamination of the sensing element and also trapping minute air-bubbles under the membrane, which can lead to lowered response and high residual current. It is best to air calibrate the DO meter in a water saturated atmosphere before it is taken to the field, and after at each sample site, for correct readings. For calibration, follow the manufacturer's instructions. The meter should be air calibrated at 4 hour intervals in the field and at the end of the sampling day. Store the electrode by cover with a wet sponge so it creates a saturated atmosphere.

Measurement by Iodometric Titration, Winkler Method

The Winkler method is not convenient for field work. The method is used to check the performance of the DO meter and the electrode. The sample is measured by the membrane electrode method, and the same sample is taken to the laboratory for analyzing by the Winkler method. The two results should show that they belong to the same sample. Improved by variations in technique and equipment and aided by instrumentation, but the iodometric titration procedure remains the most precise and reliable measurement for DO.

The test is based on the addition of divalent manganese solution, followed by a strong alkali to the sample in a glass stoppered bottle, called Winkler or BOD bottle. DO rapidly oxidizes an equivalent amount of the dispersed divalent manganous hydroxide precipitate to hydroxides of the higher valency states. In the presence of iodide ions and acidification, the oxidized manganese reverts to the divalent state with the liberation of iodine equivalent to the original DO content. The iodine is then titrated with a standard solution of sodium thiosulfate ($No_2S_2O_3$). The titration end-point can be detected with starch indicator.

The reactions involved are represented by the following formulas:

$$MnSO_4 + 2\ KOH = Mn(OH)_2 + K_2SO_4$$

Manganous sulfate reacts with potassium hydroxide (KOH) or sodium hydroxide (NaOH) to produce precipitate. If white precipitate is present, there is no dissolved oxygen in the sample and there is no need to proceed further. A brown precipitate shows that oxygen is present and reacted with the manganous hydroxide. The brown precipitate is manganic basic oxide:

$$2\ Mn(OH) + O_2 = 2\ MnO(OH)_2$$

Upon the addition of sulfuric acid (H_2SO_4), the precipitate is dissolved, forming manganic sulfate:

$$MnO(OH)_2 + 2\ H_2SO_4 = Mn(SO_4)_2 + 3\ H_2O$$

At this point, the oxygen is converted to a form ready for further analysis. The sample is ready to transport to the laboratory to determine the oxygen, as follows: manganese sulfate ($MnSO_4$) reacts with potassium iodide (KI) to liberate iodine resulting in the typical iodine coloration of the water in the presence of starch indicator:

$$Mn(SO_4)_2 + 2\ KI = MnSO_4 + K_2SO_4 + I_2$$

The quantity of I_2 liberated by these reactions is equivalent to the quantity of oxygen present in the sample. (I^- ion is oxidized by the oxygen present to I_2 molecule). The quantity of iodine is determined by titrating with a standard solution (with a known normality) of $Na_2S_2O_3$, which is a reducing agent. During the reaction, use a starch indicator. In the presence of iodide ion, the starch gives a blue color, and the blue color disappears in the presence of iodine molecule formation, indicated at the end point of the titration. (More detailed methodology and calculation is given in the laboratory manual used in laboratory practices). The working paper for determination of dissolved oxygen is on Table 2-3.

Sample Collection for the Winkler Method

Collect samples very carefully. Methods of sampling are highly dependent on the source to be sampled, and to a certain extent, on the method of analysis. Collect samples in narrow-mouth glass stoppered Winkler bottles (commonly called BOD bottles) of 300 ml capacity, with tapered and pointed ground-glass stoppers and flared mouths. Rinse the bottle with the sample. Pour the sample into the bottle and let the sample overflow over the top of the bottle (at least 1/3 of the volume of the bottle should be allowed to overflow). The bottle is then stoppered when all the air bubbles go out of the bottle. The temperature of the sample should be recorded. Remove the stopper and add 2 ml of manganese sulfate ($MnSO_4$) reagent (480 g $MnSO_4.4\ H_2O$ dissolve and dilute to 1 liter), followed by 2 ml of alkali-iodide-azide reagent (10 g of sodium azide (NaN_3) 480 g NaOH, 750 g sodium iodide (NaI) in 1 liter). Introduce both reagents beneath the surface of the sample. Replace the stopper, being careful not to trap air inside. Mix by inverting the bottle at least 15 times. Allow the floc to settle, invert again. Allow the floc to settle again, and remove the stopper. Immediately add 2 ml conc H_2SO_4 by allowing the acid to run down the neck of the bottle, restopper, and mix until the precipitate dissolves leaving a clear yellow-orange colored solution. Dissolution should be complete. Transfer samples to the laboratory to complete the test. A sample stored at this point should be protected from strong sunlight and analyzed as soon as possible.

Testing Chlorine Residual

The chlorination of water supplies and polluted waters serves primarily to destroy or deactivate disease-producing microorganisms. Chlorine also reacts with sulfide, iron, manganese, and ammonia, which is also beneficial for water quality. Reaction of chlorine with organic substances may cause adverse effects. For example, reaction of chlorine with phenolic compounds creates chlorophenols, known for their bad taste and odor. Regulated limitations of phenolic compounds are based mostly on taste resulting from chlorination and not because of toxicological concern. Reaction of chlorine with certain organics creates potentially carcinogenic compounds called trihalomethanes (THMs), which contain one carbon, one hydrogen, and three halogen atoms. These organic compounds are clearly derivatives of methane (CH_4). Methane gas is not in question, but rather the reaction of chlorine in raw or treated water with certain compounds, mainly defined as humic acids (part of organic material associated with decaying vegetation). From the listed 10 THMs, the most frequently expected in drinking water is chloroform (trichloromethane), bromodichloromethane, dibromochloromethane, and bromoform (tribromomethane). There are no fluorine-containing THMs. Chlorine can be applied as chlorine gas, chlorine hypochlorite, chlorine dioxide, and chloramine. Most commonly used is chlorine gas. In water, chlorine reacts as follows:

$$Cl_2 + H_2O \rightarrow \underset{\text{hypochlorous acid}}{HOCl} + \underset{\text{hydrochloric acid}}{HCl}$$

$$HOCl \rightarrow H^+ + \underset{\text{hypochlorite ion}}{OCl^-}$$

The relative proportion of hypochlorous acid and hypochlorite ion is pH- and temperature-dependent. These forms are called *free chlorines.* Free chlorine reacts readily with ammonia and certain nitrogenous substances to form *combined available chlorine*, called *chloramine*, as monochloramine (NH_2Cl), dichloramine ($NHCl_2$), and nitrogen trichloride (NCl_3). The production of chloramine will be the function of pH and ammonia presence. At low pH, nitrogen trichloride is produced. Chlorinated wastewater effluents, as well as certain chlorinated industrial effluents, normally contain only combined chlorine.

Field Methods for Testing Chlorine Residuals

Comparator with orthotolidine reagent disc — In the presence of chlorine, addition of orthotolidine reagent produces a yellow color. The deepness of the yellow color is proportional to the concentration of the chlorine. The concentration is read directly from the accompanied chlorine disc in terms of mg/l. The very commonly used orthotolidine method has been deleted because of the toxic nature of orthotolidine.

Comparator with DPD (N,N-diethyl-p-phenylene-diamine) reagent — Chlorine reacts with DPD by producing a red color. The deepness of the red color corresponds to the concentration of chlorine present expressed as mg/l and read from the proper disc. For the correct operation of the above comparator KITs, follow manufacturer's instructions.

Laboratory Method to Measure Chlorine Residual

Iodometric titration method — Correct operation and performance of the KIT measurement has to be checked periodically by a more accurate laboratory method. The laboratory method is a simple oxidation-reduction titrimetric method. Because chlorine is a strong oxidation agent, the method is similar to the previously discussed oxygen determination. Chlorine liberates free iodine from potassium iodide (KI) at pH 4 or less. The iodine is titrated with a standard reducing agent, such as $Na_2S_2O_3$ using starch as an indicator.

$$2\ KI + Cl_2 = I_2 + 2\ KCl$$

The result is calculated as chlorine, Cl_2 in mg/l. (More detailed methodology and calculation is given in the laboratory manual used in laboratory practices.) The working paper for chlorine determination is on Table 2-4.

Determination of Carbon Dioxide

Carbon dioxide (CO_2) is an odorless, tasteless, incombustible gas, heavier than air, and moderately soluble in water. Surface waters usually contain less than 10 mg/l free CO_2, while groundwater supply contains much higher values. It is not harmful but the presence of CO_2 contributes to corrosion problems. In water treatment, the presence of CO_2 effects coagulation and iron removal. Collection of the samples must be done carefully because of the easy loss of CO_2. Determine the free CO_2 immediately at the point of sampling. Where a field determination is impractical, fill the bottle completely, keep at a temperature lower than the sample was collected, and make the laboratory examination as soon as possible.

Titrimetric Method

The titrimetric method is a simple, rapid method that is satisfactory for field use. Free CO_2 reacts with NaOH to form sodium bicarbonate ($NaHCO_3$). Completion of the reaction is indicated by the development of the pink color characteristic of phenolphthalein indicator at the equivalence pH of 8.3.

The following reagents are needed:

- NaOH standard, 0.1 N. 4.0 grams of NaOH dissolved in DI water, and dilute to 1 L.
- Dry potassium hydrogen phthalate (also called potassium biphthalate) at 120°C for 2 hours. Weight 10 grams, dissolved and diluted to 1 liter with DI water. The concentration is 0.05 N.
- NaOH standard, 0.02 N. Dilute 200 ml 0.1 N NaOH to 1 liter with DI water. Standardize against potassium biphthalate solution as follows:
  (a) Measure 15 ml of 0.02 N potassium biphthalate solution into Erlenmeyer flask. Add 1 ml phenolphthalein indicator. Solution is colorless.
  (b) Titrate with 0.02 N NaOH until a pink color develops.
  (c) Calculate the exact normality of the NaOH:

$$N = (\text{ml biphthalate used} \times N_{prepared}) / \text{ml NaOH used or}$$

$$(A \times B)\ /\ (204.2 \times C)$$

  A = g of potassium biphthalate used
  B = ml potassium biphthalate solution measured
  C = ml NaOH used for titration
  204.2 = equivalent weight of potassium biphthalate
- Phenolphthalein indicator solution is made by dissolving 5 g of phenolphthalein in 500 ml isopropyl alcohol (also called 2-propanol) and dilute to 1 liter with DI water.

Use the following procedure:

- Measure 100 ml sample
- Add a few drops of phenolphthalein indicator (solution colorless)
- Titrate with 0.02 N NaOH solution until a pink color appears

Calculation:

$$mg/L\ CO_2 = A \times N \times 44 \times 1000/ml\ sample$$

A = ml NaOH titrant used
N = normality of the NaOH
44 = equivalent weight of CO_2

Preparation and standardization of the solutions are made by the laboratory and transported to the field. The working paper for the determination is on Table 2-5.

Table 2-5. Working Paper for Carbon Dioxide Analysis

Parameter: CO_2 Method No. 4500-CO_2C Reference : SM 17th Ed.
Method Detection Limit (MDL) : ____________
Titrant and Normality (standardized) ____________
Analyzed by ____________ **Date** ____________

ID No.	*Sample Identification*	*Sample type*	*Sample ml*	*Titrant ml*	*Result mg/L*	*Remark*
	Blank					
	Duplicate					

INFORMATIONS RELATED TO QC CHECKS

ID of duplicate samples ________
Precision as Relative Percent Deviation (RPD) ____________

Approval of Supervisor ____________ **Date** ____________

Chapter 3

Sample Collection from Different Matrices

3.1. Groundwater Sampling

Groundwater accounts for the base flow of all perennial streams, over 90% of the world fresh water resources. About 50% of the population uses groundwater as its primary source of drinking water. Groundwater is subsurface water that occurs beneath the water table in soils and geological formations that are fully saturated.

The *aquifer* is water stored in a water bearing stratum of permeable rock, sand, or gravel, and its function is to store and transit water due to its porosity and permeability. There are two main types of aquifers — unconfined and confined. An *unconfined* or *water table aquifer* contains water under atmospheric pressure. The *confined*, or *artesian aquifer* is sandwiched by layers of relatively impermeable material. The water is under greater than atmospheric pressure (Figure 3-1). Groundwater is constantly in response to gravity, pressure, and friction. Gravity and pressure drive the water, and the friction restricts its motion. Rates of the movement vary from feet per year to feet per day. The flow rates of groundwater can range from a few millimeters to several meters per day. The flow rate of groundwater is very important to determine contaminant movement. Groundwater flow is slower than surface flow. The flow of fluids through porous media is governed by *Darcy's Law*. It states that the velocity of flow of fluid through a porous medium is proportional to the hydraulic gradients.

$$V = K (h_1 + z_1) - (h_2 + z_2)/L$$

V = Darcy velocity of fluid (L/t)

K = hydraulic conductivity of medium (L/t)

h_1 = pressure head at point 1 (L)

z_1 = elevation head at point 1 (L)

h_2 = pressure head at point 2 (L)

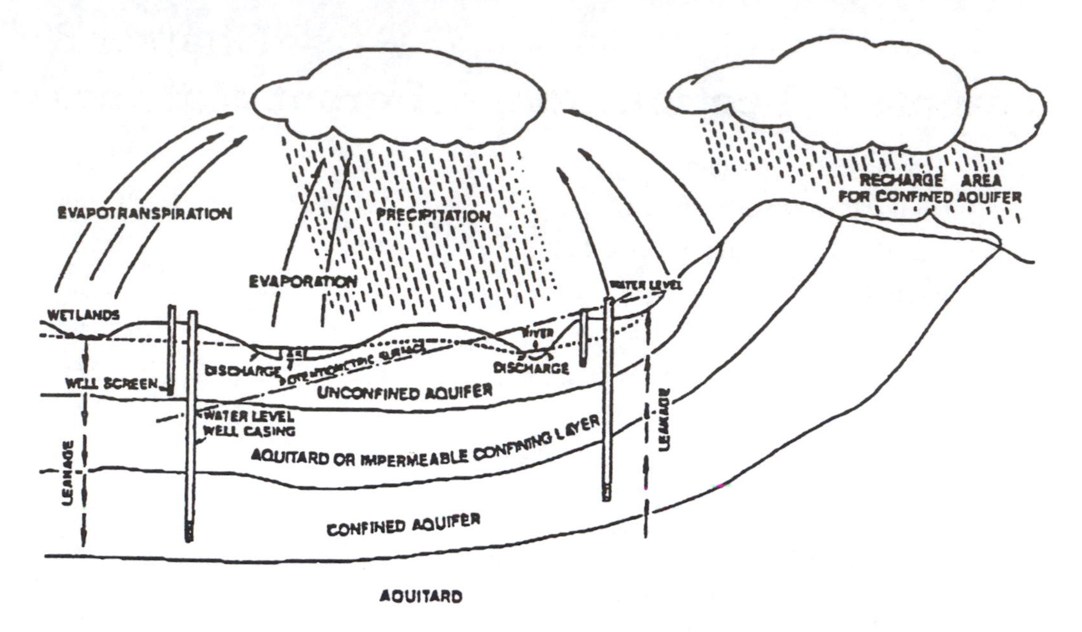

FIGURE 3-1. The hydrologic cycle, aquifer systems. The movement of the water through the environment from the atmosphere to earth and back again is called the *hydrologic* or *water cycle*. Gravity keeps water moving from the earth to the atmosphere via *evaporation* (water surfaces, land) and *transpiration* (plants). Water moves from the atmosphere to the earth in the form of *condensation* and *precipitation*. It then moves between points in streamflows and groundwater movement systems. The hydrologic cycle, according to the U.S. Geological Survey, is a natural machine, a constantly running distillation and pumping system, which is supplied by energy from the sun.
The figure shows the *characteristics of the two major types of aquifers*. Above the *unconfined aquifer*, there is no flow-restricting material, and the free water surface (water table) can rise and fall. A *confined aquifer* is between two confining layers and has no free water surface. The water level in the well represents the confining pressure at the top of the aquifer. The hydrostatic pressure within a confined aquifer is sufficient to cause the water rise in a well high enough so it flows out on the land surface (flowing artesian well).

z_2 = elevation head at point 2 (L)

L = distance of flow between points 1 and 2 along a streamline

Groundwater may have relatively high pressures of nitrogen (N) and carbon dioxide (CO_2), so water samples need to be handled very carefully. Both the sampling point and the sampling mechanism, as well as the expertise of the sampler are critical to minimize the variation of errors introduced into analytical results.

Monitoring Wells

In many instances of groundwater contamination, the ability to predict how the contaminant plume will behave in the future can only be done on the basis of extensive drilling and sampling program. The most frequently used approach in groundwater quality monitoring is to collect and analyze water samples from monitoring wells. Once the areal extent of the plume has been defined, several monitoring wells are installed in or adjacent to the plume. The purpose of a monitoring well is to determine hydrogeologic properties, provide a facility for collection water samples, and monitor the movement of the contaminant plume. The number and location of the monitoring wells and selection of the depths where samples are taken is critical. Usually one well

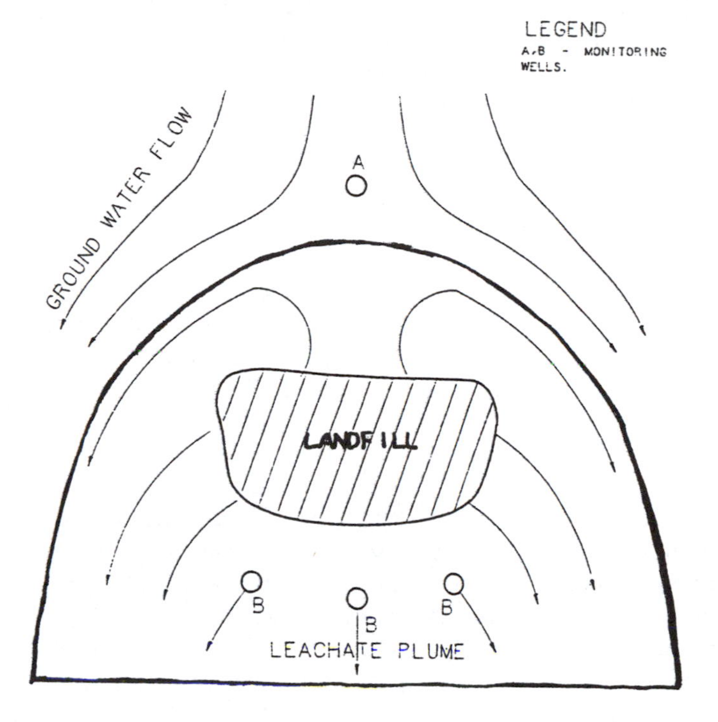

FIGURE 3-2. Idealized monitoring network. Monitoring wells "B": one well is sited near the center of the plume just downgradient from the contaminant source. Another well is installed to verify the amount of dispersion taking place. Monitor well "A" is placed upgradient of the contaminant source for ambient environmental data.

is sited near the center of the plume just downgradient from the contaminant source. Another well is installed downgradient of the contaminant source, outside the limit of the plume, and a third well is placed to monitor upgradient chemistry. Other wells may be installed to verify the amount of dispersion taking place. The design of the monitoring network is extremely important. In practice, the number of wells varies greatly depending on the local hydrogeology. Selection of the depths where samples are taken is a difficult decision. Ideally, some wells should be installed at more than one depth to verify if vertical flow is occurring or the spread of the contaminant varies at different depths. Idealized monitoring network is shown in Figure 3-2. It shows the location of monitoring wells designed to check contamination from landfill leachates. Monitoring well "A" is the background well, it serves to check the groundwater quality of an area before the source of contamination is introduced. The only direct means of sampling from groundwater is through a *borehole* that penetrates the aquifer. Boreholes are rarely left open when drilled. Most often some type of tubing (casing) is placed in the hole, to support the wall of the borehole.

The selection of the proper *casing materials* is very important, and influences the analytical data. For example, substituting Teflon with PVC is cost saving, but the adsorption of organics on PVC, or by bleeding of organic constituents in PVC cement is critical. The following materials may be used as sampling equipment and also as casing: Teflon and Stainless steel 316 and 304. Careful consideration of the materials chosen is required in each individual case. Diameter should be sufficient to allow the sampling tools to be lowered into the well to the desired depth. The part of the well where the water enters must be properly constructed and developed to avoid sampling problems.

Commercially made *well screens* are recommended. Casing and well screens should be washed with detergent and rinsed thoroughly with clean water, and some cover should be used to prevent contamination.

The importance of proper sampling of monitoring wells cannot be overemphasized. Care should be taken so that the sample collected is neither altered nor contaminated during the sampling process. In general a sample is taken only after the pH, conductivity, and temperature of water being pumped from the well have been stabilized.

Several pre-field procedures must be considered prior to sampling activities: selecting proper sampling equipment, decontamination of sampling equipment, selection of sample bottles, preservatives, preparation and calibration of field analytical instruments, documentation, such as, sample labels, chain-of-custody forms, field notebook, etc. as described under the general considerations of sampling in Chapter 1.

Selection of Materials for Sampling Equipments

Selection of materials used for well purging, sample collection, handling, and storage is a critical consideration. They should not adsorb or leach chemicals that may lead to false analytical results. Also the material selected should be easily washed, decontaminated, and transported. According to the literature, the following materials are recommended for use when sampling groundwater.

Stainless Steel 316 or 304

It is convenient to sampling for all parameter groups, except metals and is excellent for using with organic parameters. It is easy to decontaminate without rinsing with nitric acid (HNO_3). High salinity water samples may cause pitting and corrosion on the surface, especially in acidic conditions.

Teflon

Recommended for use in most groundwater and soil monitoring for all parameter groups.

Carbon Steel and Galvanized Steel

Used for soil sampling as spoon and core barrel soil sampling. High salinity, acidic environment, presence of sulfide cause corrosion, and contamination of samples.

Polypropylene, Polyethylene

Resistant to organic solvent, and resistant to corrosion, it is comparable to Teflon.

Polyvinyl Chloride (PVC), Viton, Tygon, Neoprene, Silicon

Are not recommended for organic analytes sampling.

Purging and Sampling Equipment

To ensure that you gather a representative groundwater sample from a potable well or monitor well, it is essential that the well is purged prior to the sampling. Stagnant water may undergo chemical changes, so the chemical analytical data on the sample will give false results. Purging out the stagnant water can be performed with various equipment. The selection of the equipment is based on the parameters of interest, the diameter of the well, the water level, and other site conditions. The most common equipments used in purging and collecting samples from groundwater are:

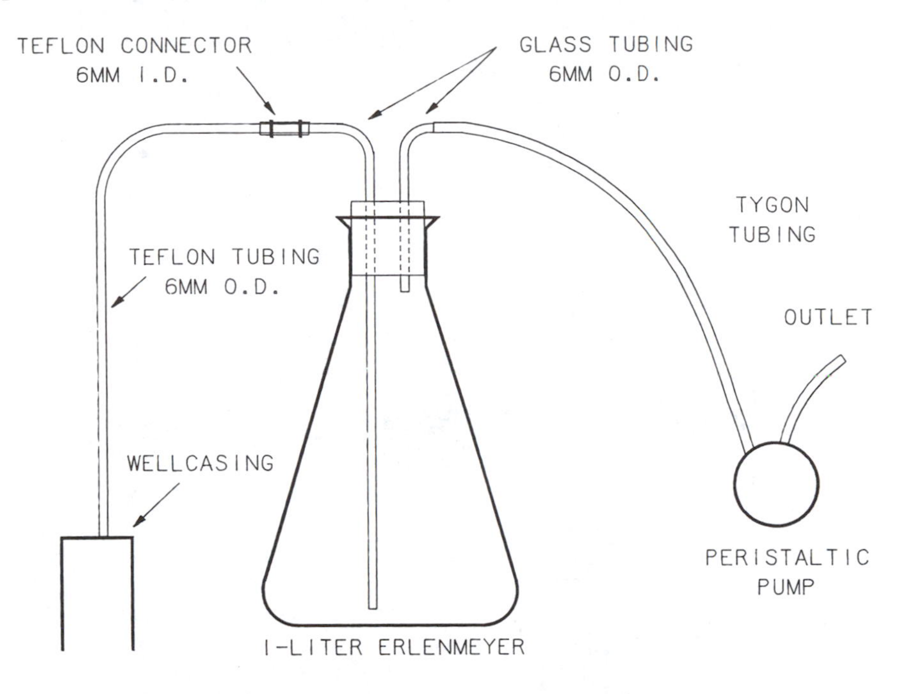

FIGURE 3-3. System for grab sampling (peristaltic pump).

- Centrifugal pumps — Used to purge 2 inch or 4 inch (5 or 10 cm) diameter wells with capacities of 2 to 10 gallon per minute (gpm) (7.6 to 37.8 liter per minute) and the water level is no deeper than 20 feet (6.5 m) below surface.
- Electric submersible pumps — Used for purging 4 inch (10 cm) or greater diameter monitor wells. The pump should be constructed of stainless steel material and the delivery hose material should be appropriate for sampling analytes.
- Hand pumps — Utilized for purging 2 inch (5 cm) diameter wells in which the static water level is too deep to use a centrifugal or peristaltic pump.
- Peristaltic pumps — Used to purge low volume, low specific capacity wells in which the static water level in the well is no greater than 20 ft (6.5 m) below land surface. The peristaltic pump is used to purge shallow wells (Figure 3-3).
- Air-lift sampler — Applying air pressure to the well to force a water sample out. Relatively portable and inexpensive, but this method is not an appropriate method for taking water samples for detailed chemical analyses due to degassing effects on the sample (Figure 3-4).
- Gas operated pumps — Can be constructed in diameters as small as 1 inch (25.4 mm). A wide variety of material may be used and is relatively easy to transport. The disadvantage is that a gas source is required, a large amount of gas is needed, and long running cycles are necessary when pumping from deep wells (Figure 3-5).
- Bailers — Can be constructed in a wide variety of diameters, and in a wide variety of materials. Bailers are easy to transport, easy to clean and inexpensive. The disadvantage of using a bailer is that atmospheric oxygen is introduced when transferring water to the sample bottle, and can cause pollution through scraping the casing wall. Bailers should be constructed of compatible materials (Teflon, stainless steel), and should be scrupulously cleaned (Figure 3-6).

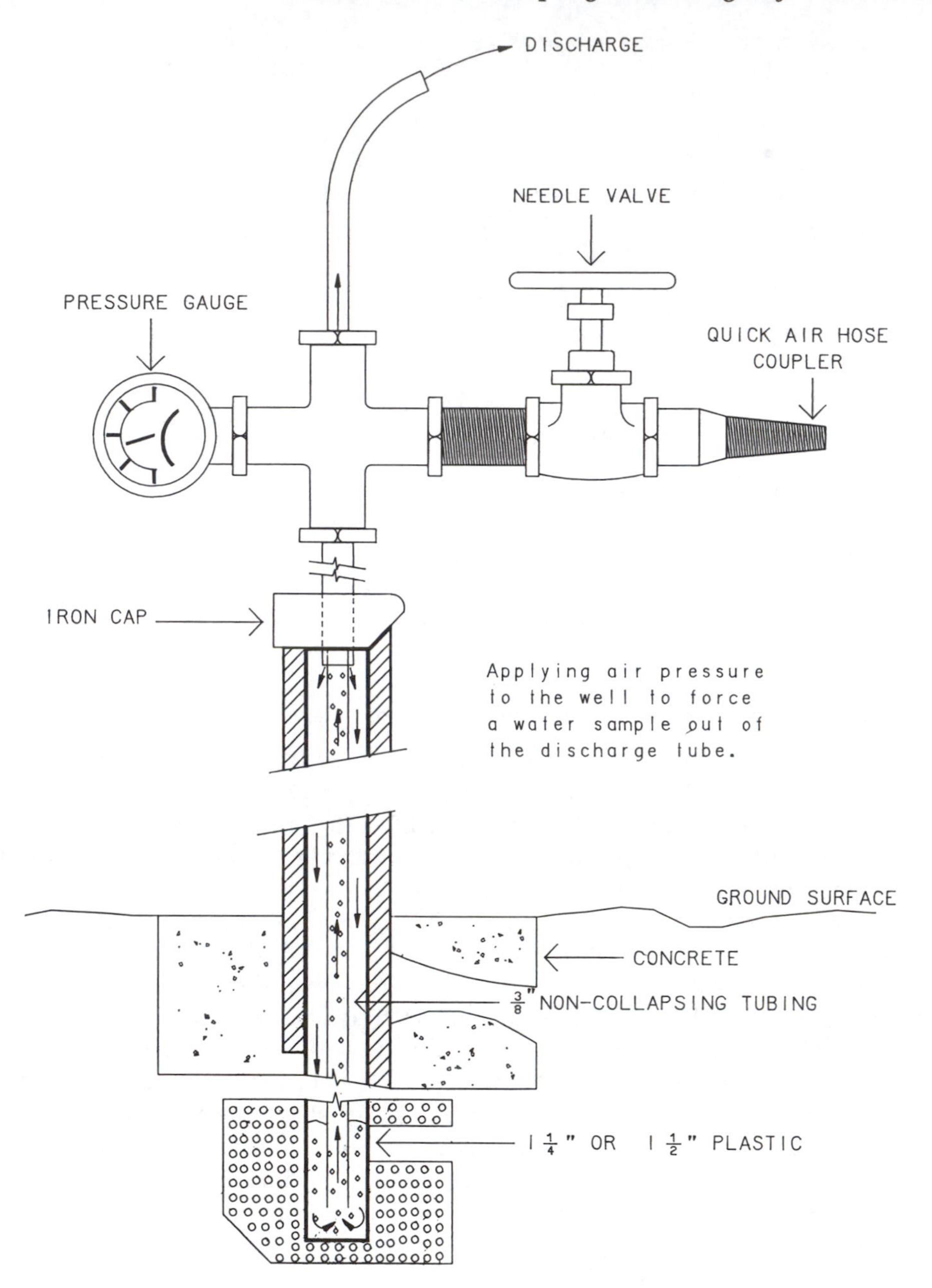

FIGURE 3-4. Air-lift sampler.

Determination of the Volume of the Purged Water Prior to Sampling

Prior to sampling, an adequate amount of stagnant well water must be removed so that the collected water sample will be representative to the groundwater conditions. Stagnant water in a well is usually chemically different from the water in the ground near the well bore. For most wells, remove three to five well volumes, or until pH, conductivity, and temperature readings of the water is stabilized. When a sample is taken, enough time should have elapsed so at least 95% of the water being pumped

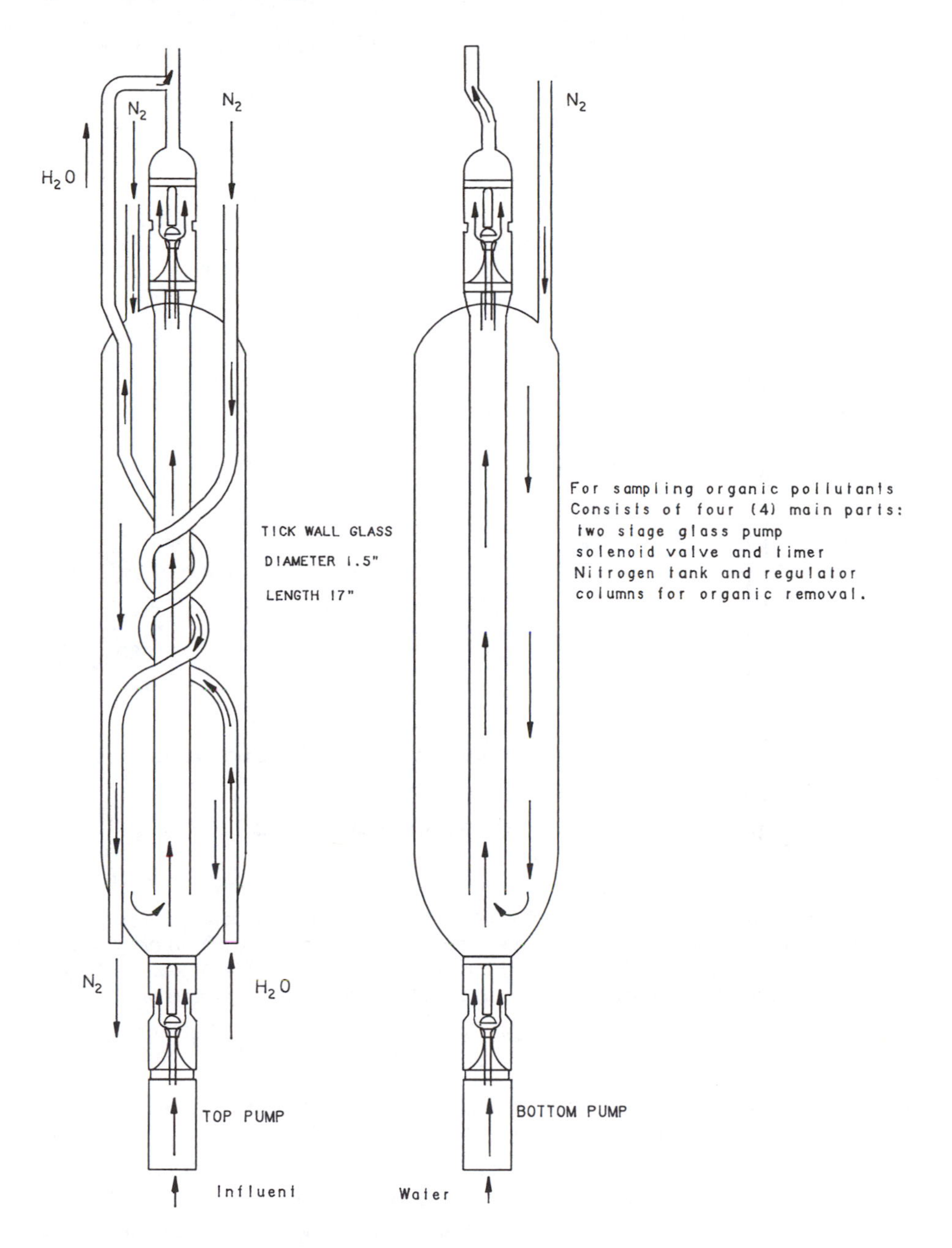

FIGURE 3-5. Nitrogen power glass-teflon pump.

comes from the aquifer. The volume of the water pumped out and the elapsed time must be documented in the field notebook. Wells should be sampled within 6 hours of purging.

Water-Level Measurement

Water level is measured by using an electronic probe, chalked tape, etc. When *electrical devices* are used, a light or ammeter indicates a closed circuit when the probe touches the water. To eliminate any errors from kinks and bends in the wires

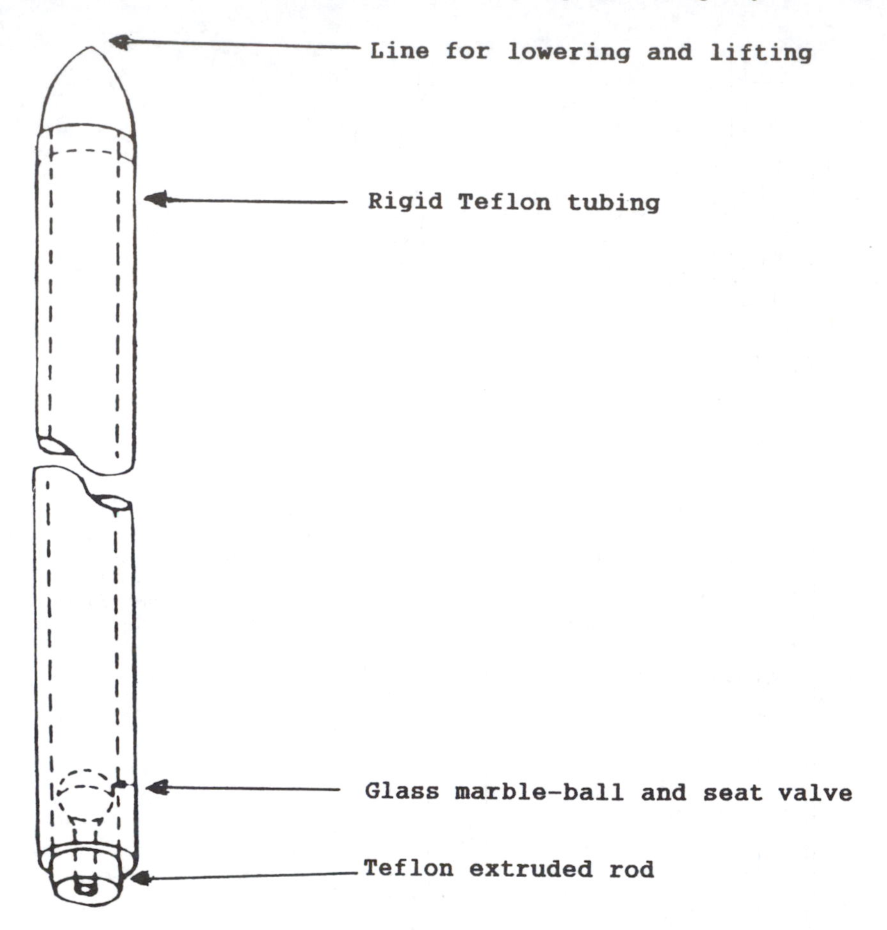

FIGURE 3-6. Teflon bailer. The top of the bailer is open and the bottom contains a simple "ball and seat" check valve arrangement. As the bailer moves down through the water in the well, the check valve remains open, and the water passes through the bailer. As the bailer is lifted, the weight of water inside the bailer causes the ball valve to seat, thus trapping the sample inside.

which may change the length slightly when the device is pulled up and let down, the cable should be left hanging in the well for a series of readings. Depth markers are commonly attached to the cable by the manufacturer at about 5 ft (1.5 m) intervals. When using a *steel tape*, a lead weight is attached to the bottom. The lower 2 to 3 ft (0.6 to 0.9 m) of the tape is wiped dry and coated with carpenter's chalk before making the measurement. The tape is let down into the well until part of the chalked section is below water and one of the foot marks is held exactly at the top of the casing. After withdrawal, the wetted line on the tape can be read on the chalked section. The reading is subtracted from the foot mark held at the measuring point; the difference is the actual depth to the water level. The total water column is obtained by subtracting the depth to the top of the water column from the total depth of the well.

$$V = (0.041)\, d^2 h$$

where V = volume in gallons, d = the well diameter in inches, h = the height of the water column in feet.

Groundwater Sampling Technique

The recommended sample collection sequence is as follows.

Volatile Organic Compounds (VOCs)
Total Organic Carbon (TOC)
Total Organic Halides (TOX)
Extractable Organics
Total Metals
Dissolved Metals
Microbiological samples
Other inorganic groups: Demands, Nutrients, Minerals, General Radionuclides

Bailers used for sampling should be properly cleaned and decontaminated prior to sampling. The material of the bailer and the purging equipment should be considered according to the parameters sampled as was discussed previously.

Using the Bailer

In most bailers, the top of the bailer is open and the bottom contains a simple check valve. In order to obtain a sample, the bailer is lowered into the well on a line. This line can be made of almost any material but for collecting specific contaminants, a noncontaminating material such as stainless steel or Teflon should be used.

- The sampler should wear latex rubber gloves to avoid sample contamination.
- Lower the bailer slowly into the well. As the bailer moves slowly down through the water in the well, the check valve remains open, allowing the water to pass through the bailer.
- At the desired depth, stop to lower the bailer.
- As the bailer is lifted, the weight of the water inside the bailer will close the valve, trapping the sample inside.
- When the bailer reaches the surface, the sample is transported to the proper sample bottle.

Groundwater Sampling for VOCs

- Inspect 40 ml glass vials to assure that they are free from breakage or leakage. Damaged vials should be discarded. A typical container for collecting samples for VOC analysis is in Figure 3-7.
- Check that vials are prepreserved, or preservation will be in the field. For proper preservation technique, see Chapter 1.
- The sample collector should wear natural latex rubber gloves to bail groundwater from the well. Remove the cap from the vial, without touching the septum.
- Fill the vial with the sample and with the least possible amount of air until a convex meniscus is formed.
- Quickly place the septum and cap over the meniscus, and close securely.
- Insert the sample and gently tap in the palm of your hand to check for air bubbles. If air bubbles are present, discard the whole sample and collect new sample.
- Collect in duplicate or triplicate samples as requested by the laboratory or by the project plan.
- Complete the sample label, wrap the vial to prevent breakage, and place into a sealed ziplock-type plastic bag.
- Place vial(s) into an ice box on wet ice at 4°C. (VOCs need separate ice box!)

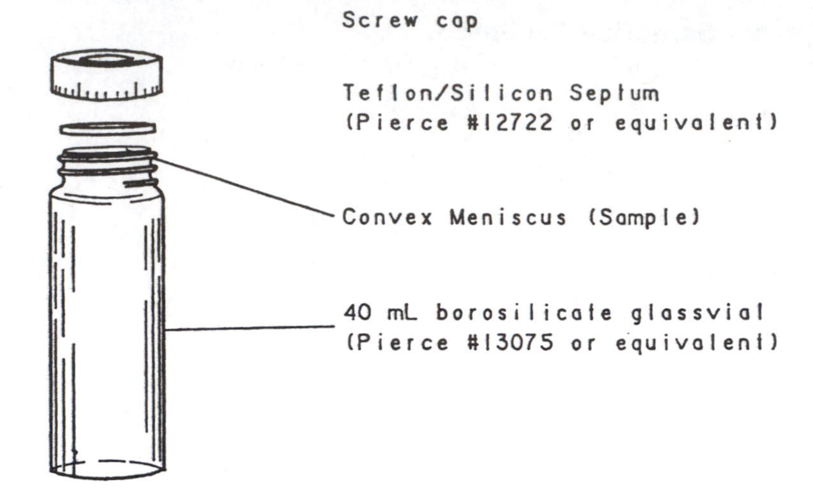

FIGURE 3-7. Sample container for purgeable organics. Grab samples obtained for analysis of purgeable organics (VOCs) are sealed to eliminate entrapped air. The sample is collected without headspace as illustrated in the above figure.

- Complete the chain-of-custody form, and be sure that all activities are documented in the field notebook with waterproof ink, as was discussed in Chapter 1.

Groundwater Sampling for Extractable Organics

- Collect samples in 1 liter glass bottles with Teflon lined caps.
- When samples are collected for pesticides, the sample must be preserved by adjusting the pH of the sample to 5 to 9 pH range with sulfuric acid (H_2SO_4) or sodium hydroxide (NaOH). After preservation, check the pH by using pH paper and documented, as described in Chapter 1. Follow the proper preservation technique per parameters as on Table 1-1.
- Wear natural latex rubber gloves and bail groundwater from the well. Remove the cap, but be careful not to touch the Teflon liner. Fill the bottle with the sample to nearly full.
- Place the cap back on and close the bottle.
- Label the bottle and complete chain-of-custody form.
- Record all sampling data in the field notebook.
- Wrap and place the sample in a plastic bag and put into an ice box on wet ice cooled to 4°C.

Groundwater Sampling for Metal Analysis

- Collect samples in a polyethylene container.
- Wear natural latex rubber gloves and bail water.
- Remove cap from the bottle and rinse the sample bottle with the water. (Do not rinse bottle when it is a prepreserved container!)
- Fill the bottle with the sample but do not fill to top. Leave space for adding preservative and for mixing.
- Add 3 ml 1+1 or 1.5 ml concentrated HNO_3 per 1 liter sample to take the pH <2. (If prepreserved bottles are used, of course, the addition of the acid is

unnecessary). Check pH by pouring the sample into a small container and then check the pH with narrow range pH paper.

- Record the preservative milliliters and the checked pH on the sample label and in the field notebook.
- The sampled equipment blank should contain the same amount of preservative as the sample! Samples with additional preservative should have separate equipment blank with the same amount of acid as in the sample.
- Affix the sample label, fill out chain-of-custody form, and record all sampling data in the field notebook.
- Samples do not need to be cooled during transportation.
- Samples for hexavalent chromium (Cr^{6+}) sampled separately from the other metals. Do not add acid preservation to this sample, and transport to the laboratory for analysis as soon as possible. The analytical request for the sample with the note of "no preservative added" should be clearly written on the label. *Holding time is 24 hours for these samples.*
- If the request is *dissolved and suspended metal* determination, the sample should be filtered prior to preservation. The acceptable filter is 0.45 u poresize membrane filter. The filter should be washed with DI water prior to filtering the sample. If the sample is filtered on the field, the water is pumped directly from the well through the filter. The filtrate will be the sample for dissolved metals and after filtration acidified the same way as for total metals (3 ml 1+1 or 1.5 ml concentrated HNO_3 per 1 liter filtrate). The filter paper with the suspended matter will be transferred to the laboratory for determination of suspended metals. If the sample is not filtered at the field, it should be transferred to the laboratory as soon as possible after collection without preservation, and filter in the laboratory. After filtration, the filtered sample will be preserved with HNO_3 to pH <2. It is recommended that all groundwater samples be filtered in the field before preservation to avoid contamination from sediment. Contaminants adhering to the sediment, especially metallic ions, may be released in the sample when dissolved by the addition of acid.

Collection Groundwater Samples for Oil and Grease and Total Recoverable Petroleum Hydrocarbons (TRPHs)

- Samples should be collected in 1 liter glass container. **Since oil and grease remain on sampling equipment, never collect composite samples for oil and grease!**
- **Never rinse the sample container with sample prior to actual sample collection!**
- Wear latex gloves during sampling. Fill the sample container almost to full capacity with the sample, and preserve by the addition of 5 ml hydrochloric acid (HCl) per 1 liter, and cap the bottle.
- Complete and affix sample label, fill up the chain-of-custody form, and the field notebook.
- Put the sample into a plastic bag and transport at 4°C.

Sampling Groundwater for Inorganic Nonmetallic Parameters and Other Miscellaneous Tests

Demands: Total Organic Carbon (TOC), Chemical Oxygen Demand (COD), and Biological Oxygen Demand (BOD)

Nutrients: Different forms of Nitrogen and Phosphorus, such as Ammonia Nitrogen (NH_3-N), Nitrite-Nitrogen (NO_2-N), Nitrate-Nitrogen (NO_3-N), Total Kjeldahl Nitrogen (TKN), Organic-Nitrogen, Ortho- and Total Phosphates (PO_4-P)

Minerals: Chloride, Sulfate, Fluoride, Hardness, Calcium, Magnesium, Alkalinity, Sodium, Potassium

Cyanide, Microbiological, Radiological and other tests

- Check for proper containers, as on Table 1-1.
- Wear natural latex rubber gloves and bail groundwater. Remove the cap from the bottle, and rinse it with water sample. Prepreserved containers and microbiological sample bottles are never rinsed with the water sample before sampling! Carefully pour sample into the container to almost full.
- Add preservatives, if required, according to the recommendation for sample containers, preservation, and holding times (Table 1-1). The same amount of preservative should be added to the equipment blanks.
- Close the container, affix sample label, complete chain-of-custody form, and record data in the field notebook.
- Place samples on wet ice and protect them from sunlight.

3.2. Drinking Water Sampling

Potable Well Water Sampling

When sampling drinking water from residential, private potable wells, they must be purged as described under groundwater sampling. If the capacity of the pressure tank is not known, purge the well about 15 to 20 minutes. After purging, reduce flow to approximately 500 ml/min. Rinse sample containers with sample water, except prepreserved sample containers, sample bottles for Oil and Grease, Total Recoverable Petroleum Hydrocarbons (TRPHs), VOCs, and TOXs. Fill sample bottles to near full capacity, except for VOCs and TOXs. Preserve samples according to the requirements, as stated in Table 1-1. To collect and preserve samples correctly, follow the guidelines as under groundwater sampling (Section 3.1) and as described in the general concepts in sampling (Chapter 1). After collecting and preserving samples, label properly, complete the chain-of-custody form, and field notebook. Place sample bottles in plastic bags, place bottles directly on wet ice, or transfer without icing and store as recommended in Table 1-1.

Sampling Drinking Water From a Distribution System

The procedures are the same as sampling from potable wells, except if there is residual chlorine present. The presence of residual chlorine in the sample may cause interference with certain types of analysis. The addition of sodium thiosulfate ($Na_2S_2O_3$) inhibits the chlorine action. 0.008% or 100 mg/L $Na_2S_2O_3$ must be added to the sample immediately after sample collection. Close bottles, label, and transfer as described above.

If samples are collected from faucets, the faucet should be clean and free from possible contamination. Samples should be collected in areas free from excessive dust, rain, snow, or other sources of contamination. When samples are collected from a faucet, it should be flushed thoroughly, generally 2 to 3 minutes, but sometimes a longer time is needed, for example, if it is a lead distribution line before the actual sample is taken. After flushing the water, adjust the flow so it does not splash against the walls of bathtubs, sinks, or other surfaces. Then collect samples. For most

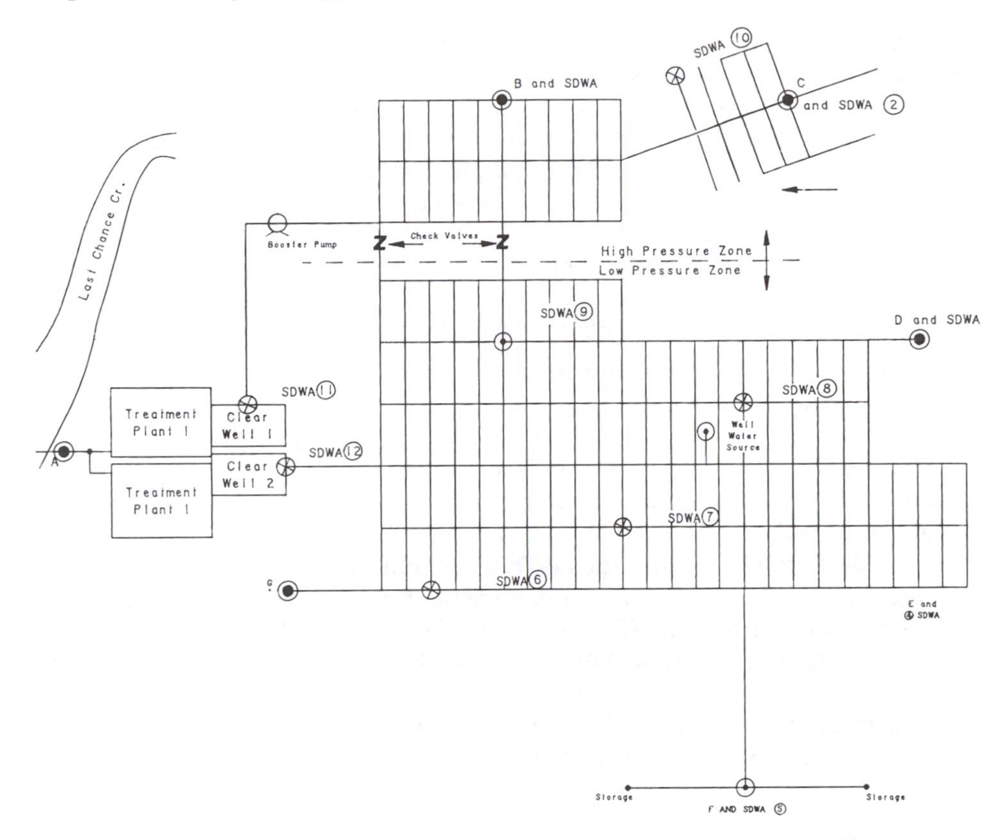

FIGURE 3-8. Combined branch and loop system. Sampling point "A" is the entry of the distribution system, required analyses are Turbidity and Trihalomethanes (THM). Sampling point "B" is the main line, sampling point "C" is the main dead end of the line, "D" and "E" are in the branch and branch dead end points. Collect samples for parameters according to the Safe Drinking Water Act (SDWA); however, sampling points "B" and "E" need complete analysis.

samples, fill the container to 1 1/2 inches from the top. Preserve, store, and transport samples according to the recommendation in Table 1-1.

Sampling Locations

Sampling point and frequency of sampling from distribution system is described by the Interim Drinking Water Regulation (IPDWR). Sampling points from a combined branch and loop distribution system is on Figure 3-8. Point "A" is not acceptable, because it is not located in the distribution system. Sampling points "B", "C", and "E" are acceptable.

Collection Samples for Microbiological Testing

Collect microbiological samples in sterile containers with caps. Handle containers and caps aseptically. Hold the cap in one hand without touching the inner surface while sampling. It is not necessary to flame the tap, just flush thoroughly before sampling. Preserve samples with $Na_2S_2O_3$ as previously mentioned, to inhibit chlorine action, and samples will represent the conditions from which they are taken.

Collecting Samples for VOCs

First run the water from the faucet until the service line is cleared and the water is representative of the main line. Normally, it takes a couple of minutes (3 to 5 minutes) with the faucet wide open. The presence of fresh water at the faucet may be recognized by the change of the water temperature. At this point, reduce the flow to a thin stream, remove the cap and septum from the sample vial, and immediately fill the container by allowing the water stream to strike the inner wall of the bottle to minimize the formation of air bubbles. **Do not rinse sample bottle!** Fill the container to the rim, but do not allow it to overflow excessively. Slide on the Teflon septum (smooth side down!). Replace the cap and tighten firmly. (Detailed outline for sampling for VOCs is in Section 3.1.) Preserve and transport samples as in Table 1-1.

Parameters for Drinking Water Analysis

Primary Drinking Water Parameters

- **Inorganics** — Arsenic, Barium, Cadmium, Chromium, Lead, Mercury, Selenium, Silver Sodium, Nitrate-Nitrogen, Fluoride, Turbidity
- **Organics** — **Pesticides-Herbicides**: Endrin, Lindane, Methoxychlor, Toxaphene, 2,4-Dichlorophenoxyacetic acid (2,4-D), 2,4,5-Trichlorophenoxyacetic acid (2,4,5-T, Silvex)

  Volatile Organic Compounds (VOCs) : Trichloroethylene, Carbon tetrachloride, Vinyl chloride, 1,2,-Dichloroethylene, Benzene p-Dichlorobenzene, 1,1-Dichlorobenzene, 1,1,1-Trichlorobenzene Ethylene Dibromide (EDB)

  Total Trihalomethanes (THMs) : Bromoform, Chloroform, Dibromochloromethane, Dichlorobromomethane
- **Bacteriology** — Total Coliform bacteria count
- **Radionuclides** — Radium 226 & 228, Gross alpha particles activity, Gross beta particles activity

Secondary Drinking Water Parameters

Color, Odor, pH, Total Dissolved Solids (TDS), Corrosivity, Iron, Copper, Manganese, Zinc, Chloride, Sulfate, Foaming agent, "p" and Total Alkalinity, as $CaCO_3$, Total Hardness, as $CaCO_3$, Carbonate, Bicarbonate, Hydroxide, Calcium, Magnesium

3.3. Surface Water Sampling

Surface water types are different, and include lakes, ponds, streams, rivers, and estuarine areas. Surface waters are under increasing pressure from stormwater runoff, sediment and nutrient loading, wastewater effluent discharges, and other impacts. The main reasons to checking the quality of surface waters are:

- To determine that the designated use of the particular surface water meets water quality limits
- To evaluate the degradation of water quality and the reduction of the diversity of the aquatic life caused by stormwater discharges, nutrient and pollutant loads, and sedimentation
- To determine the effects of a specific discharge on a certain body of water

The selection of the sample sites depends on the nature of the sampling project, the parameters of interest, and the type of samples. Sites may already be determined by EPA, DER permit, or by a permanent monitoring station. The **Surface Water Improvement and Management (SWIM) program** has been established to monitor and protect surface waters. The following guidelines are suggested in the EPA Model State Water Monitoring Program for selecting monitoring stations:

- At key locations of water, when the water used for domestic water supply, recreation, propagation, and maintenance of fish and wildlife
- In the main stream, upstream, and downstream from the confluence of major tributaries
- Near the mouths of major rivers where they enter an estuary
- At locations in major water bodies potentially subject to input of contaminants from areas of concentrated urban, industrial, or agricultural use
- At key locations in water bodies largely unaffected by human activities

Surface Waters Have Been Classified According to Their Designated Uses

Class I. Potable water supply
Class II. Shellfish propagation or harvesting
Class III. Recreation, propagation, and maintenance of a healthy, well-balanced population of fish and wildlife
Class IV. Agricultural water supplies
Class V. Navigation, utility, and industrial use

Sample volume, containers, preservation, and handling of samples follow the outline as in the General Sampling Consideration, in Chapter 1.

General Rules in Surface Water Sampling

- If sampling with a boat, samples must be taken from the bow, away and upwind from any outboard gasoline engine
- Samples, both water and sediments, should always be collected from downstream to upstream
- Care should be taken not to disturb sediments when taking the water samples
- When water and sediment samples are taken from the same area, water samples must be collected first
- Do not take samples at or near dams, piers, or bridges because the unnatural waterflow may disturb the representativeness of the sample.

Sample Types and Sampling Equipments Used to Collect Surface Waters

Grab Samples

Grab samples are taken by using **unpreserved sample containers**. Collecting grab samples are a common and accepted method because no additional equipment interferes with the procedure. Preservation should be done immediately after sample collection (within 15 minutes) as in Table 1-1. Take grab samples from surface waters as follows:

- Submerge the container into the water.
- Invert the bottle, so the neck is upright and pointing to the water flow, and return the filled container quickly to the surface.
- In the case when an intermediate sampling vessel is used, pour the sample into a sample container.
- Pour out a few milliliters of the sample to allow room for the preservative.
- Cap, close, and label the bottles.

Another sampling technique is to use a pole-mounted flask or other container. With the help of a long pole, samples can be taken from the shore, boat, or bridge. This kind

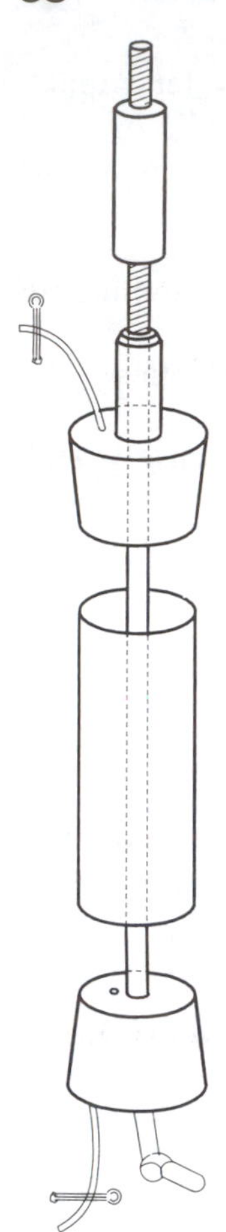

FIGURE 3-9. Modified Kemmerer sampler.

of equipment must be constructed of material and must not interfere with the sampled parameters. (If the pole is aluminum, of course, no sample is accepted for aluminum testing!).

Composite Samples
Composite samples are taken when a given depth interval is desired for the sample. Care should be taken that all of the subsamples have equal volumes! (see Chapter 1).

Using Pump
A peristaltic pump may be used to take composite samples from depths, but is also useful for grab sampling. This technique does not work for parameters such as Oil and Grease, TRPHs, and Microbiological samples.

Equipment that Takes Samples from Different Depths
Samples are taken from different depths in the same sample location: just below the surface, mid-depth, and just above the bottom. These samples may be taken by depth-specific samplers (**Kemmerer, Niskin,** etc.) or by using a pump. **Bailers** are also acceptable. Sampling devices have to be selected according to the sampling parameters. For example, many Kemmerer samplers are constructed of plastic and rubber, which prohibits their use for sampling organic parameters. Stainless steel, Teflon, or Teflon-coated equipment is acceptable for all parameters. Prior to using any equipment, carefully read and follow the manufacturer's instructions. A Kemmerer sampler is shown in Figure 3-9.

3.4. Wastewater Sampling

Sampling wastewaters, both municipal and industrial, are required by Regulatory Agencies for the **National Pollutant Discharge Elimination System (NPDES) permit program**. The permits specify the types and amounts of pollutants that may be discharged. Industries discharging pollutants into waterways or sewage systems must meet pretreatment standards to prevent these systems from potentially toxic pollution sources. The most important goal of collecting wastewater samples for quality analysis is to assure that the quality of the effluent is in compliance with its permit and the quality of the effluent does not endanger the groundwater or surface water standard. The analytical results of these effluents are also helpful when evaluating the success of the wastewater treatment.

Sample Types

Grab Samples

Grab samples represent the conditions that exist at the moment the sample is collected; therefore, the sampling time should not exceed 15 minutes. To obtain a representative sample, samples must be taken where wastewater flow is properly mixed. When taking a grab sample, the container should be inverted and submerged below the waste stream surface and filled.

Composite Samples

Composite samples give information for the average characteristics during the compositing time. The type of compositing method and the sampling point(s) depends on the project plan. The use of an automatic sampler is practical. Detailed sample collection for composite samples is in Chapter 1.

Sampling Locations

Sampling locations should follow the permit requirements and the locations necessary to establish environmental impact. If the sampling locations, described by the permit or the project plan, are not available for taking representative samples, the sampler must determine the best sampling points, but must document in detail the reason for the changes. The following areas are the most common sampling sites:

Effluents — The most representative sample may be taken downstream from the wastewater stream before it enters the disposal site (surface water, wetland, deep-well injection, etc.)

Influents — The best point for sampling is from the turbulent flow where the sample has a good mixing.

Common Parameters in Wastewater Analysis

Physical parameter — pH - Total and Suspended Solids (TS, SS) - Total Volatile Solids (TVS)

Demands — Biochemical Oxygen Demand (BOD_5) - Chemical Oxygen Demand (COD) - Total Organic Carbon (TOC)

Nutrients — Ammonia-Nitrogen (NH_3-N) - Nitrite-Nitrogen (NO_2-N). Nitrate-Nitrogen (NO_3-N) - Total Kjeldahl Nitrogen (TKN). Organic Nitrogen - Total Nitrogen - Total Phosphorus (PO_4-P)

Others — Chlorine Residual (Cl_2) - Oil and Grease - Total Recoverable Petroleum Hydrocarbons (TRPH) - Total Phenols - Cyanide - Dissolved Oxygen (DO) Volatile Organic Compounds (VOCs) - Metals (total, dissolved, filtered)

Microbiological — Fecal coliform (*Escherichia coli*) bacteria count. Sampling for additional parameters is dictated by the permit. Containers, preservation, volume, and storage of the samples are as described in Chapter 1, Table 1-1.

3.5. Agricultural Discharges Sampling

Agricultural discharges can be separated into three types: (1) concentrated animal waste or manure, (2) run-off from an agricultural watershed, and (3) irrigation return flow. The frequency and location, the number of samples, and the required parameters must be followed as given in the discharge permit. If there are no special requirements for parameters, samples must be taken for the following analytes:

- Physical parameters: Temperature - Conductivity - pH - Total Dissolved Solids (TDS) -Suspended Solids (SS) - Total Solids (TS)
- Demands: Chemical Oxygen Demand (COD) - Biochemical Oxygen Demand (BOD_5) - Total Organic Carbon (TOC)
- Minerals: Chloride (Cl) - Sulfate (SO_4) - Calcium (Ca) - Magnesium (Mg) - Carbonate (CO_3) - Bicarbonate (HCO_3) - Sodium (Na) - Potassium (K)
- Nutrients: Ammonia-Nitrogen (NH_3-N) - Nitrite-Nitrogen (NO_2-N) -Nitrate-Nitrogen (NO_3-N) - Total Kjeldahl Nitrogen (TKN) - Organic Nitrogen - Ortho-Phosphate (PO_4-P) - Total Phosphate (PO_4-P)
- Organics: Pesticides and Herbicides
- Microbiological: Total Coliform - Fecal Coliform - Fecal Streptococcus
- Metals: as required

Sampling for additional parameters may be dictated by the permit. Sample containers, preservation, storage, and holding time for these parameters is in Table 1-1.

3.6. Domestic Sludge Sampling

Sampling points, minimum sample numbers, and sampling intervals are specified in the wastewater facility operation permit. All samples for sludge classification shall be representative and taken after final sludge treatment but prior to utilization disposal. Samples must be collected in three different containers. One must have no preservation, one must be preserved with H_2SO_4, and one must contain HNO_3 preservative, according to the requested parameters as described in Table 1-1.

If no organic analytical parameters are required, the preferred containers are plastic; however, if organic constituents have to be analyzed, the sample must be collected in a glass container with a Teflon lined cap.

Analytical Parameters and Usual Reporting Units in Sludge Analysis

Total Nitrogen, as N .... % on dry weight
Total Phosphorus, as P .... % on dry weight
Total Potassium .... % on dry weight
Cadmium, as Cd. .... mg/kg on dry weight
Copper, as Cu .... mg/kg on dry weight
Lead, as Pb .... mg/kg on dry weight
Nickel, Ni .... mg/kg as dry weight
Zinc, as Zn .... mg/kg on dry weight
pH .... pH unit
Total Solids .... %

Analysis of additional parameters may be required based on changes in the quality of wastewater or sludge as a result of a new discharge to the treatment plant, changes in the treatment process, changes in the disposal of the sludge, presence of highly toxic substances or other considerations.

3.7. Soil Sampling

The goal of the soil sampling is to obtain representative portions of the soil. Locations should be selected from areas where natural vegetation is disturbed, or dead, or soil shows foreign staining indicating chemical spills. Together with the affected sample site, samples must also be collected from noncontaminated soil areas, indicated by healthy, undisturbed vegetation. Type of sample containers, preservation technique, and holding time for soil samples are listed in Table 1-1.

Sampling Equipment

The material of the sampling equipment must be selected to not contaminate the analyte group sample collected. The recommended material for organic parameters is stainless steel. Cleaning and decontamination of the equipment must follow the general rules as stated in Chapter 1.

Surface Soil Sampling

Before taking samples, remove leaves, dirt, and grass from the area where soil want to collect, and take the sample by a stainless steel spoon or scoop.

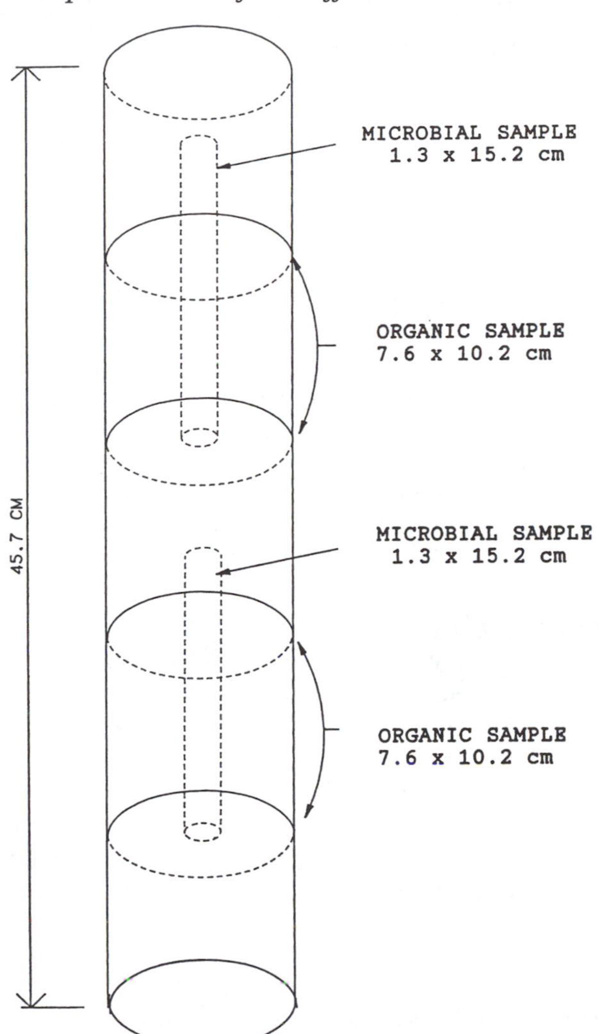

FIGURE 3-10. Typical location of subsamples. Successful sampling of subsurface earth solids required both acquisition of subsurface solids at desired depths. After a subcore for microbial analysis has been removed from core, a 10 cm (4 in) length of core material for chemical analysis is obtained. For organic analysis, the sample is extruded directly into a clean disposable aluminum baking pan and lightly covered with clean aluminum foil. For other analysis, the sample container that is used is dependent on the type of analysis to be performed. Polyethylene bags, which allow the passage of air, but not water vapor, are good sample holders because the samples have access to air and yet are kept from drying.

Shallow Subsurface Soil Sampling

To obtain a shallow subsurface soil sample, dig a hole with a stainless steel shovel to the desired depth or, as an alternative, use a stainless steel bucket auger. Some soil has low cohesion, and the hole may collapse. To avoid this problem, insert a rigid PVC support into the hole. After the sample has been collected, remove the support and the hole filled with the excavated soil. Typical locations of subsamples after sampling by an auger are shown in Figure 3-10.

Deeper Subsurface Soil Sampling

Deeper soil samples are taken from holes greater than 15 feet below land surface. There are different sampling devices available for this type of sampling. For rocks and hard soil surfaces, the head of the sampling device has a small diamond to cut through the hard surfaces as the drill rod is rotated.

General Rules in Soil Sample Collection

- Wear natural latex rubber gloves.
- Select the appropriate sampling device.

- Select sample containers for the required analysis.
- Take the soil sample, and cut portions from the sample.
- For VOC analysis, the sliced portion should be immediately transferred into a 40 ml glass vial without headspace. This can be done by tapping the sample into the vial with a glass rod.
- For other analyses, mix soil well in a stainless steel plate and transfer to the appropriate sample containers with minimal headspace.
- Clean sample containers outside, if necessary. Label the bottles, and complete chain-of-custody form and field notebook notes.
- Place the sample container into a plastic bag and put on wet ice for transportation.
- When samples have been collected from a large area, to reduce the number of samples, composition of soil samples is recommended. Composition of soil samples should be handled in stainless steel or glass containers. The origin and sample size of each subsample has to be documented in the field notebook. Soil samples for VOC analysis are never composited or homogenized to avoid volatilization.

3.8. Collecting Sediment Samples

Sediment samples are usually taken as a part of surface water samples. All surface water samples should be taken prior to the sediment collection. Equipment used for sediment sampling differs according to the sample location, depth of the water, sediment grain size, water velocity, and the type of the analytes.

Scoops

Scoops are generally most useful around the shore line. The scoops may be attached to a pole for taking samples from a boat or from shore.

Corers

Corers are used for softer sediments. They can be constructed from different materials according to the need. The corer is rotated as it is pushed into the sediment. Upon withdrawal from the water surface, the core is extruded out into a pan or tray and homogenized; then the appropriate sample containers are filled up.

Other Equipments

The three pieces of equipment most commonly used are the Peterson, Eckman, and Ponar samplers. Peterson and Ponar are suitable for hard rocky sediments, while Eckman is mostly used for sand, silt, or mud type sediments (see Figure 3-11).

3.9. Hazardous Waste Sampling

Hazardous wastes are waste products of industries that, if not disposed of properly or destroyed, pose a threat to the environment. Even a small leak of hazardous waste can contaminate large quantities of groundwater. For example, a leak of only 15 cups of hazardous liquid per hour can contaminate nearly 3.8 million liters (1 million gallons) of groundwater in a single day. Since most environmental samples taken in hazardous waste fields are heterogenous, it is important that representative sampling be performed. To ensure representative sampling, a sample plan or protocol is required. This plan must ensure that the correct number of samples will be taken at an appropriate frequency and must minimize sample loss or degradation. Hazardous

FIGURE 3-11. Ekman bottom grab sampler.

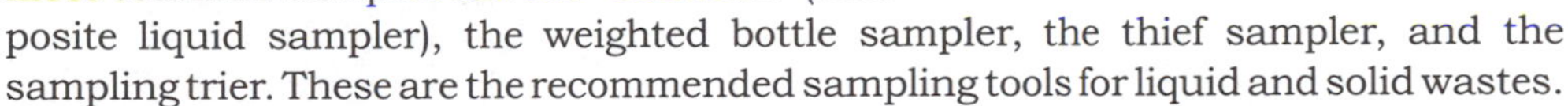

material samples can be gas, liquid, solid, paste, sludge, or a combination; therefore, different methods must be used for sampling. It is very important to consider **phase changes** that may occur in addition to the number of phases found while sampling (volatile gas losses). For example, a sample may contain volatile organic compounds in a solid phase or in the spaces between the solid particles. Once the sample is placed into the container and closed, the volatile organics will move into the gaseous space at the top of the container. When the sample is removed from the container, some volatile will be lost. Reaction of the sample with sunlight or temperature extremes are also important considerations during sampling hazardous wastes.

Sampling Equipment

Equipments used for taking hazardous materials are shown in Figure 3-12 and Figure 3-13. The most common samples are the "Coliwasa" (composite liquid sampler), the weighted bottle sampler, the thief sampler, and the sampling trier. These are the recommended sampling tools for liquid and solid wastes.

Coliwasa

Designed to take free-flowing liquids and slurries from drums, shallow open tanks, pits, etc., the Coliwasa consists of a glass, plastic, or metal tube, equipped with an end-closure, which can be opened or closed. Open, lower into the waste, and let the tube fill. Lock the stopper and withdraw from the waste. Wipe the outside with a disposable cloth. (Figure 3-12).

Trier

Used for sticky solids and loosened soils, the trier is a tube, cut in half lengthwise with a sharpened tip, which allows the sampler to cut the solid sample (Figure 3-13). Insert the clean trier into the waste sample, cut the core, remove with concave side up, and transfer the sample into a container.

Sampling Thief

Useful for any bulk material, but especially useful for grain and grain-like samples, the sampling thief consists of two slotted concentric tubes, usually made from brass or stainless steel (Figure 3-12). Insert the clean closed thief into the material to be sampled. Wiggle the sampler to let the material enter into the thief. Close, withdraw, and remove inner tube. Transfer the sample into sample container.

Auger

Convenient to sampling hard and packed solid wastes and soils, the auger consists of a sharpened spiral blade attached to a hard metal central column. Auger until the desired sampling depth, withdraw the auger, and transfer the sample in a pan.

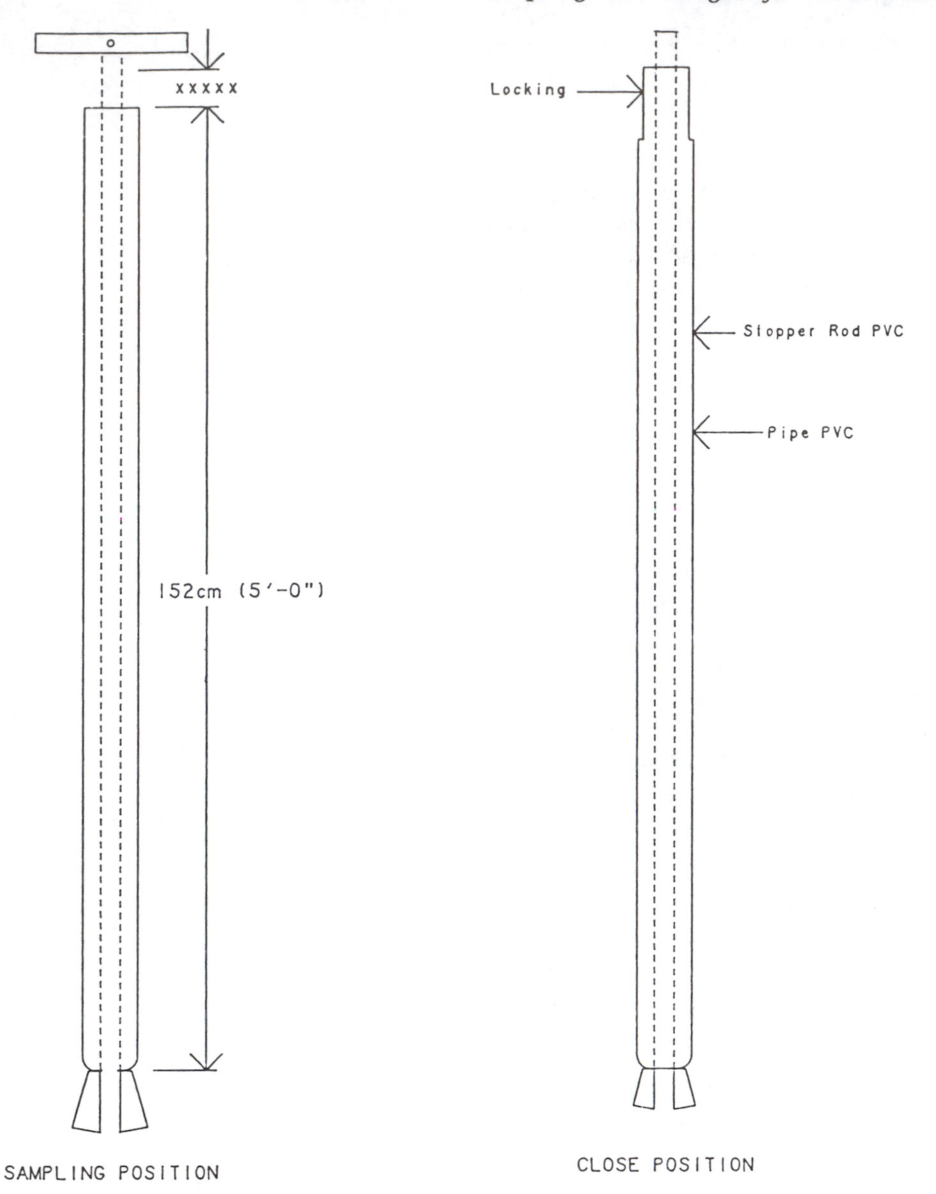

FIGURE 3-12. Composite liquid waste sampler, coliwasa. Sampling free flowing liquids and slurries from drums, shallow open tanks, pits, etc. To be sure, the sampler is clean, and the stopper provides a light closure. Open, lower into the sample, and let fill the tube. Lock the stopper, and withdraw from the sample. Wipe the outside with a disposable cloth.

Weighed Bottle

The weighed bottle samples liquids and free-flowing slurries. It is a glass or plastic bottle with a sinker, stopper, and a line that is used to lower, raise, and open the bottle. Lower the bottle into the sample, open, and allow to fill completely. (It is filled, when it ceases bubbling.) Raise and use the bottle itself as a sample container.

Dipper

The dipper samples liquids and free-flowing slurries. It is simply a beaker on the end of a long pole. Dip the beaker facing in the down position into the sample, turn right

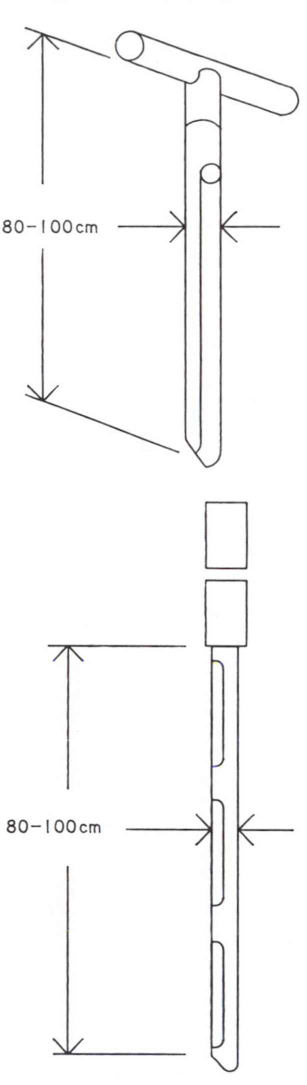

FIGURE 3-13. Sampling trier. For sticky solids and loose soils, it is a tube cut in half lengthwise with a sharpened top, which allows the sampler to cut the solid sample. Insert the clean trier into the sample, cut the core, remove with concave side up, and transfer sample into the container. Thief sampler. For any bulk material, but especially useful for grain-like material. Consists of two slotted concentric tubes, usually made from brass or stainless steel. Insert the clean, closed thief to sample. Wiggle the sampler to let the material enter into the thief. Close, withdraw, and remove inner tube; transfer sample into sample container.

side up to allow it to fill with the sample. Raise and transfer the sample into a sample container.

Scoops and Shovels

For sampling granular or powdered materials, use a scoop or shovel. **Be sure that the material of the sampling equipments and sample containers do not react with the waste sample!** Plastic is never used for organic materials. Metal devices are never used for sampling acidic or basic wastes. Glass is best for avoiding any reaction or contamination, but never used if the sample contains hydrofluoric acid (HF) or Fluorine (F). Teflon is the best material to use for sampling for trace organic analysis.

Sample Containers

The material of the sample container should be chosen so it will not react with the sample. It should be resistant to leakage and breakage and should be the appropriate size to collect your sample. Wide mouth containers with tight, screw-type lids are more desirable for hazardous waste samples. Glass and Teflon containers are recommended for organic analytes, and plastic containers are recommended for inorganic parameters. Samples for VOC analysis must be collected in 40 ml glass vials with Teflon septums. After they are closed and sealed, they must be stored in an inverted position with as little air space as possible.

Random Sampling

To get the most accurate and representative sample, random sampling is recommended. (A random sample is taken by chance.) There are three types of random sampling which may be applied to sampling hazardous materials.

Simple random sampling — When a suitable number of samples are randomly selected.

Stratified random sampling — When strata are identified and simple random sampling is done within each stratum.

Systematic random sampling — When the first sample is randomly selected and subsequent samples are chosen at fixed times or space intervals. For example, a waste site might be divided into 1000 grid squares. The first sampling site randomly selected, square 5. Then every 10th square will be sampled, square 15, square 25, etc. Various sampling strategies must be employed depending on the size and location of the hazardous waste site. The wastes may be in a container, barrel, tank, lagoon, pond, lake, or traveling through a pipe. If the waste is in several containers, every container should be sampled. Containers can also be randomly chosen.

Three dimensional simple random sampling — When a surface is divided into grids, and sampling points are randomly chosen. For example, landfills and lagoons are often sampled by this method.

Two dimensional simple random sampling — When the chosen sampling points on the surface are sampled along the entire length of the vertical column.

Safety in Collecting Hazardous Waste Samples

The person collecting samples must realize that these samples are hazardous materials and must be handled with extreme care. The material can be flammable, toxic, or reactive. There are also physical hazards from heavy equipment used for sampling or waste removal. In many cases, there is information about the wastes and proper safety precautions that should be followed.

In cases where nothing is known about the material, full protection is necessary. Wear protective gloves, face aspirators, and hazard hat!

In the case of flammable and combustible material, an appropriate fire extinguisher should be available, if the site is small in size. If a large amount of flammable material is present, a fire truck should be present.

When sampling flammable materials with high vapor pressure or low flashpoint, all equipment should be grounded, and all sources of ignition should be prohibited.

WHEN HANDLING EXPLOSIVE, VOLATILE, OR TOXIC MATERIALS, THE FIRMS SHOULD NOTIFY THE SAMPLING CREW OF THE DANGER OF THE AREA BEFORE THE SAMPLING. NO SMOKING OR EATING DURING SAMPLING! WHEN SAMPLING HAZARDOUS MATERIAL, SPECIAL SAFETY CLOTHES, SHOES, AND HAT SHOULD BE WORN FOR FULL PROTECTION! AFTER SAMPLING, SAMPLE CLOTHES, SHOES, AND HAT SHOULD BE REMOVED. IN SOME CASES A FULL SHOWER IS NECESSARY (for example sampling asbestos). THE SAMPLER MUST ALWAYS WASH HANDS AND EXPOSED PORTIONS OF THE BODY AFTER SAMPLING.

Transportation and Shipping Hazardous Materials

According to the Department of Transportation (DOT), hazardous wastes should be categorized as follows.

Hazardous material — A substance or material, including a hazardous substance which has been determined to be capable of posing an unreasonable risk to health, safety, and property.

Hazardous substance — A material and its mixtures or solutions that is listed as a hazardous substance and is in a quantity, in one package, which equals or exceeds the reportable quantity.

Hazardous waste — Any material that is subject to the hazardous waste manifest requirement of the EPA specified in the CFR, Title 40, Part 262, or subject to these requirements. (Manifest the attached document to the waste, that gives information as the type, quantity, and chemical concentration of the waste. It also identifies who is handling the waste and who will receive it).

Multiple hazard material — A material has more than one hazard class definition.

Packaging and Labeling of Hazardous Materials

Correct packaging and labeling of hazardous materials are specified and required by the federal law.

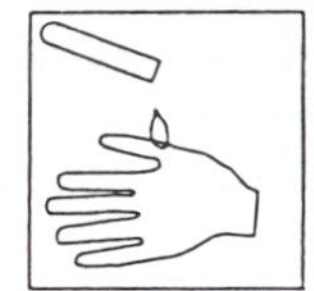

FIGURE 3-14. Safety labels.

Combustive and flammable materials — Flame symbol with a red background.
Oxidizing materials — Flame surrounding an "O" on a yellow background.
Poisons — A skull and cross-bone symbol with a white background.
Explosives and reactants — A blast pattern, or the words "Blasting Agent" with an orange background.
Corrosives — A test tube with either a hand, a brick, or some other object being eaten away on a black and white label, along with the word "corrosive".
Biomedical materials — Infectious substances.
Radioactive materials — Have their own distinctive symbol as well as a variety of different colors.
Hazardous wastes — Yellow background with red lettering and borders with space for specific written information.
Common safety labels are shown in Figure 3-14.

Hazardous Waste Verification

For laboratory identification of hazardous wastes, samples must be collected without preservatives. The identification process is based on the following laboratory tests.

Corrosivity

Corrosivity is measured by the general ability of a substance to erode the surface of a metal disc. It is generally measured by an extremely low or high pH. Criteria of corrosivity is if the pH is <2 or >12 pH unit. Corrosive wastes may destroy containers, contaminate soils and groundwater, or react with other materials by releasing toxic gases. Corrosive materials are dangerous to human tissues and also to aquatic life.

Reactivity

Reactivity is measured by the determination of cyanide (CN) and sulfide (S) content of the waste. Samples should be collected with a minimum of aeration. The sample bottle should be filled completely and stoppered. Analysis should commence as soon as possible. Samples should be kept in a cool, dark place until analysis begins. Any quantity of cyanide and sulfide present designates the waste as hazardous.

Ignitability or Flammability

This is determined by flashpoint measurement. The waste is characterized to be hazardous if the flashpoint is less than 60°C (140°F).

Chromium Hexavalent (Cr^{6+})

The sample should contain no **preservative**, as stated in the general sampling procedures for metal samples. The presence of Cr^{6+} gives the waste the hazardous character.

Toxicity

It is the measure of the effect of the waste materials when they contact water or air, and are leached into the groundwater or disposed of into the environment. The basic concept of the leachate test is that it should simulate the conditions in the field. The two most common leachate tests are the **Extraction Procedure Toxicity Test (EP Toxicity)** and the **Toxicity Characteristic Leachate Procedure (TCLP)**. In the EP Toxicity test, by the end of the 24 hours extraction the extract is filtered through a 0.45 u membrane filter, and the filtrate is analyzed for toxic metals (arsenic, barium, cadmium, chromium, lead, mercury, selenium, and silver), insecticides (endrin, lindane, methoxychlor, toxaphene), and herbicides (2,4-D and 2,4,5-T). The Maximum Contaminant Levels (MCLs) criteria is 100 times higher than drinking water standards, because the generic dilution/attenuation factor is 100. Maximum concentration of contaminants for EP Toxicity is on Table18-13.

The TCLP test has been developed from the EP Toxicity test, by the addition of 38 organic substances. Also, the TCLP test applies compound-specific dilution/attenuation factors rather than the uniformly used 100. EP Toxicity identifies wastes that create hazard to the environment by leach significant concentrations of specific toxic substances. The TCLP test is designed to determine the mobility of the contaminants present in the waste. The analyzed parameter with the regulatory levels in the TCLP extracts are in Table 18-14.

Statistical Approach of Sampling Hazardous Wastes

Calculate the Proper Number of the Samples

To determine the proper number of samples to be taken and the frequency of the sampling, a preliminary determination with a relatively few number of samples should be made. First, obtain a mean $(\overline{x})$ and the variance (s^2) of the chemical contaminant of interest. Next, calculate the appropriate number of samples (N) to be taken. Mean $(\overline{x})$ = sum the results of the measurements divided by the total number of measurements (n).

Calculate the variance (s^2) according to Equation (1):

$$\text{Variance } (s^2) = [Ex^2 - (Ex)^2/n]/n - 1 \qquad (1)$$

a/ sum the values for the measurement (Ex)
b/ squared all the measurements, and sum the squared measurements (Ex^2)
c/ square the sum of the measurements $(Ex)^2$ and divide by the number of the measurements $(Ex)^2/n$
d/ subtract $(Ex)^2/n$ from (Ex^2) and divide by the number of measurements minus 1 (n - 1)

E = sum
n = number of measurement
x = results of the measurements
Ex = sum of the measurements
$(Ex)^2$ = square the sum of the measurements
Ex^2 = sum of the squared measurements

Calculate the number of the samples (N) according to Equation (2)

$$\text{Number of the samples (N)} = (t^2 s^2)/(RV - \bar{x})^2 \quad (2)$$

t = "t" value taken from the Students "t" variate statistical table (Table 9-1)
s^2 = variance
RV = Regulated Value
$\bar{x}$ = mean

For example: Calculate the number of the samples to be taken from a pond probably contaminated by barium (Ba). The regulated value for Ba in hazardous wastes is 100 ppm (mg/L). Preliminary samples were analyzed, and the Ba levels reported as 86, 90, 98, and 104 ppm.

x	x^2
86	7,396
90	8,100
98	9,604
104	10,816
E 378	35,916

n = 4 Ex = 378
n - 1 = 3 Ex^2 = 35,916
$\bar{x}$ = 378/4 = 94.50 s^2 = 65

To calculate the number of samples required to be collected for the next sampling program, calculate s^2 by using Equation (1) and calculate the number of the samples need to collect (N) by using Equation (2). Substitute the values and using the students "t" table with the degree of freedom (df) value of 4 - 1 = 3 (Sample number is 4).

$$N = [(1.638)^2 \, 65]/(100 - 94.5)^2$$

$$N = 174.4/30.25 \qquad N = 5.76$$

Therefore, at least six samples must be taken based on preliminary results. Generally three additional samples are taken in case the preliminary sampling was inaccurate. So, a total number of nine samples are taken for Ba determination.

Tentative Determination of Contamination

The calculated number of samples are collected and analyzed for the contaminant level of the interested parameter. If the mean $(\overline{x})$ of the analytical data is equal or greater than the applicable. Regulated Value (RV), use the Confidence Interval (CI) calculation. If the upper limit of the CI is less than the RV, the sample contains an unhazardous level of the contaminant. If the upper limit of the CI is greater than RV, the sample is tentatively determined to be hazardous.

Calculation of Confidence Intervals (CI)

The CI for the calculated mean depends on the number of measurements (n), the standard deviation (s), and the level of the confidence desired. To calculate the confidence interval, use Equation (3):

$$CI = \overline{x} \pm (ts/\sqrt{n}) \qquad (3)$$

$\overline{x}$ = mean
t = value from the student "t" table, depends on the level of confidence desired and the number of degree of freedom (n - 1)
s = standard deviation
n = number of measurements

Standard deviation is determined by using a calculator or can be converted from the calculated variant (s^2) as described above in the sample number calculation. For example, in the previous sample, the number of calculated and collected sample number was 6. The results of the analyses give the following data: 89, 90, 87, 96, 93, 113 ppm.

x	x^2
89	7,921
90	8,100
87	7,569
96	9,216
93	8,649
113	12,769
E 568	54,224

$\overline{x}$ = 94.66
s^2 = $[54{,}222 - (568^2/6)]/6 - 1 = 90.8$
s = $\sqrt{90.8} = 9.53$
t = 2.571 (for 95% confidence)

$$CI = 94.66 \pm (2.571)(9.53/\sqrt{6})$$
$$CI = 94.66 \pm (2.571)(3.89)$$
$$CI = 94.66 \pm 10.00$$
$$CI = 104.7 - 84.7 \text{ ppm}$$

Upper Confidence Limit (UCL) = 104.7 = 105
Lower Confidence Limit (LCL) = 84.7 = 85

The UCL is greater as RV (100 ppm); therefore, the Ba level indicates a possible hazardous effect. If this tentative conclusion is reached, recalculate the number of samples for a new collection and analysis to prove contamination. The standard deviation should be computed to three significant figures when used in calculations, and rounded to two when reported as a data. The calculated confidence limits are rounded up.

3.10. Fish Tissue Sampling

Fish tissues are usually analyzed for metals and organic pollutants. All equipment used for sampling must be decontaminated properly by washing with laboratory detergent, followed with DI water, isopropyl alcohol, and a final analyte-free water rinse. After drying, wrap the equipment in aluminum foil for storage until use.

Fish may be captured by electroshock, net, hook, etc., and immediately placed in wet ice in a cooler. Weigh the fish and prepare the tissue. The balance pan must be cleaned after each fish. The fillet should not be skinned, because numerous organic compounds are fat soluble (for example, halogenated pesticides) and a layer of fat tissue lies directly under the skin. Duplicate samples from the same region are recommended. All samples taken should be wrapped in clean aluminum foil, preserved by keeping samples on wet ice (maximum 24 hours), and should be frozen for longer storage in the laboratory.

3.11. Collecting Air Samples

The primary concern of the sampler must be directed to the collection of representative samples and the homogeneity of the air mixtures employed to calibrate both the collection and the analytical systems. The source of contaminant, air flow direction, and velocity (whether due to wind or thermal gradients), density of the contaminants, intensity of sunlight, time of day, and presence of obstructions such as trees, buildings, machinery, etc. (which produce turbulence and humidity) influence the concentration of contaminant at any given location.

Sampling and Storage of Particles

There is a great variety of methods available, and the method selected depends on the purpose of the sampling. For example, if information on individual particles is desired, particles should be collected for *microscopic examination* which determine the shape and size of the particles. Particle size distribution can be detected from a larger amount of sample. For *chemical analysis*, a Hi-Vol sampler is employed. It is necessary that all parts of the stream be sampled and properly weighted to be representative of the whole stream. The size and type of the filter paper is usually dictated by the instrument and sample site. Particles can react with filter paper, evaporate, and sublimate, depending on the nature of the sample. The sampler must always consider the method prior to the sampling. For many purposes, particulate samples tend to keep well for a long period of time. Site selection is important in any air sampling, but especially for particles because they are much less uniformly dispersed in the ambient air as well as in process equipment. Particles of all sizes are continually emitted into the atmosphere. The *larger particles* fall rapidly, while the *smaller sizes* fall more slowly. The height of the source, the wind velocity and turbulence, and particle size distribution will determine how fast the particles settle out. The *very small particles (Aitken nuclei)* tend to become attached to larger particles. *Aerosol samples* are the dispersion of any material in the solid or liquid phase in a gas stream or the atmosphere. According to their sizes, particles may be divided into the following groups:

Settleable particles, larger than 30 um diameter
Suspended particles, smaller than 30 um diameter
Condensation or Aitken Nuclei, 0.01 to 0.1 um diameter
Agglomerates, several small particles, attracted by a large particle, or attracted to each other

Fine particles, particles with <2.5 um size
Coarse particles, particles with >2.5 um size

Isokinetic Sampling of Particles

According to the physical law, momentum of the particle is mass × velocity. Different sized particles will be displaced by different amounts. Isokinetic sampling refers to taking a sample under such conditions that there is no change in momentum. This is accomplished by using a thin, walled tube aligned with the stream flow and drawing the sample into it, at the same linear velocity as the stream flow at that point. With this condition, all particle sizes may be collected efficiently. An illustration of the isokinetic sampling is in Figure 3-15.

General Rules for Sampling Particulates

A 24 hour sampling has become a standard practice. The general rules in sampling particulates are:

- Sample should be taken at the point of major interest.
- Sampler should not be placed directly downwind from a major point source.
- Locate the sampler about 1.5 m above ground level.
- Locate downwind from major obstacles a distance of about 10 times their height.

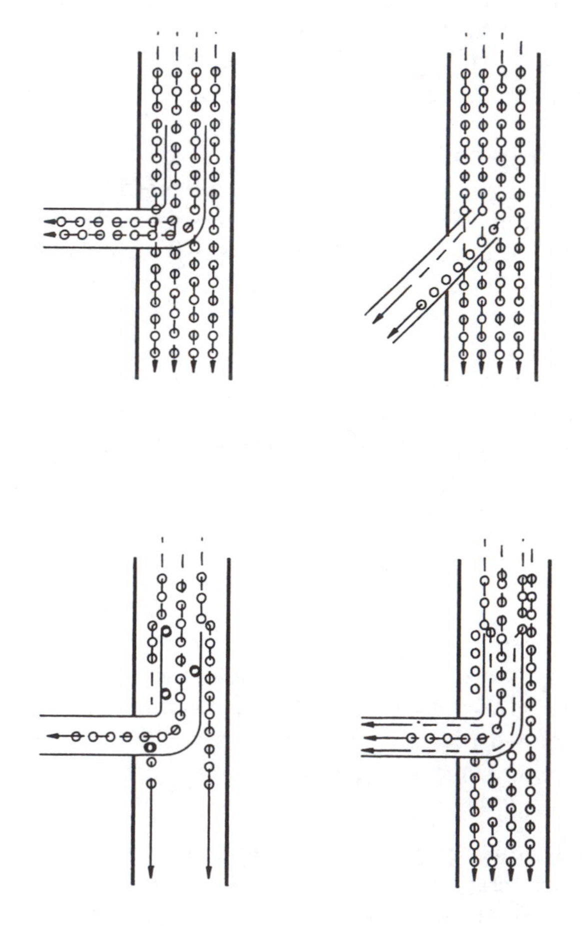

FIGURE 3-15. Schematic illustration of isokinetic and non-isokinetic air supply.

- Take several samples at different locations in the area of interest.
- Sample the time of day of greatest interest, or make 24 hour sampling.

The objective is to collect a sample that is representative of the material that is emitted. The specific points of sampling are generally determined by discussion with plant engineers or others who understand the process or the source of emission. A site visit is generally required for final selection. Particulate sampling should be carried out with probes inserted in the duct at each end of the points of flow measurement. Care must be exercised to be sure that particles do not settle out in the duct. If the temperature is too hot, particles may be vaporized, and if the temperature is too low, water or other vapors form mists that will collect with the solids and plug up the filter, giving bad results. For each sample, the following data must be attached: date and time of collection, sample location, sample flow rate, sample pressure, sample temperature, dew point, plant operating condition, sampler's name, date, and time of sampling.

Sampling and Storage of Gases and Vapors

Gases and vapors follow the normal laws of diffusion and mix freely with the surrounding atmosphere. The appropriate sampling procedures should be selected according to the parameters of interest.

Sampling Equipments

Gas pumps — They have adequate capacity and perform uniformly. For sampling reactive gases, such as ozone, or trace organics, Teflon lined pumps are essential.

Gas metering device — Measure the total volume of air sampled. For this purpose, two types of meters are available, volume and rate meters. *Volume meters* record the total volume of sample that is passed through the sampling train. *Rate meters* indicate the flow rate. This rate multiplied by the sampling time gives the total volume sampled. *Sample probes* must be constructed of inert materials such as glass and Teflon, to avoid loss of gas and vapor samples.

Sampling Techniques, Without Concentration of Gases and Vapors

Sampling techniques are given for low boiling point compounds, such as methane, ethane, propane, and fixed gases as N_2, O_2, and CO.

Plastic bags — Different trade marks are available, such as Scotch pack, Aluminized Scotch pack, Mylar, Saran, and Tedlar. Bags can be reused after purging with clear air and checking for any gaseous residues.

Glass containers — Excellent for inert gases such as O_2, N_2, methane (CH_4), CO, CO_2. They are not recommended for reactive gases such as H_2S, nitrogen oxides, and SO_3. The pump is attached to one end of the container and a sampling probe to the other end. The pump is activated and approximately 10 times the sampling tube volume is passed through the system.

Metal containers — Stainless steel containers are good for collecting inert gases, but are not suitable for reactive gases. Size is from 1 to 34 liters.

Sampling Techniques that Extract and Concentrate Gases and Vapors

Wet Collection System

It is composed from bubblers or bubblers with diffusers. The *impinger* contains the absorbent, sometimes called *scrubbing solution.* The quality, concentration, and quantity of the absorbent depends on method used (Figure 3-16).

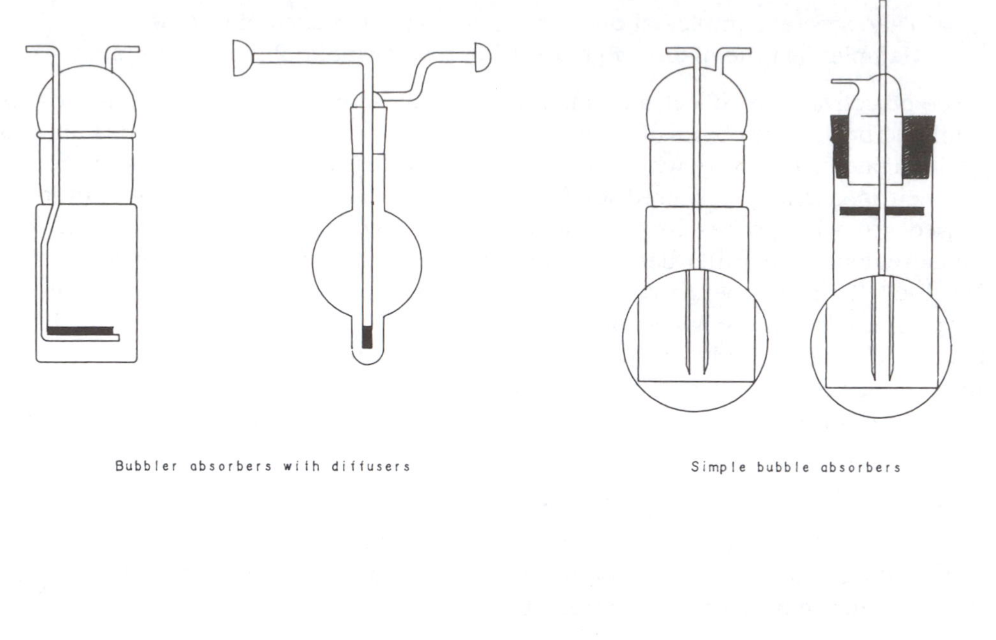

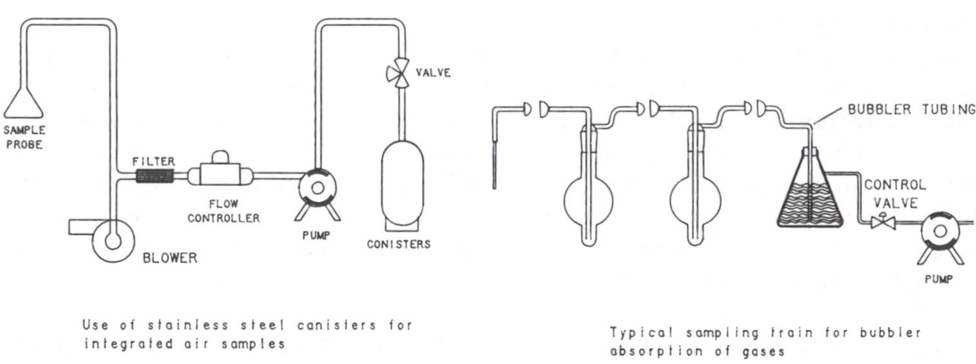

FIGURE 3-16. Wet collection systems in air sampling.

Dry Collection System

It is based on the practical value of porous solids, as adsorbents. Activated carbon, silica gel, activated alumina, and various active earths used as adsorbents. Synthetic polymers have also been used for adsorption of gases, particularly for volatile organic gases.

Most widely used materials are silica gel, charcoal, and tenax. After the gases are collected, the adsorbent tubes should be sealed immediately and properly stored. Storage in a sealed plastic or glass vial is most satisfactory. The adsorbent gases must be solvent extracted and analyzed by gas chromatographic (GC) or Infrared Spectrophotometric (IR) methods. Solvents used for extraction are usually polar substances, such as alcohols, dimethyl sulfoxide, or water in combination with carbon disulfide. Activated charcoal is the most commonly used. Tenax, the polymer material, is used in parts per billion (ppb) or lower levels of volatile compounds. In this case, the gases are not solvent extracted, but are thermally desorbed, by heating to 240°C for 10 to 15 seconds while purging with helium gas, and introduced to the GC as a single injection. (Permeation tube methods are shown in Figure 3-17).

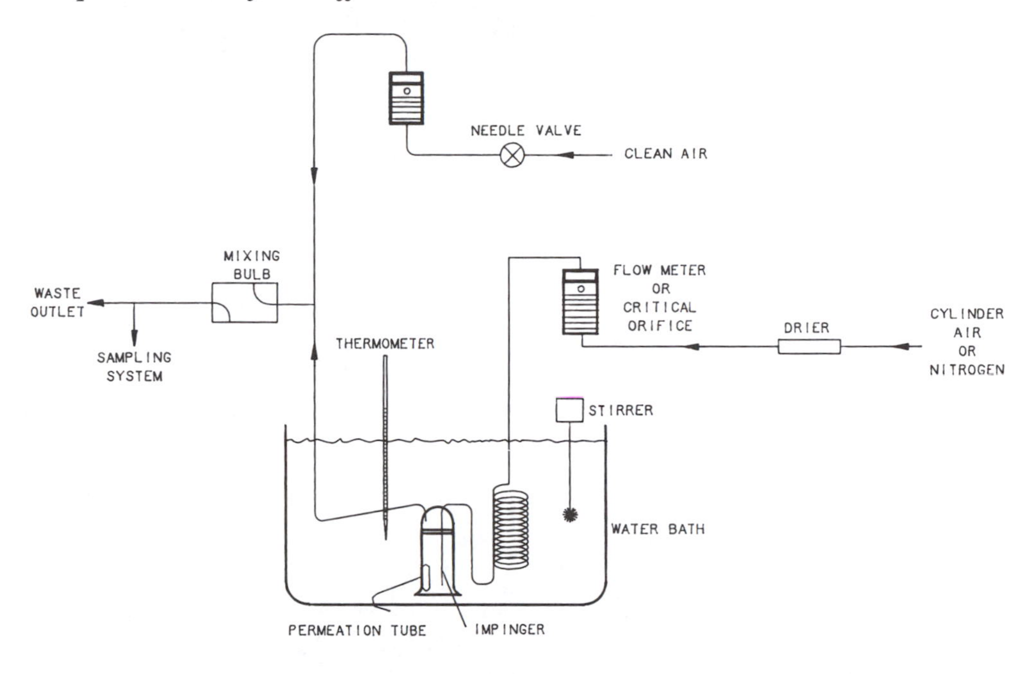

–Gas dilution system for preparation of standard concentrations of sulfur dioxide for field use by the permeation tube method.

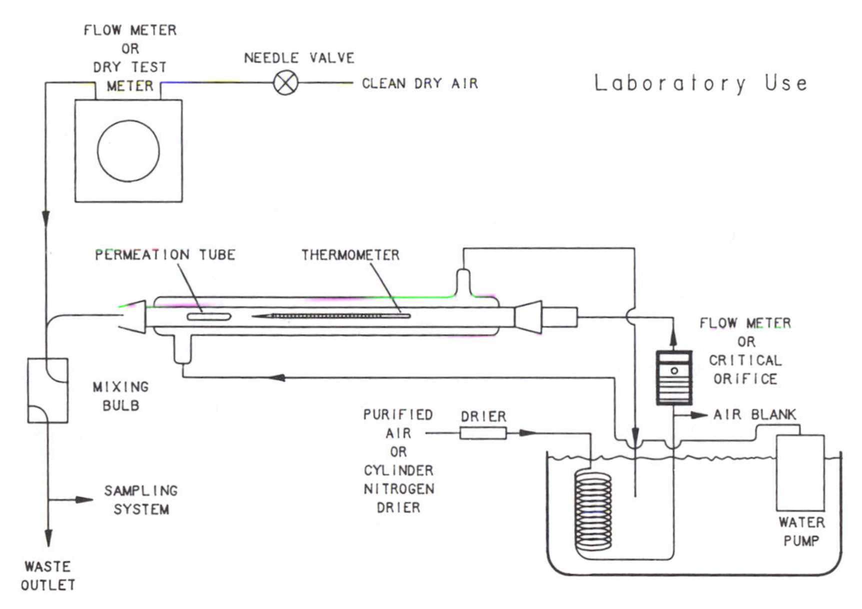

–Gas dilution system for preparation of standard concentrations of sulfur dioxide for laboratory use by the permeation tube method.

FIGURE 3-17. Permeation tube methods.

Part 2

Quality Assurance and Quality Control

Chapter 4

Introduction to Quality Assurance and Quality Control (QA/QC)

For collecting and analyzing environmental samples, it is necessary to use EPA approved methods, but that does not guaranteed the correctness of the analytical values. Matrix interferences, failure of equipments, and errors by the analysts may cause inaccuracies. Therefore, quality control procedures are necessary to approve that the obtained analytical data and the reported values are correct. The validation and approval of analytical data is based on a well-designed and regularly applied Quality Assurance/Quality Control (QA/QC) program. This program supports the criteria that "data must be technically proven and legally defendable".

This chapter defines the QC procedures and components that are mandatory in the performance of analyses, and indicates the QC information that must be generated with the analytical data. The components of a QA/QC program can be classified as management (Quality Assurance, QA) and as functional (Quality Control, QC). The differentiation of such programs are the following:

Quality Assurance (QA) — Quality assurance is a definite plan for laboratory operation that specifies standard procedures that help to produce data with defensible quality, and reported results with a high level of confidence. QA includes quality control and quality assessment activities.

Quality Control (QC) — Quality control is a set of measures within a sample analysis methodology to assure that the process is in control.

Quality Assessment — Quality assessment is a process to determine the quality of the laboratory measurements through internal and external QC evaluations. It includes performance evaluation samples, laboratory intercomparison samples and performance audits.

4.1. Quality Assurance Programs

QA is a necessary part of data production, and it serves as a guide for the operation of the laboratory for production quality data. Each laboratory involved in environmental analytical processes have to develop their own QA/QC program. The program should be delineated in a QA/QC manual, and should be sufficiently comprehensive to apply to most of the operations. The laboratory should have only one comprehensive QA program plan, but would have a QA project plan for each of the projects. These programs should be strictly enforced, but be flexible to deviate occasionally as necessary. The comprehensive QA/QC (CompQA/QC) program should be continually reviewed and updated as needed.

The following are items that must be included in a CompQA/QC plan.

- Introduction to laboratory organization
- QC targets for precision and accuracy
- Sampling procedures
- Sample custody
- Analytical methodology
- Calibration procedures and frequency
- Preventive maintenance schedule and frequency
- QC activities
- Specific routine procedure used to assess data precision and accuracy
- Data reduction, validation, and reporting
- Performance and system audit
- QA reports

There are several items that should be included in a QA/QC project plan. QA/QC project plan should be prepared and approved in advance. It should address the following points:

- Project description
- Project organization and responsibilities
- QA objectives for measurement data in terms of precision and accuracy
- Sampling procedures
- Sample custody
- Calibration procedures and frequency
- Analytical procedures
- Data reduction, validation, reporting
- Internal QC checks and frequency
- Performance and system audit and frequency
- Specific routine procedures used to assess data precision and accuracy
- Corrective actions
- QA report and management

The goal of this part is to introduce and explain these requirements according to the request of EPA and DEP. Environmental technicians and analysts should have a knowledge of the methods and instruments used, and understand the principles of the analysis. Above all, analytical chemists should recognize and solve the problems during the analysis. It has been said that if you can state a problem clearly, it can be solved. The knowledge of the QA/QC program helps the analysts to be a responsible participant of the laboratory system, one who produces defendable, precise, and accurate analytical data.

4.2. Responsibilities of the QA/QC Officer

The responsibility of ensuring that the QA/QC measures are properly employed must be assigned to a knowledgeable person who is not directly involved in the sampling and analysis. Responsibilities of such an officer may range from periodic reviewing of the program to conducting system audits and internal performance audits on a routine and continuing basis. The basic duties are to conduct the QA program and taking and recommending corrective actions, when required. The following are also included.

- Develops and carries out QC programs, including statistical procedures and techniques.
- Monitors QA activities to determine its cooperation.
- Conducts system audits, and makes recommendations for corrective actions if necessary.
- Advises management by reviewing technology, methods, equipment, facilities with aspect of QA.
- Coordinates all of the calibration and performance checks of the instruments.
- Evaluates data quality, checks records, control charts, calibrations, and other documentation.
- Conducts investigations related to quality problems.
- Advises and trains staff in QA matters.
- Writes QA reports.

The quality assurance officer should be independent of daily operation and report directly to the management.

4.3. Duties of the Laboratory Supervisor

The duties of the laboratory supervisor are as follows.

- Direct supervision of the QA/QC program
- Train technical staff for QA/QC activities
- Identification of inaccurate data
- Advises and conducts corrective actions

4.4. Duties of the Analyst

Analysts must be well-trained for both technical and QA/QC responsibilities. By doing their job correctly, they have the first possibility to detect default data and work on the correction. They are responsible for:

- all of the reported data
- the overall analytical operation, including complete QA/QC requirements

Chapter 5
Sample Custody

Sample custody is the process of protecting the samples collected and analyzed by the laboratory. All the paperwork involved in the sample custody process defends and secures the quality of the reported data. It shows how samples are collected, preserved, stored, transported to the laboratory, treated, numbered, and tracked during the analytical processes. Sample custody documentation has to involve all of the records that have been composed during the sample collection and through the laboratory analytical processes. All records have to be maintained to be easy to find and ready at all times for immediate inspection. They must be clear and easy to understand without interpretation or explanation. They should be easy to follow and should cover all of the historical data for the sample from the collection until reporting of the data.

All documentation must be signed or initialed by the responsible person, and recorded with waterproof ink. Never use any erasures or markings. All corrections must be made with one line marked through the error, accompanied with a signature or initial, date and the corrected form. The procedures used must be reported, as well as where the sample was taken (i.e., found in the field or laboratory SOPs). The method numbers and references must also be given. Copies and originals of all documentation for sample collection, sample preparation, and sample analysis must be kept in order.

Documented information related to the sample include the following:

- Sample collection, field activities
- Sample receipt
- Sample distribution
- Sample preparations prior analysis
- Analytical methods
- Reagents and standards preparation
- Calibration procedures and frequency
- Analytical data and calculations
- Detection limits
- Data validation and reduction
- Data reporting

According to these requirements, sample custody is divided into field and laboratory custody.

5.1. Field Custody

Detailed discussion of these requirements are in Part I.

- Preparation for field work
- Collect and preserve samples properly
- Record data in field notebook
- Label samples, complete chain-of-custody and field records

5.2. Laboratory Custody

Sample Bottles and Sample Identification Checklist

After samples arrive in the laboratory, the sample custodian department checks the samples and their identifications. Analyses will be performed only on correctly collected and preserved samples. The check list for the sample custodian is as follows:

- Check sample bottles for leakage or breakage. If any are found, samples in the same ice-box are neglected or marked as "suspect for contamination".
- Check the sample bottle labels for any damage or mistakes.
- Compare field ID numbers on sample labels with the ID numbers of the field records. Arrange samples according to the field ID numbers.
- Check the field sample log for each sample and count the number of the sample bottles for each Field Identification Number (FID) and for each sample site.
- Check for correct temperature preservation. Only "wet ice" preservation is accepted! Otherwise, samples should be discarded.
- Check for correct pH preservation and the accompanying records.
- If any sample shows incorrect identification, reject it with a written statement, which is signed by the sampler and the sample custodian and dated accordingly.

Sample Log-In

If all of the samples and documentation are checked and accepted, samples have to be logged into a laboratory log book.(Table 5-1). Each sample receives a Laboratory Identification Number (LID) given to the sample label. It is very helpful if the ID number is written on a colored label that is applied to the sample container. The colored labels provide easy identification of preservation. For example: red label — HNO_3 preserved; yellow label — H_2SO_4 preserved; blue label — NaOH used for pH adjustment; white label — no preservation, etc. Numbering of the log book starts with one and the year, for example 1-93, and closes with sample numbers on December 30 of the year. Each laboratory has its own computer system, so sample logs are computerized.

Holding Time and Storage of the Samples

An important part of the laboratory custodian's work is the proper storage of the samples and the documentation of the holding time. Each parameter or parameter group has a specified holding time and storage requirements, as monitored on Table 1-1.

Samples should be stored separately from all standards, reagents, cleaning supplies, food, etc. VOC samples must be stored separately from all other samples.

The person with this specific duty is responsible for the proper storage for both the original and the prepared samples, and record on the sample holding time log form on Table 5-2.

Sample Preparation for Analysis

Sample pretreatment prior to the actual analysis, such as digestion, extraction, distillation, filtration, etc. has to be documented on a sample preparation log form, as on Table 5-3. Standard Operation Procedures for sample preparation should be available, with method numbers and references and should be followed as described.

All numerical data relating to the preparation processes and that need further calculations should be briefly documented. (Volume or weight of the samples used for preparation, dilution factors, concentrations, reagents preparations, pH checks, etc.) Each container carrying the pretreated sample should be properly identified with sample ID, date of preparation, and all the pertinent information (volume, weight, dilution, concentration, etc.) related to the preparation procedure.

Documentation During Analytical Processes

All records and information must be documented during sample analyses. These records include:

- Sample identification
- Date of the analysis
- Method number for the analytical method used
- Generated raw data
- Calculations
- Data validation and reduction
- Corrected, reportable data
- Electronic data documentation
- Preparation of reagents, standards
- Analytical calibration and standardization and frequency
- Determination of Method Detection Limits
- QC check samples preparation, QC requirements, and QC routine checks related to the analysis

Disposal of the Samples, Extracts, Digestates, and Analytical Endproducts

Samples and pretreated samples as extracts and digestates should be stored properly until the end of the effective holding time. Refrigerators and separate storage areas have to be designated for this purpose. Regular water samples may be disposed of into the sewer system, with the exception of hazardous wastes. Hazardous laboratory wastes are stored in specified containers until collected and transported by a professional waste disposal company. Such vessels should be marked clearly according to the nature of the waste. For example: " Acid wastes", "Organic solvents", "Cyanide containing wastes", "Mercury wastes", etc. All waste from a microbiological laboratory is collected in special "hazard" orange plastic bags and, prior to the disposal into a regular garbage bag, should be autoclaved at 15psi and 120°C for one hour. Disposal of the samples and treated products should be documented as on Table 5-4 and Table 5-5.

Table 5-1. Laboratory Sample Log Book Page

Date Collected	Date Received	Lab I.D. Number	Field I.D.	Sample Source	Sample Location	Number of Bottles	Analysis Required	Initial of Logger

Table 5-2. Sample Holding Time Log

			Holding Time (days)			Date of Preparation				Storage	
Sample ID	**Matrix**	**Analysis Required**	**prep**	**anal**	**dispo**	**rec.**	**prep**	**anal**	**dispo**	**Sample Prepared**	**Sign**

Sample ID = Sample Identification Number
Analysis Required = Analysis required
prep. = Prepared
anal. = Analysis
dispo. = Disposal

rec. = Received
Sign. = Signature of Logger

Holding Time Explanation:

prep. = number of days between the date sample received and the date sample prepared
anal. = number of days between the date sample prepared and the date of actual analysis
dispo. = number of days between the date sample received and the date sample disposed

Storage Designations:

R.T. = Room Temperature in designated area
Ref. O. = Refrigerator, designated for organic samples
Ref. I. = Refrigerator, designated for inorganic samples
Fr. = Freezer, designated for special samples

Table 5-3. Sample Preparation Log Sheet (Per Analyte Group)

Sample I.D. No.	Matrix	Test for	Method No. Analysis	Method No. Prep.	Date Sample Rec'd	Date Sample Prep.	Sample Size		Final Volume ml.	Signature
							ml.	g		

Method References:

Table 5-4. Sample Disposal Log

Samp ID	*Matrix*	*Analysis required*	*Preser vation*	*Holding time*	*Date rec.*	*Date disp*	*Mode of dispo*	*Sign*

Sample ID = Sample Identification Number
Analysis requ. = Analysis required
Date rec. = Date sample received
Date disp. = Date sample disposed
Mode of dispo = Mode of disposal
Sign. = Signature of logger

Mode of sample disposal :

Samples flush to sewer system, exception of hazardous wastes

Table 5-5. Disposal Log Form for Digestates and Extracts

Sample ID	*Date sample rec.*	*prep.*	*Mode of prep*	*dispo*	*Date of dispo*	*Sign*

Sample ID = Sample Identification Number
rec. = received
prep. = prepared
dispo. = disposed
Sign. = Signature of logger

Mode of preparation :

D = digested **E** = extracted **Dist** = distilled

Mode of disposal :

		Storage:
Ac.W.	= Acid Waste container	designated area
B.W.	= Basic Waste container	designated area
Org.S.	= Organic solvent container	"Flammable" cabinet
Hg.W.	= Mercury Waste container	designated area
CN	= Cyanide Waste container	designated area

Chapter 6
Approved Analytical Methods and References

All analyses must be performed by EPA or DEP approved methods. These methods are specified according to the sample matrices. If there is no EPA or DEP method for a specified component, it is possible to use other sources to establish the method. Sources of such methods are:

Standard Methods for the Examination of Water and Wastewater, 17th Ed., 1989. Prepared and published jointly by American Public Health Organization (APHA), American Water Works Association (AWWA), and Water Pollution Control Federation (WPCF)

Annual Book of ASTM Standards, American Society for Testing and Materials (ASTM), 1989, Vol 11.01 and 11.02 (Water I and II).

Methods for Determination of Inorganic Substances in Water and Fluvial Sediments, U.S. Geological Survey (USGS) 1979, Book 5, Chapter A1.

Official Methods of Analysis of the Association of Official Analytical Chemists (AOAC), 2nd Ed., 1989.

Before applying any method that is different from the approved method, it should be reviewed and accepted by DEP. Permission, issued by the regulatory agency, should be available for acception. When any modification of a method has been suggested, the modification also needs the approval of DEP prior to use.

The recommended methods are listed by sample matrices. The groups for environmental sample matrices are identified as:

- Drinking water
- Surface water
- Groundwater
- Wastewater effluents (domestic and industrial)
- Soils and sediments
- Domestic and industrial sludges
- Solid and hazardous wastes

6.1. EPA Approved Methods and References for Analyzing Water Samples

Drinking Water

Methods for Chemical Analysis of Water and Wastes, EPA 600/4-79-020, Revised March 1983

Methods for the Determination of Organic Compounds in Drinking Water, EPA 600/4-88-039, December 1988

Standard Methods for the Examination of Water and Wastewater, 17th Ed., 1989. (APHA-AWWA-WPCF)

Manual for Certification of Laboratories Analyzing Drinking Water, EPA 570/9-90/008, April 1990

CFR Part 141, Subpart C and Subpart E

Monitoring and Analytical Requirements

Note: "500" series methods shall be used for drinking water related analyses unless approved by DEP for use in other matrices.

Surface Water, Groundwater, and Wastewater Effluents

Methods for Chemical Analysis of Water and Wastes, EPA 600/4/79/020 Revised, March 1983.

40 CFR Part 136, Tables IA, IB, IC, ID, and IE, July 1989.

Test Methods for Evaluating Solid Waste, Physical Chemical Methods, EPA SW-846, 3rd Ed., 1986, and its Revision I, dated, December 1987.

Water Sources (Surface and Groundwater) Analyzed Pursuant to 40 CFR Part 261 (Resource Conservation and Recovery Act, RCRA) and the DER Chapter 17-700 Series

Test Methods for Evaluating Solid Waste, Physical and Chemical Methods, EPA SW-846, 3rd Ed., 1986, and its Revision I, 1987.

Methods listed in 40 CFR Part 261, Appendix III, 1989.

USEPA Contract Laboratory Program Statement of Work for Inorganic Analyses, EPA SOW #7/88, July 1988.

USEPA Contract Laboratory Program Statement of Work for Organic Analyses, EPA SOW 2/88, February 1988.

EPA Approved Methods for Biological Analyses of Water Samples

Microbiological Tests

40 CFR Part 141, Subpart C (July 1989) Monitoring and Analytical Requirements

40 CFR Part 136, Table IA (July 1989).

Microbiological Methods for Monitoring the Environment, EPA 600/8-78-017, 1987.

Bioassays

Methods for Measuring the Acute Toxicity of Effluents to Freshwater and Marine Organisms, EPA 600/4-85-013, 1985, 3rd Edition.

Short Term Methods for Estimating the Chronic Toxicity of Effluents and Receiving Waters to Freshwater Organisms, EPA 600/4-89-001, 2nd Ed.

Short Term Methods for Estimating the Chronic Toxicity of Effluents and Receiving Waters to Marine and Estuarine Organisms, EPA 600/4-87-028, 1988.

6.2. EPA Approved Methods and References for Analyzing Soils, Sediments, and Residuals

Soils and Sediments

Test Methods for Evaluating Solid Waste, Physical and Chemical Methods, EPA SW-846 3rd ed., 1986; Revision, December 1987.

Procedures for Handling and Chemical Analysis of Sediments and Water Samples, EPA/Corps of Engineers CE-81-1, 1981.

USEPA Contract Laboratory Program Statement of Work for Inorganic Analysis, EPA SOW #7/88, July 1988.

USEPA Contract Laboratory Program Statement of Work for Organic *Analysis*, EPA SOW #2/88, February 1988.

Domestic and Industrial Sludges

Test Methods for Evaluating Solid Waste, Physical and Chemical Methods, EPA SW-846 3rd ed., 1986; Revision, December 1987.

USEPA Contract Laboratory Program Statement of Work for Inorganic Analysis, EPA SOW #7/88, July 1988.

USEPA Contract Laboratory Program Statement of Work for Organic Analysis, EPA SOW #2/88, February 1988.

POTW Sludge Sampling and Analysis Guidance Document, EPA Permits Division, August 1989.

Solid and Hazardous Wastes

Test Methods for Evaluating Solid Waste, Physical and Chemical Methods, EPA SW-846 3rd ed., 1986; Revision, December 1987.

40 CFR Part 261, Appendix III, July 1989.

USEPA Contract Laboratory Program Statement of Work for Inorganic Analysis, EPA SOW #7/88, July 1988.

USEPA Contract Laboratory Program Statement of Work for Organic Analysis, EPA SOW # 2/88, February 1988.

6.3. DEP Approved Methods and References for Analyzing Water and Groundwater Sources

Bromide SM 4500-Br^- B. (Phenol Red Colorimetric Method), Standard Methods, APHA-AWWA-WPCF 17th Ed., 1989.

Bromates EPA-SOP.

Determination of Inorganic Disinfection Byproducts by Ion Chromatography, Jack D. Pfaff and Carol A. Brockoff, U.S EPA Cincinnati, OH.

Chlorophyll SM 10200 H

Standard Methods, APHA-AWWA-WPCF 17th Ed., 1989 Spectrophotometric method.

Corrosivity, $CaCO_3$ stability, Langelier's index SM 203 Calcium Carbonate Saturation, Standard Methods, APHA-AWWA-WPCF 16th Ed., 1985.

ASTM D513-82, Annual Book of ASTM Standards, 1989.

Ethylene Dibromide, EDB in Groundwater EPA 601 Modified, EPA 504

Using one Electrone Captured Detector instead of Electrolytic Conductivity Detector

HRS (Florida Health and Rehabilitative Services) "Method for the Analysis of EDB" EPA 8011 EPA SW-486, 1986, Revised 1987.

Odor SM 2150 B Threshold Odor Test

Standard Methods, APHA-AWWA-WPCF, 17th Edition 1989.

Salinity SM 2520

Standard Methods 17th Edition, 1989.

Un-Ionized Ammonia DER -SOP

DER Central Analytical Laboratory, Tallahassee, FL Revision No.1, October 3, 1983. (Florida DER, QA Section)

6.4. DEP Approved Methods and References for Analyzing Solid Matrices

Total Recoverable Petroleum Hydrocarbons SW-846, 9073 Draft

Total Halides, TOX EPA SW 846, 5050/9056 Draft
EPA SW 846, 5050/9252 Draft
EPA SW 846, 5050/9253 Draft

Draft methods are available from Florida DER QA Section, Tallahassee.

Estuarine Sample Preparation and Analysis

"Deepwater Ports Maintenance Dredging and Disposal Manual" DER Coastal Zone Management, Revision 4, 1984.

6.5. Approved Modifications of EPA Methods

EPA Method 300

This method may be used for the analysis of specified ions in groundwater and surface water, except fluoride.

EPA Methods 601, 602 , 624, and 625

Capillary columns may be used instead of the specified packed columns if the laboratory meets the accuracy and precision criteria and detection limit with this modification.

EPA Methods 601 and 602

The photoionization detector and electrolytic conductivity detector may be used in a series if the laboratory can meet the performance criteria of the modified method.

EPA Methods 602, 8020, 8021

May include the analysis of xylene and methyl-tert-buthyl-ether (MTBT).

EPA Methods 610, 625, 8100, 8310, 8250, 8270

May include the analysis for methylnaphtalenes.

EPA Method 5030/8010

Must be modified to analyze EDB in soils. An electron captured detector instead of an electrolytic conductivity detector must be used. Tables 6-1, 6-2, 6-3, 6-4, and 6-5 are summaries of the most common methods and references both for aqueous and solid matrices.

Table 6-1. Methods for Determination of Physical Properties and Inorganic Nonmetallic Constituents

Parameter	Method	Method no.	Ref.
Color	Colorimetric, Pt-Co	110.2.	R-1
Conductance	Specific conductance	120.1.	R-1
Odor	Threshold odor	140.1.	R-1
pH	Potentiometric	150.1.	R-1
Total dissolved solids, TDS	Gravimetric, 180°C	160.1.	R-1
Suspended solid, SS	Gravimetric, 103°C	160.2.	R-1
Total solids, TS	Gravimetric, 103°C	160.3.	R-1
Volatile solid, VS	Gravimetric, 550°C	160.4.	R-1
Temperature	Thermometric	170.1.	R-1
Turbidity	Nephelometric	180.1.	R-1
Acidity, as $CaCO_3$	Titrimetric	305.1.	R-1
Alkalinity, as $CaCO_3$	Titrimetric	310.1.	R-1
Carbon dioxide	Titrimetric	4500-Co_2C	R-2
Carbon dioxide	Nomographic	4500-Co_2B	R-2
Chloride	Mercuric nitrate	325.3.	R-1
Chlorine residual	Iodometric method I	330.3.	R-1
Cyanide, amenable to chlorination	Colorimetric after distillation	335.1.	R-1
Cyanide total	Colorimetric after distillation	335.2.	R-1
Fluoride	Ion selective electrode (ISE)	340.2.	R-1
Hardness, total as $CaCO_3$	Titrimetric, EDTA	130.2.	R-1
Nitrogen-Ammonia	Ion selective electrode (ISE)	350.3.	R-1
Nitrogen-Ammonia	Nesslerization	4500-NH_3C	R-2
Nitrogen-Nitrite	Colorimetric	354.14.	R-1
Nitrogen-Nitrate	Ion selective electrode (ISE)	4500-NO_3D	R-2
Nitrogen-Nitrate	Cadmium reduction	4500-NO_3E	R-2
Nitrogen-Organic	Macro-Kjeldahl	351.4.	R-1
Oxygen dissolved	Membrane electrode	360.1.	R-1
Oxygen dissolved	Iodometric	4500-O-B	R-2
Phosphorus	Ascorbic acid	365.2.	R-1
Silica	Molybdosilicate	370.1.	R-1
Sulfate	Turbidimetric	375.4.	R-1
Sulfide	Iodometric	376.1.	R-1
Sulfite	Iodometric	377.1.	R-1

R-1 = Methods for Chemical Analysis of Water and Wastes, *EPA-600/4-79-020, Revised March 1983.*
R-2 = Standard Methods for the Examination of Water and Wastewater, *AWWA, 17th Ed., 1989.*

Table 6-2. Methods for Determination of Metals

Parameter	FL	GR	Other	Method no.	Ref.	Method no.	Ref.
Aluminum	+	+	—	202.1&2	R-1	7020	R-3
Antimony	+	+	—	204.1&2	R-1	7040	R-3
Arsenic	–	+	—	206.2.	R-1	7060	R-3
Barium	+	+	—	208.1&2	R-1	7080	R-3
Beryllium	+	+	—	210.1&2	R-1	7090	R-3
Boron	–	–	Curcumin	4500-BB	R-2	—	—
Boron	–	–	Carmine	4500-BC	R-2	—	—
Cadmium	+	+	—	213.1&2	R-1	7130	R-3
Calcium	+	–	—	215.1.	R-1	7140	R-3
Calcium	–	–	EDTA titrimetric	215.2.	R-1	—	—
Chromium	+	+	—	218.1&2	R-1	7190	R-3
$Chromium^{6+}$	–	–	Colorimetric	3500CrD	R-2	7196	R-3
Cobalt	+	+	—	219.1&2	R-1	7200	R-3
Copper	+	+	—	220.1&2	R-1	7210	R-3
Iron	+	+	—	236.1&2	R-1	7380	R-3
Lead	+	+	—	239.1&2	R-1	7420	R-3
Magnesium	+	+	—	242.1&2	R-1	7450	R-3
Manganese	+	+	—	243.1&2	R-1	7460	R-3
Mercury	–	–	Cold vapor	245.1.	R-1	7470, 7471	R-3
Molybdenum	+	+	—	246.1&2	R-1	7480	R-3
Nickel	+	+	—	249.1&2	R-1	7520	R-3
Potassium	+	–	—	258.1.	R-1	7610	R-3
Selenium	–	+	—	270.2.	R-1	7740	R-3
Silver	+	+	—	272.1&2	R-1	7760	R-3
Sodium	+	–	—	273.1.	R-1	7770	R-3
Thallium	+	+	—	279.1&2	R-1	7840	R-3
Tin	+	+	—	282.1&2	R-1	7870	R-3
Titanium	+	+	—	283.1&2	R-1	—	—
Vanadium	+	+	—	286.1&2	R-1	7910	R-3
Zinc	+	+	—	289.1&2	R-1	7950	R-3

Metal analysis by Inductively Coupled Plasma (ICP) method is widely used according to the method 6010, with reference of R-3.

Sample Preparation Methods for Metal Analyses

Acid Digestion of Waters for Total and Dissolved Metal Analysis by Flame AA and ICP method, *Method No.: 3005 Reference: R-3.* Acid Digestion of Aqueous Samples and Extracts for Total Metals by Flame AA and ICP, *Method No.: 3010 Reference: R-3.* Acid Digestion of Aqueous Samples and Extracts for Total Metals by Graphite Furnace AA, *Method No.: 3020 Reference: R-3.* Dissolution Procedure for Oils, Greases, and Waxes, *Method No.: 3040 Reference: R-3.* Acid Digestion of Sediments, Sludges, and Soils, *Method No.: 3050, Reference: R-3*

Fl = Flame atomic absorption technique; Gr = Graphite furnace atomic absorption technique; R-1 = Methods for Chemical Analysis of Water and Wastes, *EPA-600/4-79-020, Revised March 1983; R-2* = Standard Methods for the Examination of Water and Wastewater, *AWWA, 17th Ed., 1989; R-3* = Test Methods for Evaluating Solid Wastes, *EPA SW-846 EPA SW-846, 3rd Ed., 1986.*

Table 6-3. Methods for Determination of Organic Analytes (Total Organic Matter Present)

Parameter	Method	Method no.	Ref.	Method no.	Ref.
Biochemical oxygen demand (BOD)	5-day test, dilution technique	405.1.	R-1	—	—
Chemical oxygen demand (COD)	Open reflux,	410.4.	R-1		
	closed reflux	5220.D.	R-2	—	—
Total organic carbon (TOC)	Combustion infrared	415.1.	R-1	9060	R-3
Oil & grease	Gravimetric	413.1.	R-1	9070	R-3
	infrared	5520.C.	R-3		
Total recoverable petroleum hydrocarbons (TRPH)	Gravimetric	418.1.	R-1	—	—
	infrared	5520.C.	R-3	—	—
Oil & grease in sludge	Gravimetric	—	—	9071	R-3
Phenols, total	Chloroform extraction	420.1.	R-1	9065	R-3
Surfactants	Anionic surfactants as MBAS (methylene blue active substances)	425.1.	R-1	—	—

R-1 = Methods for Chemical Analysis of Water and Wastes, *EPA-600/4-79-020, Revised March 1983.*
R-2 = Standard Methods for the Examination of Water and Wastewater, *AWWA, 17th Ed., 1989.*
R-3 = Test Methods for Evaluating Solid Wastes, *EPA SW-846, 3rd Ed., 1986.*

Table 6-4. Methods for Determination of Organic Analytes (Individual Organic Compounds or Group of Compounds)

Name of compound or group	Analytical technique	Method no.	Ref.	Sample prep.
Purgeable halocarbons	GC-HALL	601	R-1	P&T
Purgeable aromatics	GC-PID	602	R-1	P&T
Acrolein and acrylonitrile	GC-FID	603	R-1	P&T
Phenols	GC-FID or ECD	604	R-1	XTN
Benzidines	HPLC	605	R-1	XTN
Phthalate esters	GC-ECD	606	R-1	XTN
Nitrosamines	GC-NPD	607	R-1	XTN
Organochlorine pesticides/PCBs	GC-ECD	608	R-1	XTN
Nitroaromatics and isophorone	GC-FID + ECD	609	R-1	XTN
Polynuclear aromatic hydrocarbons, PAHs	GC-FID	610	R-1	XTN
Haloethers	GC-HALL	611	R-1	XTN
Chlorinated hydrocarbons	GC-ECD	612	R-1	XTN
2,3,7,8-Tetrachlorodibenzo p-Dioxin (TCDD)	GC/MS	613	R-1	XTN
Organohalide pesticides	GC-ECD	614	R-1	XTN
Halogenated volatile organics	GC-HALL	8010	R-2	P&T
Nonhalogenated volatile organics	GC-FID	8020	R-2	P&T
Aromatic volatile organics	GC-PID	8020	R-2	P&T
Acrolein, acrylonitrile, acetonitrile	GC-FID	8030	R-2	P&T
Phenols	GC-FID	8040	R-2	XTN
Phthalate esters	GC-ECD	8060	R-2	XTN
Organochlorine pesticides/PCBs	GC-ECD	8080	R-2	XTN
Nitroaromatics and cyclic ketones	GC-FID	8090	R-2	XTN
Chlorinated hydrocarbons	GC-ECD	8120	R-2	XTN
Organophosphorus pesticides	GC-FPD or GC-NPD	8140	R-2	XTN
Chlorinated herbicides	GC-HALL or GC-ECD	8150	R-2	XTN

R-1 = 40 CRF Part 136 (Vol. 49, No. 209/Friday, October 26, 1984); R-2 = Test Methods for Evaluating Solid Wastes, *EPA SW-846, 3rd Ed., 1986; GC = Gas Chromatographic Technique; HALL = Halogen Sensitive Detector (Electrolytic Conductivity); PID = Photoionization Detector; FID = Flame Ionization Detector; ECD = Electron Capture Detector; NPD = Nitrogen Phosphorus Detector; FPD = Flame Photometric Detector; HPLC = High Performance Liquid Chromatography; P&T = Purge and Trap; XTN = Extraction methods; Sample Preparation Methods (Ref.: EPA SW-846, 3rd Ed., 1986); 5030 = Purge & trap, direct injection of liquid samples. Solid samples extracted prior injection; 3510 = Separatory funnel extraction of liquid samples; 3520 = Continuous liquid-liquid extraction; 3540 = Soxhlet extraction for solid samples; 3550 = Sonication extraction for solid sample.*

Table 6-5. Methods for Microbiological and Miscellaneous Tests

Parameter	Technique	Method no.	Ref.
Heterotrophic plate count (HPC)	Pour plate method	9215B	R-1
Total Coliform bacteria	Multiple tube fermentation	9221B	R-1
	membrane filter	9222B	R-1
Fecal Coliform bacteria	Multiple tube fermentation	9221C	R-1
	membrane filter	9222C	R-1
Fecal Streptococcus	Membrane filter	9230C	R-1
Iron and sulfur bacteria	Microscopic examination	9240B,C	R-1
Chlorophyll	Spectrophotometric	10200H	R-1
Calcium carbonate saturation (corrosivity)	Calculation of Langelier's saturation index	203	R-2
Soil pH	Electrometric	9045	R-3
Cation exchange capacity	Ammonium acetate	9080	R-3
	Sodium acetate	9081	R-3
Ignitability (flashpoint)	Closed cup	1010	R-3
EP (extraction procedure) toxicity test	—	1310	R-3
TCLP (toxicity characteristics leachate procedure)	—	1310	R-3

R-1 = Standard Methods for the Examination of Water and Wastewater, *AWWA 17th Ed., 1989; R-2* = Standard Methods for the Examination of Water and Wastewater, *AWWA 16th Ed., 1985; R-3* = Test Methods for Evaluating Solid Waste, *EPA SW-846, 3rd Ed., 1986.*

Chapter 7
Analytical System Calibration and Performance Check

Analytical chemistry is the branch of chemistry dealing with the separation and analysis of chemical substances. Analytical chemistry includes both qualitative and quantitative analysis. Qualitative analysis is concerned with what is present and quantitative analysis is concerned with how much. The *qualitative identification* of the substances presented in an analytical sample is of great importance. Sometimes chemical reactions are based on precipitation or color change and can be used for identification. Emission spectrophotometry, chromatography, and infrared spectra are the most common techniques of modern qualitative analysis. Methods of *quantitative analysis* are based on chemical reactions, on the measurement of certain chemical and physical properties, or on the measurement of a combination of chemical and physical properties. Quantitative analytical techniques may be divided into two main groups: *wet chemistry* and *instrumentation analytical techniques*. The so called "wet chemistry" procedures include gravimetric analysis (based on the weight of the substance) and titrimetric or volumetric analysis (based on the measurement of the volume). It is a quick, accurate, and widely used way of measuring the amount of substances in a solution. Both wet chemistry and instrumentation chemical analytical techniques are involved in environmental testing. Although more and more instrumentation methods are developed, chemical methods are still vital and used. Properly used instrumental and chemical methods supplement each other. The knowledge of chemical methods provides the background needed for a real understanding of instrumental analysis. Calibration and standardization of analytical systems are necessary to ensure that the produced data are acceptable or, in other words, successful calibration demonstrates that the data produced during the analytical performance is correct.

7.1. Calibration of Instruments

Initial and Continuing Calibration

Calibrations are performed at the beginning of the analysis to make sure that the instrument is working properly. This initial calibration has to be proven during the

Name and concentration : ____________________

Test for used : ____________________

Source : ____________________

Lot number ____________________

Date received : ____________________

Date opened : ____________________

Expiration date : ____________________

Manufacturer's certification : ____________________

Storage : ____________________

Date of disposal : ____________________

Mode of disposal : ____________________

Remarks: ____________________

____

____

____________________ Signature of the logger

____________________ Date

FIGURE 7-1. Documentation log form for purchased calibration stock and standard solutions.

analytical process by continuing calibration. When continuing calibration fails to meet acceptance of the criteria, initial calibration should repeated. Calibrations must be performed according to the direction of the analytical methods. Initial calibration is determined for each parameter tested and it is based on the instrument responses for different concentrations of standards, called calibration standards. The number and the optimum concentration range of the calibration standards used for each particular method is given by the approved methodology and described by the laboratory Standard Operation Procedures (SOP). If this information is not in the method, then a minimum of a blank and three standards must be utilized for calibration. Calibration of the instruments vary with the type and model of the equipment. Detailed operation and calibration procedures for each instrument are available in the laboratory SOP and in the manufacturer's instruction.

7.2. Calibration Stock and Standard Solutions

Calibration stock solutions are either commercially available, or "house prepared" by the laboratory, according to the SOP.

Name and Concentration:____

Test for use:____

Source Of Chemical Used:

Name and Formula:____

Grade:____

Lot Number:____

Source:____

Date Received:____

Date Opened:____

Preparation Procedure:

Temperature and drying time of chemical prior to preparaton:_____________________________________

Volume prepared:____

Quantity of chemical used:_____________________________________

Remarks:

Date:____________ **Signature:**________________________________

Date:____________ **Approval by Supervisor:**________________________

FIGURE 7-2. Documentation log form for preparation of calibration stock solution.

Commercially Available Calibration Stock Solutions

Records should be maintained with information such as name, concentration, chemical grade or purity, lot number, test for use, source, date received, date of expiration, date of logging, and the signature of the logger, as on Figure 7-1.

Laboratory Prepared Calibration Stock Solutions

Record each time when prepared. The log form contains the name and concentration of the calibration stock, name, formula, chemical grade, source, lot number, date received, date opened, and expiration date of the chemical used for preparation of the stock. Information about preparations such as the volume of the stock prepared, quantity of chemical used, and detailed preparation steps are also included in this log form, as on Figure 7-2.

Test:____

Optimum Calibration Range:____

Stock Solution, Concentration:____

Final Volume of Standards, ml:____

Concentration of Standards	Volume Stock Used
____________	____________
____________	____________
____________	____________
____________	____________
____________	____________

Concentration of Continuing Cal. St.____

____

Date of Preparation:____

Expiration Date:____

Storage Description (if applicable):____

____

____

Date:______________ Signature:____

Date:______________ Approval by Supervisor:____

FIGURE 7-3. Documentation log form for preparation of calibration standards.

Calibration and Continuing Calibration Standards (CCS)

Preparation of the log form must include the concentration of the stock, the dilution technique for the desired concentration, date and signature of the preparer, and, if applicable, holding time and mode of storage of the standard. Log form is shown in Figure 7-3.

Frequency of Standard Preparation and Storage of the Standards

If no direction is available from the manufacturer, calibration standards must be renewed when the response of the instrument is not satisfactory. It is recommended that primary standards change yearly, or as indicated by the expiration date, and working standards are to be prepared fresh each time when calibration is performed.

Table 7-1. Storage and Preparation Frequency of Selected Stock and Standard Solutions

Test	Calibration stocks and standards	Storage	Preparation frequency
pH	pH 4.00, 7.00, 10.00 buffers	Room temp.	Expiration date indicates
Conductance	0.01 M KCl	Room temp. glass stop bottle	6 months
Turbidity	400 NTU stock	Refrigerate	1 month
	Dil. standards	Refrigerate	1 week
Bromide Br^-	500 ppm stock	Room temp.	3 months
Cyanide CN^-	1000 ppm stock	Refrigerate	1 month check weekly
Fluoride F^-	100 ppm stock	Room temp.	3 months
Ammonia-Nitrogen NH_3-N	1000 ppm stock	Refrigerate	3 months
Nitrate-Nitrogen NO_3-N	1000 ppm stock	Refrigerate preserve with chloroform	6 months
Nitrite-Nitrogen NO_2-N	250 ppm	Refrigerate preserve with chloroform	3 months
Phosphorus PO_4-P	50 ppm	Refrigerate	3 months
Silica SiO_2	10 ppm	Refrigerate in tightly stoppered plastic bottle	1 month
Sulfate SO_4^{2-}	100 ppm	Room temp.	6 months
Metals	1000 ppm	Room temp.	Expiration date indicates
	10 to 100 ppm standards	Room temp. preserve with 0.5% HNO_3	1 month
COD	500 ppm	N/A	Prepare fresh for each
TOC	1000 ppm and 10 to 100 ppm standards	Refrigerate brown glass bottles	3 months
Oil and grease	"Reference oil" calibrate each time used	Freezed in sealed container	3 months
Total phenols	1000 ppm	Refrigerate in glass bottle	3 months
Trace organics	Concentrate depends on methods and analytes	Freeze in individually packaged vials	Expiration date indicates
Tannin and lignin	1000 ppm stock and 10 ppm standard	Room temp.	6 months

Standards must be stored according to the recommendation of the method or according to the statement of the supplier. If there is no guidance for the storage time, change the standards when the reading of the values are decreased. Documentation for storage and preparation of the standards must be available. Table 7-1 contains the recommended storage and preparation frequency for selected calibrations stock and standard solutions.

7.3. Calibration Requirements for Field Instruments

Calibration of field instruments must be performed on a regular basis. The records must include the method used, date and time of calibration, number of the standards, standard concentrations and the responded readings of the instrument. The name

Name and Concentration ____________________
Storage ____________________

Outside source

Manufacturer: ____________________

Certification Available : ____________________

Date received: ____________________

Expiration date : ____________________

Storage : ____________________

In House Prepared

Name, formula, and grade of the chemical used

____

____

Lot number ____________________

Date received ____________________

Date opened ____________________

Expiration date ____________________

Preparation procedure:

____

____

____

Signature of logger ____________________ Date ____________

Approval by supervisor ____________________ Date ____________

FIGURE 7-4. Documentation log form for preparation of calibration verification standards (CVS) or quality control (QC) check standards.

and model of the instrument is also important. Detailed calibration of field instruments are in Part I, Chapter 2. Initial calibration should be performed at the beginning of the field tests and continuing calibration is recommended for every four hours during the sampling day.

7.4. Calibration of Laboratory Instruments

Initial Calibration

Initial calibration is based on the response of the instrument on different concentrations of calibration standards against calibration blanks. The response of the instrument should be linear with the concentration of the introduced standards. Standards are prepared by the analysts or by suppliers of the highest reliability. The analyst preparing the standards should have the capability to evaluate them, and if the source of the standards is commercial, the analyst should require proof of the correctness based on the supplier's quality assurance program. Every standard should be accompanied by the expiration date and should not be used beyond such date. The number of the standards is recommended by the method, or by the manufacturer of the instrument. The concentration of the standards should be bracketed in the optimum concentration range given by the analytical method. When the number of the standards are not described, a 3 standards calibration is satisfactory. The concentration of the "high standard" is the upper level of the optimum range, the concentration of the "middle point standard" is half of the highest standard, and the value of the "low level standard" is five times lower than the highest standard. The concentration of the standards and the response of the instrument should plot on the calibration curve, or the instrument software should automatically prepare the curve. After the calibration curve is prepared, it should be approved with the calculation of the corresponding *correlation coefficient* by using the linear regression calculation. Its value should be >0.9998; otherwise, new calibration has to be made. Calculations and preparation of the calibration curves may vary by instrumentation. It may be constructed manually or by using computer software, and should be stored in a manner to be available at any time for inspection.

Some general rules in the preparation of calibration curves follow.

- Ordinary rectangular-coordinate paper is satisfactory for most purposes. For some graphs (for example, measurement of MV response, by using ion selective electrodes), semilogarithmic paper is preferable.
- Plot the independent and dependent variables on abscissa and ordinate in a manner that can be easily comprehended.
- Choose the scales so that the value of either coordinate can be found quickly and easily.
- Plot the curve to cover as much of the graph paper as possible.
- Choose the scales so that the slope of the curve approaches unity as nearly as possible.
- Other things being equal, choose the variables to give a plot that will be as nearly a straight line as possible.
- Entitle the graph to describe adequately what the plot is intended to show. Present legends on the graphs to clarify possible ambiguities. Include in the legend complete information about the conditions under which the data were obtained, volume of the sample, wavelength used, time between addition of the reagent and the reading, data obtained from the linear regression calculation, parameter determined, method, date of preparation, and the name and signature of the analyst. Typical calibration curve is on Figure 7-5.

Continuing Calibration Standard (CCS) — Used to assure calibration accuracy during each analytical run. It represents the value of the mid-point initial calibration standard, and used to assure calibration accuracy during each analytical run. It must run immediately after the standard curve was established, and during the analytical batch analysis at the frequency of 5% and after the last sample was analyzed. The deviation from the original value should be within ±5%.

Calibration Verification Standard (CVS) — It is a known value standard, used to verify that the standards and the calibrations are accurate and also confirm the

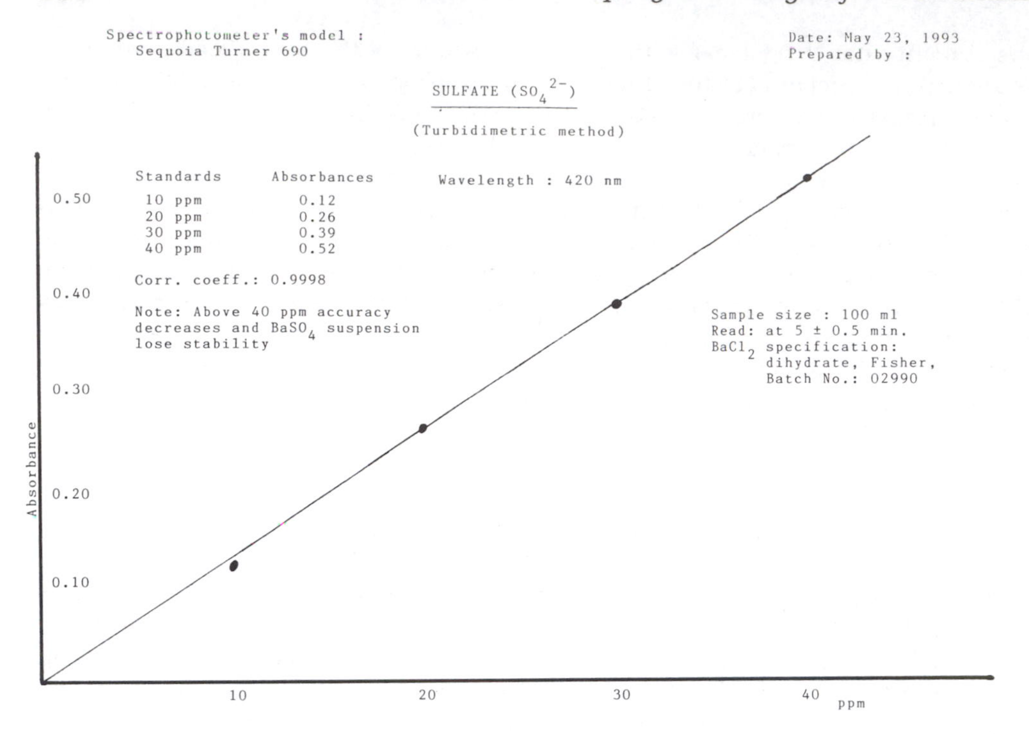

FIGURE 7-5. Typical calibration curve.

calibration curve. It should be a certified (purchased from EPA or from other sources), or independently prepared standard from a source other than the calibration standards. The value is accepted within ±10% deviation from the 100% recovery. Record for the preparation of CVS is on Figure 7-4.

Preparation blank (Prep blank) and Laboratory Control Standard (LCS)—When samples go through any pretreatment (digestion, distillation, extraction, filtration, etc.) prior to the actual analysis, it should be verified and the effects from the sample preparation should be monitored. To support these measures, blank and standards should be prepared and analyzed together with the sample. The preparation blank is a volume of analyte-free water processed through the sample preparation procedure. Laboratory Control Standard (LCS) is the same concentration as the Calibration Verification Standard (CVS) except it is carried through the preparation and analysis procedure the same as the samples. Recovery is a ±15% deviation from the 100% recovery.

Accepted Calibration

Preparation of the calibration curve with the accepted correlation coefficient (>0.9998), CCS within the proper range, and CVS with the correct percent recovery is the criteria of the appropriate *initial calibration* for the instrument and the analytical system. With a 5% frequency, or once in one analytical batch the initial calibration has to be approved by *continuing calibration.* (Analytical batch is defined as samples that are analyzed together with the same method sequence and the same lots of reagents. Samples in each batch should be similar matrices.) Continuing calibration includes analysis of the calibration blank, CCS, and CVS, with the frequency and acceptance mentioned above. If this criteria fails, initial calibration should take place again and samples analyzed before this failure must be analyzed again.

Calibration Frequency

The frequency of calibration curve preparation depends on the instrumentation. If the accepted calibration curve is available, for example by using VIS/UV spectrophotometer (utilizing lights in the visible and ultraviolet wavelength range) use the calibration curve until the correct calibration is approved. Initial calibration is based on a 4 to 6 points standard curve in the optimum linear range, stated in each particular methodology. It must be performed at least every 6 months or on failure of any continuing calibration standard. Curve should be checked to assure linear correlation coefficient of >0.9998. Each time a calibration curve is established or checked, records on the origin of the calibration stock solution and the prepared calibration standards should be attached, as on Figure 7-1, 7-2, and 7-3. Daily calibration is made by zeroing the instrument with a calibration blank, checking original calibration curve with the ±5% recovery of the CCS, and with the ±10% recovery of the CVS. After analyzing 10 samples, or one analytical batch, start again with CCS, CVS, blank. If CCS or CVS fail the calibration criteria, the run must be stopped, and a new initial calibration must be performed. Samples analyzed before the failed standard must be analyzed again. Sample values are calculated according to the linear regression calculation (using calculator or computer software). Detailed calibration procedures for selected instrumentation are available in the next section.

7.5. Summary of Calibration Procedures

Initial Calibration

- Establish the calibration curve by using a blank and a number of calibration standards in the optimum range according to the method.
- Determine acceptance or rejection of the curve by calculating correlation coefficient.
- Run CCS to verify the curve. Must be within 5% of the verified value.
- Run CVS. It has to be from other source as the calibration standards, as was stated above. Must be within 10% of true value to accept initial calibration.
- If the analytical method includes sample pretreatment, such as distillation (for example cyanide, phenols), digestion (for example total phosphorus, TKN, metals), a treated blank called "prep blank" and a treated CVS, now called Laboratory Control Standard (LCS) should be incorporated in the analytical run. The result must be within 15% of the true value to accept.
- Analyze 10 samples (including reagent blanks, spikes, and duplicates).
- After the analysis of these samples, check the curve again with CCS. It must be within 5% of the verified value for satisfaction.
- The accepted curve should be certified again with CVS. If the result is within 10% of the true value, continue the measurement of the samples.
- If CCS or CVS fails criteria, run must be stopped, and new initial calibration performed. Samples analyzed before to failed standards must be analyzed again. If the verification of the initial calibration is satisfying, follow the analysis, and repeat the same verification with 5% frequency.

Daily Calibration

UV/VIS Spectrophotometer

- Zeroed the instrument with blank.
- Check curve with one CCS. It must be within 5% of true value.
- Verify curve by analyzing CVS. It must be within 10% of true value.

- When the curve was checked and verified, analyze 10 samples, including reagent blank, spikes and duplicates.
- Analyze CCS again to check any drifting of the curve.
- Continue analysis scheme as mentioned above in initial calibration procedure.

Atomic Absorption Spectrophotometer (AA)

The calibration is based on a 3 point (low, middle, and high level) standard curve in the optimum linear range as specified per analyte in the methodology. This is performed each time the instrument is used, or in the failure of any continuing calibration standards. Calculation of the acceptance of the calibration curve is by computer software using linear regression. The curve is checked to assure linear correlation of >0.9998. Complete the initial calibration by checking the curve with CCS and CVS and, using continuing calibration, check in the same manner as VIS/UV Spectrophotometer.

Inductively Coupled Plasma Analyzer (ICP)

By monitoring several wavelengths, either all at once or in a programmed sequence, many elements can be determined in one automated analysis with ICP. It offers significant speed advantage for determination of metals in environmental samples. Calibration is made by mixed, multi-element calibration standards. Number of the calibration standards and calibration technique is defined by the instrument's manufacturer. Initial calibration and checking with continuing calibration are the same as in the general discussion above.

Ion Analyzer, Ion Selective Electrodes (ISE)

The specific calibration procedure, as calibration of ion analyzer, uses ISE by establishing the proper slope for the curve, which is based on two standards. Calibration acceptance criteria of certain instruments are based on manufacturer's instrument specifications, and the verification of the calibration depends on the acceptable recoveries specified by QC check sample suppliers (For example, *Gas Chromatograph*). Calibrations are similar as described in the initial and continuing calibration procedures in the concentrations of the detector linear range.

7.6. Standardization of Titrating Solutions

In the titrimetric method, the determination of the exact concentration of the titrating solution, also called titrant, is critical. Detailed descriptions of preparing and standardizing the titrant solutions are in the test method. All of the data related to this activity should be documented in a solution preparation log (Figure 7-6) and in the standardization log form (Figure 7-7). The preparation log form includes the information on the chemical used and a detailed outline of the work.

The log form in Figure 7-7 is designed for standardization of titrant solutions by two different procedures. The first method is using primary standard solid chemicals. (Detailed standardization procedures according to the approved methodology are in the laboratory SOP). By registering the exact weight of the chemical and the amount of titrant used, the correct normality (N) of the solution is calculated by following the formula (1) on the log form. The final normality is based on the average value of three parallel determinations.

The second part of the log form is used when a standard solution with exact concentration is used to establish the normality of the titrant. By knowing the amount of standard solution used, the normality of the standard and the amount of titrant

Titrant, Normality:____________________

Test used for:____________________

Date of Preparation:____________________

Chemical Used For Preparaton:

Name, Formulas, Grade:____________________

Lot Number:____________________

Date Received:____________________

Date Opened:____________________

Preparation Of Titrant:

g or ml Used:____________________

Final Volume:____________________

Date:__________ **Signature:**____________________

Date:__________ **Approval by Supervisor:**____________________

FIGURE 7-6. Preparation of titrants.

consumed, the exact normality is calculated by using the formula (2) on the log form. The exact normality is calculated by averaging the results of three determinations.

7.7. Performance Check of Laboratory Instruments

Performance Check of pH and Conductometer

After calibration, but before measuring samples, a known value QC Check Sample (sample obtained from an independent source with a known value) is measured. If the measured value of this check sample is satisfactory, the accurate and reliable performance of the meter is approved. Calibration and performance check of pH meter is in Table 2-1 (Chapter 2).

Performance Check of Dissolved Oxygen Meter

The performance of the DO meter is annually checked against the Winkler titrimetric method, as described in Chapter 2. DO content of the sample is measured by the

Name and Normality of the titrant: ______________________
Test used for: ______________________
Date of preparation: ______________________
Date of standardization: ______________________

STANDARDIZATION BY PRIMARY STANDARD CHEMICAL

Name and formula of the primary standard ______________________

Manufacturer's name ______________________
Date received ____________ Date opened ____________

Weight of flask (W_1)	Weight of fl & chem. (W_2)	Weight of chemical (W_2–W_1)	Titrant used	Normality calculated
g	g	g	ml	N

CALCULATION: $N = g \times 1000 / \text{ml titrant} \times Eqw_{\text{primary standard}}$ (1)

FINAL NORMALITY (N) : ____________
(based on the average value of three parallel determinations)

STANDADIZATION WITH STANDARD SOLUTIONS

Name and concentration of the standard: ______________________
Date of preparation : __________ Log book __________ Page ________

ml Standard solution used			
ml Titrant solution used			
Calculated normality (N)			

CALCULATION: $N = \text{ml standard} \times N \text{ of standard} / \text{ml titrant}$ (2)

FINAL NORMALITY (N) : ____________
(based on the average value of three parallel determinations)

Signature of technician ______________ Date ______________

Approval of Supervisor ______________ Date ______________

FIGURE 7-7. Documentation form for standardization of titrant solutions.

membrane electrode method in the field. From the same sample source, collect samples for the Winkler titrimetric method by carefully following the method described in Chapter 2. If there is serious deviation between the two results, properly conducted corrective action should verify and solve the problem. Documentation and records of the tests are in Table 2-3 (Chapter 2).

Performance Check of Residual Chlorine Kit

Chlorine KIT performance should be checked annually against laboratory methods, such as iodometric titration, or by other approved laboratory tests, as in Chapter 2. Compare the two results for reliability of the correct work for KIT. If the two results disagree, document the action needed for correction. Documentation and records of the two methods are in Table 2-4 (Chapter 2).

Performance Check of Turbidimeter

The meter is calibrated each time before measuring samples by Reference Standards, available from the manufacturer. The performance of the meter is checked periodically (at least semiannually). Any evidence of the malfunction of the meter is checked by measuring the results of the formazine standards prepared by the laboratory.

- Formazine stock solution:
  Always prepare one day prior to performing the check!
  Solution 1. 1.00 g Hydrazine sulfate $(NH_2)_2.H_2SO_4$ dissolve and dilute to 100 ml with DI water.
  Solution 2. 10 g Hexamethylene tetramine $(CH_2)_6N_4$ dissolve and dilute to 100 ml with DI water.
  These solutions are stored in a refrigerator and are usable for 6 months.
  In a 100 ml volumetric flask, mix 5 ml of Solution 1 and 5 ml of Solution 2. Mix and allow to stand for 24 hours at room temperature. Then dilute to 100 ml with DI water, and mix well. The value of this "stock solution" is 400 NTU (Nephelometric Turbidity Unit), and it remains good for one month if stored in a refrigerator. Wait until it has warmed up to room temperature and mix well prior to use.
- Dilute this stock solution for different ranges and check the performance of the meter. In case of deviation, conduct corrective action until a satisfying performance is achieved. Special care should be taken to use scrupulously clean and scratch-free sample cuvette. The performance of the meter should be documented, as on Table 7-2.

Performance Check of VIS/IV Spectrophotometer

Spectrophotometers should retain their wavelength accuracy for the life of the instrument under normal operating conditions. To confirm the performance of the spectrophotometer, the wavelength accuracy must be periodically checked. Superior results will be obtained by using a DIDYMIUM calibration filter for this measurement.

Wavelength Calibration Check

If the didymium filter is not available, good results may be produced by measuring the absorbance of a cobalt chloride solution (22 to 23 g of cobalt chloride, $CoCl_2$, dissolve and dilute to 1 liter with 1% HCl solution) on 500, 505, 510, 515, and 520 nm wavelengths. The wavelength calibration check is satisfying when maximum absorbance (or minimum transmittance) occurs between 505 and 515 nm. The specific absorbance values are not interesting. Documentation for the wavelength calibration check is on Table 7-3.

Linearity Check

Linearity check of the instrument is given by the measurement of the absorbance at 510 nm of the stock and the 1:1 diluted cobalt chloride solution. The absorbance of the 1:1 dilution solution should be half of the stock solution value in correct operation. Documentation of the linearity check is on Table 7-4.

Performance Check of Atomic Absorption Spectrophotometer (AA) Flame and Graphite

Performance of the AA Spectrophotometer is checked each time a different metal is analyzed as part of the analytical procedure. The performance check is an indicator

Table 7-2. Calibration and Performance Check of Turbidimeter

Model : ______________________________

	Calibration				Performance				
Date	0.4 NTU	4.0 NTU	40 NTU	100 NTU	0.2 NTU	2.0 NTU	20 NTU	100 NTU	Sign.

NTU = Nephelometric Turbidity Unit

of any deterioration of either the lamps or the spectrophotometer and shows the optimal condition of the instrument. It is measured by a "sensitivity check standard", with which a concentration is given by the method for each metal. The absorbance of this standard should be 0.200. If it differs by more than ±10%, the instrument is not performing correctly and has to be corrected. The concentration of the sensitivity standards for flame and graphite techniques are in Table 7-5 and Table 7-6.

Performance Check of Ion Analyzer and Ion Selective Electrodes (ISE)

Performance and calibration are checked each time a test is performed as part of the analytical procedure. The calibration is based on two standards that differ in concentration by a factor of ten. Correct operation of the electrodes are indicated by the proper slope reading. Correct slope values are included in the analytical procedures (SOP). Satisfying slope value and the accuracy of a known value QC check standard indicates the accepted performance of the analyzer and the electrodes. Otherwise, corrective action must be taken to find out the problem and set it right for operation. Records on slope values are on the daily workpapers, designed for using ISE with autoanalyzer.

Table 7-3. Spectrophotometer Wavelength Calibration Check

Model ______________________________

Method for check ______________________________

Absorbance read at wavelength

Date	500 nm	505 nm	510 nm	515 nm	520 nm	Remarks	Sign.

nm = nanometer, unit of the wavelength

The calibration check is satisfying when maximum absorbance (or minimum transmittance) occurs between 505 nm and 515 nm wavelengths.

Performance Check of Infrared (IR) Spectrophotometer

A 0.05 mm thick film of polystyrene (commercially available) is used according to the following procedure:

- Using the polystyrene film provided, record a spectrum using SCAN TIME 12.
- Compare the spectrum with the one in Figure 7-8.

Resolution — Shoulders "A" and "B" should remain quantitatively the same in their depth resolution from the main bands.

Frequency positions — The bands shown in Figure 7-8 should appear at the frequencies indicated and should not deviate from the designated frequency by more than the tolerance indicated.

Ordinate accuracy — The bands whose depth range between 80 and 100% transmittance should vary by no more than 2% transmittance between tests.

General contour — The test spectrum should have the same general shape and appearance as illustrated on Figure 7-8. If the test spectrum is not within the tolerance indicated, adjustment is necessary, probably by service representatives.

Table 7-4. Spectrophotometer Linearity Check

Model: ____________________

Date	Checking solution	Absorbance at 510 nm	Remark	Sign.
	Stock Cobalt Soln. 1:1 Cobalt Soln.			
	Stock cobalt soln. 1:1 Cobalt Soln.			
	Stock Cobalt Soln. 1:1 Cobalt Soln.			
	Stock cobalt Soln. 1:1 Cobalt Soln.			
	Stock Cobalt soln. 1:1 Cobalt Soln.			
	Stock Cobalt Soln. 1:1 Cobalt Soln.			
	Stock Cobalt Soln. 1:1 Cobalt Soln.			
	Stock Cobalt Soln. 1:1 Cobalt Soln.			
	Stock Cobalt Soln. 1:1 Cobalt Soln.			

The absorbance of the 1:1 Diluted Cobalt solution should be half of reading produced by the Stock Cobalt solution

Band no.	Frequency cm^{-1}	Tolerance cm^{-1}
1	3027	±6
2	2851	±6
3	1944	±6
4	1601	±6
5	1181	±6
6	1028	±6
7	907	±6
8	699	±6
9	540	±6

Table 7-5. Performance Check of Flame AA

Model: ______________________________

Element	Concentration of Sensitivity Check Standards, in mg/L
Aluminum (Al)	50
Antimony (Sb)	25
Barium (Ba)	20
Beryllium (Be)	1.5
Calcium (Ca) at 422.7 nm	4.0
Calcium (Ca) at 287.4 nm	60
Cadmium (Cd) at 228.8 nm	1.5
Cadmium (Cd) at 368.4 nm	850
Cobalt (Co)	7.0
Chromium (Cr)	4.0
Copper (Cu)	4.0
Iron (Fe)	5.0
Lead (Pb)	20
Potassium (K)	2.0
Magnesium (Mg)	0.3
Manganese (Mn)	2.5
Molybdenum (Mo)	30
Nickel (Ni)	7.0
Silicon (Si)	100
Sodium (Na)	0.5
Strontium (Sr)	5.0
Tin (Sn)	150
Titanium (Ti)	80
Thallium (Tl)	30
Tungsten (W)	450
Zinc (Zn)	1.0
Zirconium (Zr)	300

The performance of the Flame Atomic Absorption Spectrophotometer (Flame AA) should b checked each time a metal is analyzed by using sensitivity check standard. The sensitivity check data given above are the metal concentration (mg/L) in aqueous solution which will give a reading of approximately 0.2 absorbance unit.

Performance Check and Calibration of Balances

Balances are delicate instruments. The proper use and care of the balances is imperative. The following comprehensive rules should be followed to protect and keep this important laboratory equipment in excellent condition.

- Balance should be on a heavy, shockproof table, with adequate working area, and a drawer for the balance accessories.
- Should be located away from laboratory traffic.
- Should be protected from sudden drafts and humidity changes.
- Keep desiccant inside the balance area to protect from humidity.
- Temperature should be room temperature.
- Special precautions should be taken to avoid spillage of chemicals on the pan or inside the balance.
- Check to make sure the balance is level.
- Check to make sure the balance is zero. Adjust to zero with zero adjustment prior to use.
- Inspect the balance to make sure that it is working properly by using certified Class S weights for monthly monitoring.

Table 7-6. Performance Check for AA Graphite Furnace

Model: ______

Element	Concentration of Sensivity Check Standards, in mg/L
Aluminum (Al)	0.05
Antimony (Sb)	0.04
Arsenic (As)	0.04
Barium (Ba)	0.03
Cadmium (Cd)	0.003
Cobalt (Co)	0.03
Chromium (Cr)	0.007
Copper (Cu)	0.01
Iron (Fe)	0.01
Manganese (Mn)	0.005
Molybdenum (Mo)	0.025
Nickel (Ni)	0.04
Lead (Pb) at 283 nm	0.025
Lead (Pb) at 217 nm	0.016
Selenium (Se)	0.06
Silicon (Si)	0.14
Tin (Sn) at 286 nm	0.14
Tin (Sn) at 224.6 nm	0.08
Titanium (Ti)	0.21
Vanadium (V)	0.15
Zinc (Zn)	0.0007

The sensitivity check data given above are the metal concentrations in mg/L in aqueous solution which will give a reading of approximately 0.2 absorbance unit when 20 μl (microliter) are used.

- When the balance is not in use, the beam should be raised, the weights returned to the beam, objects removed from the pan, and the weighing compartment closed.
- All balances should be checked and adjusted yearly.

Documentation of routine balance check is on Table 7-7.

Performance Check of Temperature Related Devices

All temperature related equipment must have their temperature checked on a daily basis. Such devices are refrigerators, ovens, incubators, etc. Temperature control log sheets are on Tables 7-8, 7-9, 7-10, 7-11, 7-12, and 7-13. Thermometers used for monitoring temperature should be checked periodically (at least annually) against precision thermometers certified by the National Bureau of Standards. Documentation is on Table 7-14.

Performance Check of Gas Chromatograph

The gas chromatograph is a multipurpose instrument; however, specific detectors and columns are needed to analyze certain compounds. The detector response to different analytical parameters and standards may vary due to the condition of the analytical column, detector, or the chromatograph. Therefore, in order to insure the performance of the chromatograph, integrated areas and response factors should be recorded in performance check sheets for the standards of specified analytes. If these results vary by more than ±20%, corrective action is needed.

Performance Check of Gas Chromatograph/Mass Spectrotometry

At the beginning of each day that analyses are to be performed, the GC/MS system should be checked to see that acceptable performance criteria are achieved by

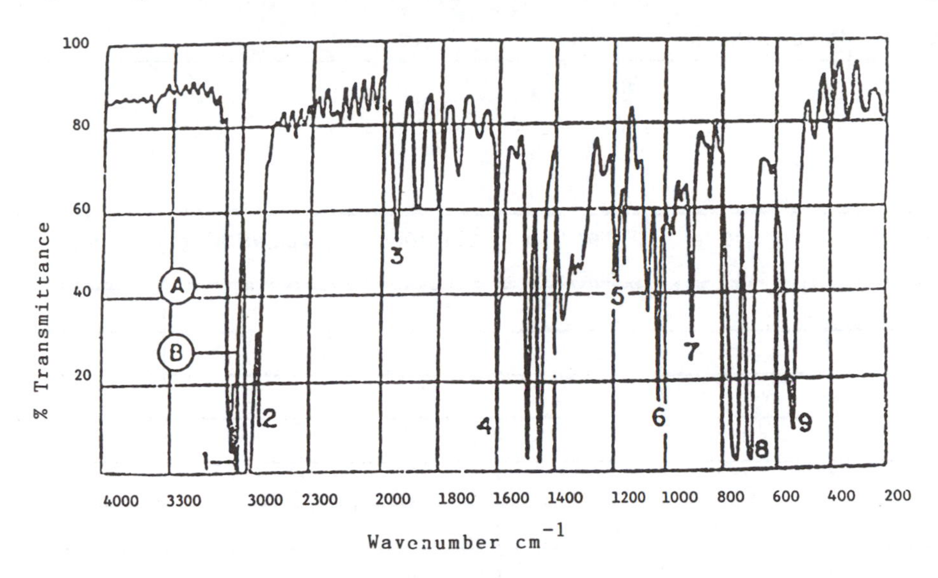

Infrared Spectra of Polystyrene

"A" and "B" = shoulders **1-9 = band numbers**

- **Shoulders "A" and "B" should remain quantitatively the same in their depth resolution from the main band.**
- **The bands shown at the spectrum of the polystyrene film should appear at the frequencies indicated.**
- **The bands whose depth range between 80% and 100% transmittance should vary by no more tha 2% transmittance between tests.**
- **The test spectrum should have the same general shape and appearance as illustrated above.**

FIGURE 7-8. Performance check of infrared (IR) spectrophotometer.

injecting calibration fluid into the column. The mass spectra obtained is checked to assure that ion abundance criteria for each mass in question is achieved. If volatile organics are to be analyzed, p-Bromofluorobenzene (BFB) is used as calibration fluid. If base-neutrals or acids are to be analyzed Decafluoro-triphenyl-phosphine (DFTPP) is used. If some problem arises, the mass spectrometer is retuned and the test is repeated until all criteria are met.

Performance Check of Autoclave

Sterilization temperature and pressure, with the sterility indicator strip should be documented each time the equipment is used.

7.8. Definitions Related to Calibration

Optimum concentration range — The concentration range where the calibration curve remains linear. It is defined by the method and varies with the instrumentation.

Calibration stock solution — Calibration stock solution is a high concentration solution of the analyte. It is commercially available or prepared by the laboratory according to the SOP.

Table 7-7. Balance Check

Model ______________________________

Date	0.0100 g	0.1000 g	1.0000 g	10.0000 g	Sign.

Intermediate standard — Standard is diluted from the stock solution to obtain concentration suitable to dilute the calibration standards.

Calibration standards — The number of exact concentration solutions, prepared by diluting the calibration stock or the intermediate standard solution. The number of the standards is determined by the method and by the instrument, while the concentration is determined by the given optimum range.

Mid-range standard — A standard in the middle of the linear range of the established calibration curve or a standard concentration in the middle of the expected sample concentration range depending on the type of determination to be performed.

Calibration blank — A volume of analyte-free water used to zero the instrument. It is also run at the end of the analytical work to check if any contamination or drifting occurs.

Preparation blank — A volume of analyte-free water processed through sample preparation procedure.

Continuing calibration standard (CCS) — Used to assure calibration accuracy during each analytical run. Analyzed at the beginning and end of the analysis, and after each 10 samples. The concentration must be at or near the mid-range level.

Table 7-8. Oven Temperature Control Log

Temperature should be 100°C - 105°C

Instrument Description ______________________________

	Jan.	Feb.	Mar.	Apr.	May	June	July	Sept	Oct.	Nov.	Dec.
1.											
2.											
3.											
4.											
5.											
6.											
7.											
8.											
9.											
10.											
11.											
12.											
13.											
14.											
15.											
16.											
17.											
18.											
19.											
20.											
21.											
22.											
23.											
24.											
25.											
26.											
27.											
28.											
29.											
30.											
31.											

Calibration verification standard (CVS) or QC check standard — An independently prepared standard, with a concentration near the mid-level standard. Prepared from a source other than the calibration standards. It is analyzed at the beginning and the end and in 10% frequencies during the analysis.

Laboratory control standard (LCS) — It is the CVS carried through the complete processes as the samples are prepared and analyzed. It is used to monitor the effects from sample preparation.

Sensitivity check standard — Used to optimization of the AA spectrophotometer. Concentration of the standard is given by each method and is also stated in the manufacturer's instructions. With this standard, the instrument must give 0.200 or very close absorbance reading.

QC check sample — Samples obtained from an independent source, with a known value of the analyte. These samples are analyzed with a sample set of similar matrix, and the obtained results are used to determined the accuracy of the laboratory performance.

Table 7-9. Oven Temperature Control Log

Temperature should be 180° ± 2°C

Instrument Description____________________________________

	Jan.	Feb.	Mar.	Apr.	May	June	July	Aug.	Sept	Oct.	Nov.	Dec.
1.												
2.												
3.												
4.												
5.												
6.												
7.												
8.												
9.												
10.												
11.												
12.												
13.												
14.												
15.												
16.												
17.												
18.												
19.												
20.												
21.												
22.												
23.												
24.												
25.												
26.												
27.												
28.												
29.												
30.												
31.												

Table 7-10. Incubator Temperature Control Log

Temperature should be 20°C ± 2°C

Instrument Description____

	Jan.	Feb.	Mar.	Apr.	May	June	July	Aug.	Sept	Oct.	Nov.	Dec.
1.												
2.												
3.												
4.												
5.												
6.												
7.												
8.												
9.												
10.												
11.												
12.												
13.												
14.												
15.												
16.												
17.												
18.												
19.												
20.												
21.												
22.												
23.												
34.												
25.												
26.												
27.												
28.												
29.												
30.												
31.												

Table 7-11. Incubator Temperature Control Log

Temperature should be 37°C ± 2°C

Instrument Description ______________________________

	Jan.	Feb.	Mar.	Apr.	May	June	July	Sept	Oct.	Nov.	Dec.
1.											
2.											
3.											
4.											
5.											
6.											
7.											
8.											
9.											
10.											
11.											
12.											
13.											
14.											
15.											
16.											
17.											
18.											
19.											
20.											
21.											
22.											
23.											
24.											
25.											
26.											
27.											
28.											
29.											
30.											
31.											

Table 7-12. Incubator Temperature Control Log

Temperature should be $44^0C \pm 2^0C$

Instrument Description ______________________________

	Jan.	Feb.	Mar.	Apr.	May	June	July	Sept	Oct.	Nov.	Dec.
1.											
2.											
3.											
4.											
5.											
6.											
7.											
8.											
9.											
10.											
11.											
12.											
13.											
14.											
15.											
16.											
17.											
18.											
19.											
20.											
21.											
22.											
23.											
24.											
25.											
26.											
27.											
28.											
29.											
30.											
31.											

Table 7-13. Refrigerator Temperature Control Log

Temperature should be 4°C ± 1°C

Instrument Description______________________________________

	Jan.	Feb.	Mar.	Apr.	May	June	July	Aug.	Sept	Oct.	Nov.	Dec.
1.												
2.												
3.												
4.												
5.												
6.												
7.												
8.												
9.												
10.												
11.												
12.												
13.												
14.												
15.												
16.												
17.												
18.												
19.												
20.												
21.												
22.												
23.												
24.												
25.												
26.												
27.												
28.												
29.												
30.												
31.												

Table 7-14. Calibration Chart for Thermometer

Date	Reading on working thermometer °C	Reading on certified thermometer °C	Difference between readings °C	Remark	Sign.

Chapter 8
Quality Management of Laboratory Instruments and Supplies

Considerate selection, care, and maintenance of laboratory instruments and supplies (chemicals, glassware, etc.) are significant factors of the laboratory management and important parts of the precise and accurate analytical performance.

8.1. Maintenance of Laboratory Instruments

Constant care and routine maintenance is the secret for the proper working condition of the laboratory instruments. Maintenance activities for each instrument are found in the manufacturer's instruction manual that accompanied the instrument. A written maintenance schedule per instrument must be available. The laboratory must have one maintenance expert or have contracted the vendor specialist for regular maintenance and simple repairs. Table 8-1 contains the recommended maintenance activities and frequencies for the laboratory instruments. Routine maintenance activities per instrument are based on the recommendation of the manufacturer per each different type and model. The frequency of the maintenance may also change according to the workload and the type of samples analyzed.

8.2. Quality Requirements for Laboratory Supplies

Laboratory Glassware

There are many grades and types of laboratory glassware from which to choose. The mainstay of the modern analytical laboratory is a highly resistant borosilicate glass, such as "Pyrex" or "Kimax". Corning brand glass is claimed to be 50 times more resistant to alkalies and is practically boron-free. Raysorb or Lo-actinic brand glass is used when the reagents or materials are light sensitive.

Stoppers and caps should be carefully selected. Use the following guidelines.

- Do not use metal caps for corrosive materials and metal solutions.
- Do not use glass-stoppered bottles for alkalines.
- Do not use rubber-stoppered bottles for alkalines, for organic solvents, and metal solutions.
- Use Teflon lined caps for organic solvents.

Volumetric Glassware

Volumetric glassware should be used for accurate volume measurement. This glassware should meet the federal specifications as Class "A". Glassware is permanently marked as "A" and also with the temperature at which calibration was made. Carefully check the glassware for "TC" (To Contains) or "TD" (To Deliver) marks, and use them according to these designations. Carefully select glassware and use them professionally. Not suitable, incorrectly used, or not properly cleaned glassware

Table 8-1. Maintenance of Laboratory Instruments

Instruments	Maintenance Activity	Frequency
pH meter	Clean electrodes	D
	Refill electrodes	W or AN
	Change battery	AN
Conductometer	Clean electrode	D
	Change battery	AN
DO meter	Clean electrode	D
	Change membrane	AN
	Change battery	AN
Ion selective electrodes	Clean electrode	D
	Store electrodes	short term
	as described	long term
Reference electrodes	Clean electrodes	D
	Use proper filling solutions and storage	D
Balances	Clean pan	D
	Replace light bulb	A (I)(C)
	Adjust scale deflection	A (I)(C)
Spectrophotometer VIS/UV	Check lamp alignment	W
	Replace lamp	AN
	Clean windows	Q (I)(C)
	Clean sample compartment	D
	Clean cuvettes after use	D
Spectrophotometer IR	Clean sample cell	D
	Clean windows	M
	Change desiccant	Q
	Check gas leakage	D
Spectrophotometer AA, flame	Clean nebulizer	D
	Clean burner head	D
	Check tubing, pump, and lamps	D
	Clean quartz windows	W
	Check electronics	SA (I)(C)
	Check optics	A (I)(C)
Spectrophotometer AA, graphite	Check graphite tube	D
	Flush autosampler tubing	D
	Clean furnace housing and injector tip	W
	Check electronics	SA (I)(C)
ICP	Clean realing torch	W
	Clean nebulizer and spray chamber	W
	Check tubing and vacuum pump oil	W
	Check water lines, gases torch compartment	D
	Check electronics	SA (I)(C)
	Check wavelength calibration, adjust as needed	SA (I)(C)
	Check interelement interference standard	SA
TOC analyzer	Clean injection port	M
	Change catalyst	M
	Inspect combustion tube	SA
Gas chromatographs	Check septa, gas flow	D
	Clean GC syringes	D
	Check for leaks	D
	Replace column	Q

Table 8-1. Maintenance of Laboratory Instruments (continued)

Instruments	Maintenance Activity	Frequency
	Clean injection port	M
	Check electronics	Q (I)(C)
	Check temperature cal.	Q (I)(C)
Purge and trap	check for leaks	M
	Clean sparger	W
	Change trap	A
	Check purge flow	M
Autoanalyzers	Check for leaks, flush system	D
	Clean spill after use	D
	Clean sample probe	M
	Check tubing	M
	Clean optics	Q (I)(C)
	Clean pump rollers, platens, colorimeter filter	M
	Clean flow cells, check oil, lubricate gears	SA (I)(C)
Refrigerators, ovens, incubators	Clean interior	M
	Check temperature against certified thermometer	A
Autoclaves	Check gaskets	W
	Clean interior	M
	Sterilization indicator tape	D
	Timing mechanism check	SA (I)(C)
Turbidimeter	Clean instrument housing	M
	Clean cells	D
Thermometers	Check for cracks and gaps in the mercury	D
Autosampler	Check needles and tubing	D
	Clean	M

D = daily; W = weekly; M = monthly; A = annually; SA = semiannually; I = instrumentation specialist; C = on contract; AN = as needed; Q = quarterly.

endangers the quality of the analytical results! Laboratory glassware must be kept spotlessly clean. Volumetric, Class A glassware should not be dried by heating!

Cleaning Procedures of Laboratory Glassware

Laboratory glassware cleaning procedures depend on the parameters determined. Detailed glassware cleaning procedures should be available for each laboratory as a part of their quality QA/QC program. Store glassware to protect from contamination, dust, breaking, or chipping.

It is best to keep glassware separated by analytical groups to avoid contamination. Always use suitable detergent!

Organics: Liquinox, Alconox, or equivalent
Inorganics:
anion: Liquinox or equivalent
cations: Liquinox, Alconox, or equivalent
Microbiology: Must pass an inhibitory residue test

When cleaning glassware for metal analysis, use the following guidelines:

- Remove all labels or marks from the glassware (may use acetone).
- Wash with hot soapy water. Use Alconox, Acationox, or equivalent detergent. Use a brush to scrub inside of glassware.
- Rinse thoroughly with hot tap water.

- Rinse thoroughly with deionized (DI) water.
- Rinse with 1+1 HCl.
- Rinse with 10% Nitric acid (HNO_3).
- Rinse with DI water.

When cleaning glassware for analyzing nutrients and minerals, use the following guidelines:

- Remove all labels or marks from glassware.
- Wash with hot soapy water, use Liquinox or equivalent detergent. Use a brush to scrub inside of glassware.
- Rinse thoroughly with hot tap water.
- Rinse thoroughly with DI water.
- Rinse with 1+1 HCl.
- Rinse with DI water.

When preparing glassware for the determination of solids, use the following guidelines:

- Remove all labels or marks from glassware.
- Wash with hot soapy water, using Liquinox or equivalent detergent. Use a brush to scrub inside the glassware.
- Rinse thoroughly with hot tap water.
- Rinse with DI water.
- Bake at 180°C prior to use, as described by the method.
- For volatile solids, instead of baking at 180°C, heat in muffle furnace for 1 hour at 550°C.

When cleaning glassware for analysis of inorganic nonmetals and physical properties, use the following guidelines:

- Remove all labels or marks from glassware.
- Wash with hot soapy water using Liquinox or equivalent detergent. Use a brush to scrub inside the glassware.
- Rinse thoroughly with hot tap water.
- Rinse thoroughly with DI water.

When cleaning glassware for volatiles or purgeables (VOCs), ethylene dibromide (EDB), and trihalomethanes (THMs), use the following guidelines:

- Remove label or marks from glassware.
- Wash with hot soapy water, using Liquinox, Alconox, or equivalent detergent.
- Do not use brush with any rubber or plastic parts on it!
- Do not use detergent stored in plastic container!
- Do not use disposable plastic gloves to wash glassware!
- Rinse thoroughly with hot tap water.
- Rinse thoroughly with DI water.
- Rinse thoroughly with pesticide grade methanol.
- Bake at 105°C for 1 hour.

When cleaning glassware for analysis of pesticides, herbicides, oil and grease, total recoverable petroleum hydrocarbons (TRPH), and total phenolics, use the following guidelines:

- Rinse thoroughly with pesticide grade acetone.
- Remove all labels or marks from glassware.
- Wash with hot soapy water, using Liquinox, Alconox, or equivalent detergent.
- Never use brush with rubber or plastic parts on it!
- Never use detergent stored in plastic container!
- Never wear plastic gloves when washing the glassware!
- Rinse with hot tap water.
- Rinse with DI water.
- Rinse thoroughly with pesticide grade acetone.
- Cleaned and dried glassware should be sealed and stored in a dust-free environment.
- After the DI water rinse, drain, then heat in muffle furnace at 400°C for 30 to 60 minutes, cool, seal, and store in dust free environment.
- Another alternative to prepare the glassware is after the tap water rinse, soak in strong oxidizing agent such as Chromic acid at 40 to 50°C. Then rinse with hot tap water, then with DI water, and finally with pesticide grade acetone.
- Prior to use, glassware should be rinsed with the solvent used in the analytical method.

When cleaning glassware for microbiology testing, use the following guidelines:

- Remove all labels or marks from the glassware.
- Wash with hot soapy water using laboratory quality detergent. Use a brush to scrub inside the glassware.
- Rinse with hot tap water.
- Rinse with DI water.
- Sterilize according to the appropriate method.

Chemicals

Chemicals are manufactured in varying degrees of purity. Carefully select the grade of the chemical that meets the need of the work to be done. Always recheck the label of the chemical that you are using! The use of a wrong chemical can cause an explosion or ruin the analytical work. Check the information carefully on the chemical container: name, formula, formula weight, percent impurities, analytical grade, health hazards, and safety codes.

Grades and Purity of Chemicals

AR Primary Standard: A specially manufactured analytical reagent of exceptional purity for standardizing volumetric solutions and preparing reference standards.
AR: Analytical Reagent Grade, for all general laboratory work
ACS: The chemical meets the requirements of the American Chemical Society Committee on analytical reagent.
USP: A grade meeting the requirements of the United States Pharmacopeia.
NF: A grade meeting the requirements of the National Formulary.
TAC/FCC: Tested Additive Chemical/Food Chemical Codex. Meets the requirement for food chemical codex and satisfactory for approved food uses.
Purified: A grade of higher quality than technical.
Technical: A grade suitable for general industrial uses.
AR Select: High purity acids for trace element analysis.

OR: Organic laboratory chemicals of suitable purity for most research work and for general laboratory purposes. It is the highest grade of the particular chemical generally available.

Certified: Applies to stains certified by the Biological Stain commission and bears their label of certification.

ChromAR: Solvents specially purified for use of chromatography.

GenAR: Used for biotechnology or genetic laboratories.

Nanograde: Specially controlled for Electron Captured Gas Chromatographic (GC) techniques, such as pesticides residue analysis, etc.

ScintillAR: Used in liquid scintillometry.

SilicAR: For column and thin layer chromatography procedures.

SpectrAR: Used for spectrophotometry.

StandAR: A line of prepared solutions including various APHA and other titrants. Also includes a complete group of atomic absorption (AA) standards.

All chemicals used for mercury (Hg) analysis should be mercury-free. All chemicals used for nitrogen analysis should be nitrogen-free. The proper storage of the chemicals, reagents, and standards is summarized on Table 8-2.

Reagent Grade Water

One of the most important aspects of analysis is the quality of reagent grade water to be used for preparation of standard solutions, reagents, dilutions, and blank analysis.

Type I Water

Type I water has no detectable concentration of the compound or element to be analyzed at the detection limit (DL) of the analytical method. Use Type I water in test methods requiring minimum interference and bias and maximum precision.

It is prepared by distillation, deionization, or reverse osmosis treatment of feedwater followed by polishing with a mixed-bed deionizer and passage through a 0.2 um poresize membrane filter. Mixed-bed deionizer typically adds small amounts of organic matter to water. The quality check by measurement of the conductivity (resistivity) does not show organic, nonionized contaminants. Thus, make separate measurements of contaminants by Total Organic Carbon (TOC), silica (SiO_2) detection, and bacterial count. It cannot be stored without significant degradation. Produce it continuously and use it immediately after processing.

Type II Water

Type II water is intended to provide the user with water in which the presence of bacteria can be tolerated. It is used to prepare reagents, dyes, or staining. It is produced by distillation or deionization. Check the same contaminants as for Type I water. It may be stored, but keep storage to a minimum and provide quality consistent with the intended use. Store only in materials that protect the water from contamination, such as Tetrafluoro-ethene (TFE), well known as Teflon, or glass for organic analysis, and plastic for metals.

Type III Water

Type III water may be used for glassware washing, preliminary rinsing of glassware, and as a feedwater for production of higher quality grade waters. Select the material of the storage container to protect from contamination.

Table 8-2. Storage of Chemicals, Reagents and Standards

Chemical	Method of storage
Acids (HCl, H_2SO_4, HNO_3, Acetic acid)	Stored in original containers in special designed and "Acids" labeled cabinet, grouping by safety color code bottles
Flammable solvents	Stored in original containers, in special designed and "Flammable" labeled cabinet. Large quantities should be stored in metal safety cans, and outside the laboratory in marked "Flammable" storage area.
Solvents	Stored in separate solvent cabinet, in original containers, and in a well-ventilated area.
Chemicals used for analysis of VOCs	Stored in separate solvent cabinet, in original containers. No other chemicals stored in this area!
Chemicals	Stored in alphabetical order in chemical storage room, with records of "date of arrival" and "date opened" on each container.
Phenol Hydrogen peroxide (H_2O_2)	Stored in "Chemical Storage" marked refrigerator with closely capped and sealed containers.
Turbidity standard, ammonia, nitrate, nitrite, phosphorus stock and standard solutions. Silica stock solution: it should be stored in a plastic bottle!	Stored in refrigerator, marked as "Refrigerator for Inorganics".
Oil & grease standard	Refrigerate in sealed containers, refrigerator marked as "Refrigerator for Organics".
Stocks and standards for trace organics	Stored in vials in special marked freezer.
Metals, stock solutions	Stored in room temperature.
Metals, standards (100 to 10 ppm)	Stored in room temperature, (each contains 0.5% acid as described in the method) in designated storage area in the laboratory.
pH, conductivity standards	Stored in room temperature in properly marked cabinet.
Microbiology (sample, media, reagents)	Stored in separate, marked refrigerator in the microbiology laboratory.

Note: All stocks and standards have to be marked by the date received or prepared, expiration date, and signature.

The American Society for Testing and Materials (ASTM) specifies four different grades of laboratory water as follows.

Grade of water	Maximum total solids mg/l	Maximum conductivity umhos/cm	pH unit at 25°C
Type I	0.1	0.06	—
Type II	1.0	1.00	—
Type III	1.0	1.00	6.2–7.0
Type IV	2.0	5.00	5.0–8.0

The pH of Type I and II water cannot be measured accurately without contaminating the water. Measure other constituents as required for individual tests.

Methods for Preparation of Reagent Grade Water

Distilled water — Distillation is the procedure in which the liquid is vaporized, re-condensed, and collected. Distilled water quality depends on the type of still and quality of feedwater. Deionized feedwater is preferred.

Demineralized or deionized (DI) water — Water purified by a mixed-bed ion exchanger. Commercial resin purification trains can produce superior water quality.

Redistilled water — Prepared by redistilling single distilled water from an all-borosilicate-glass apparatus.

Reagent water — A sample of water that conforms to ASTM grades II, III, or IV.

Analyte-free water — Water is free from the substance analyzed.

Reagent grade water — This is the highest quality laboratory pure water. It is even purer than triple distilled water. It is prepared by distilled water passed through an activated carbon cartridge to remove the dissolved organic materials. It is then passed through two deionizing cartridges to remove the dissolved inorganic substances. Finally, it is passed through membrane filters to remove microorganisms and any particulate matter with the diameter as large as 0.22 um. This kind of high quality water is commercially available and used for AA, gas chromatographic (GC) work, tissue culturing, etc. The quality of the laboratory pure water should be checked regularly. Parameters should be checked and documented as on Table 8-3. The frequency of the checking is as on Table 8-4.

Table 8-3. Quality Check of Laboratory Pure Water

Parameter	Monitoring frequency	Limit
Conductivity	D	1–2 umhos/cm
pH	D	5.5–7.5 unit
Total organic carbon	A	<1.0 mg/L
Trace metal single Cd, Cr, Cu, Ni, Pb, Zn	A	<0.05 mg/L
Trace metal total Cd, Cr, Cu, Ni, Pb, Zn	A	<1.0 mg/L
Ammonia, as NH_3-N	M	<0.1 mg/L
Free chlorine, Cl_2	M	<0.1 mg/L
Heterotrophic count		
Fresh water	M	<1000 cnt/ml
Stored water	M	<10,000 cnt/ml
Water suitability test	A	Ratio: 0.8 - 3.0

A = annually; M = monthly; D = daily; cnt = count (bacterial).

Table 8-4. Documentation of Laboratory Pure Water Quality

Date	*pH*	*Cond. mhos/ cm*	*TOC*	*Cd*	*Cr*	*Cu*	*Ni*	*Pb*	*Zn*	*NH_3 as N*	*Het. Bact. c/ml*	*In.*

Parameters are reported in mg/L, except as noted

Cond. = Conductivity, umhos/cm
TOC = Total Organic Carbon
Cd = Cadmium
Cr = Chromium
Cu = Copper
Ni = Nickel
Pb = Lead
Zn = Zinc
NH_3-N = Nitrogen Ammonia
Het.bact. = Heterotrophic Bacteria Count, count/ml
In. = Initial of the logger

Chapter 9
Quality Control Requirements

A good Quality Control (QC) program of one analytical laboratory involves a number of practices. QC is a set of measures within a sample analysis methodology to assure that the process is in control. The data acquired from QC procedures are used to evaluate analytical data and to determine the necessity or the effect of correction action procedures. As all parts of a Quality Assurance/Quality Control (QA/QC) program, QC requirements are also divided into two groups: field and laboratory QC checks.

9.1. Field Quality Control Checks

Correct and usable chemical measurements should meet both field and laboratory quality check criteria. The goal of the QC of the field operation is to check for contamination during sample collection. This control is accomplished by the collection and submittal for analysis of appropriate blanks.

Equipment Blank

It is analyte-free water taken from the precleaned equipment rinsate on sampling site, *before* sample collection. Equipment blanks are collected, preserved, documented, and transported in the same manner as the samples.

Field Blank

It is analyte-free water taken from equipment rinsate at the sampling site at the *end* of the sampling event, after equipment has been cleaned in the field. A field blank is collected, preserved, documented, and transported in the same manner as the samples.

Trip Blank

Analyte-free water blank is prepared in the laboratory, transported to the field, unopened, and submitted to the laboratory with the samples. Trip blank check is recommended when collecting samples for Volatile Organic Compounds (VOCs). Each VOCs cooler should have one trip blank.

Duplicate Samples

Duplicate samples are collected at the same time from the same source. When collecting less than 5 samples, no duplicates are needed. With 5 to 10 samples, one duplicate sample must be collected, and when the number of the collected samples is more than 10, the frequency of the duplicate samples is 10% per each parameter group.

Calibration of Field Equipments

All field instruments are calibrated at the beginning of the measurements (initial calibration). Calibration check is recommended at the beginning of the measurements and in every 4 hours (continuing calibration). If the continuing calibration fails the calibration criteria, initial calibration must be repeated. All calibration data must be documented. Detailed initial and continuing calibration performance is in Chapter 7, and calibration procedures for field equipments are in Chapter 2. Detailed field QA/QC requirements are discussed in Chapter 1.

9.2. Quality Control Checks in Chemical Measurements

Laboratory QC checks include all practices and activities that provide the accuracy and the precision of the analytical measurements. It serves as a tool for the analysts and for the laboratory to approve the correctness of the generated analytical data. The laboratory should follow all of the QC requirements specified by each method. The following items are given as guidelines for the essential control of a chemical measurement.

Reagent blank — It is analyte-free water analyzed with the samples, one per sample set.

Preparation blank or method blank — If the method includes any pretreatment (digestion, distillation, etc.), one analyte-free water blank must be processed and analyzed the same way as the samples. It is used to detect any contamination during the sample preparation process.

Spiked sample or matrix spike — Spiked samples are samples that have a specific concentration of various parameters of interest added. They are used to measure the performance of the complete analytical system including any chemical interferences from the sample matrix. A small quantity of a known concentration of analyte stock solution is added to the sample (or aliquot of the sample). Always use high concentration stock solution and a small quantity of spike addition so the volume change of the sample becomes negligible. Water samples are thoroughly mixed by shaking the sample container before measuring out the sample aliquot. Soil, sediment, or other solid matrices are mixed with a glass rod or by pouring the sample into a glass mixing bowl and mixing before removing the aliquot. *The spike addition should produce a minimum level of 10 times and a maximum level of 100 times of the Instrument Detection Limit (IDL).* According to DER QA/QC requirements, the concentration of each analyte in the spiking solution should be approximately 3 to 5 times the level expected in the sample. *The spike solution must be added before the sample preparation!*

Use the following formula to calculate the volume of the spike stock solution that should be added to the sample:

$$c_1 v_1 = c_2 v_2$$

c_1 = concentration of the spike stock solution
v_1 = ? (it is the unknown)

c_2 = desired spike concentration
v_2 = volume of the sample you want to be spiked

For example, it is available at an 11,000 mg/L chloride spike stock solution. How many milliliters of stock solution are needed to add to a 50 ml sample for a 110 mg/L chloride spike value? By using the above equation:

$$11{,}000\ v_1 = 110 \times 50$$

$$v_1 = 5500/11{,}000$$

$$v_1 = 0.5 \text{ ml or } 500 \text{ ul}$$

The same formula may be used to calculate the value of the added spike, when the concentration and the added volume of the spike stock solution and spiked sample volume are known. If more than one matrix is in the analytical set, matrix spike should be prepared for all matrix types. Analyze once per set and at the rate equivalent to 5% of all samples.

Reagent water spike — Analyte-free water is spiked with the analyte and is prepared in the same way as the samples. It monitors the effectiveness of the method. Run with 5% frequency. Concentration and preparation is the same as described for a matrix spike. The level of the spike addition is recommended to be 5 times the method quantification limit.

Surrogate spikes — Surrogates are organic compounds that are similar to analytes of interest in chemical composition, extraction, and chromatography, but that are not normally found in environmental samples. Surrogate spikes are used in trace organic determination, GC and Gas Chromatograph/Mass Spectrophotometric (GC/MS) methods. These compounds are spiked into all blanks, standards, samples, and spiked samples prior to purging or extraction according to the appropriate methods.

Blind QC check samples — QC check samples (also known as Reference Material) are obtained from an independent source with known analytical level(s). The analyst is not told the true concentration value of the sample. It measures and validates the analytical system and the performance of the analyst. The frequency of the analysis is semiannual and it is a part of the QA report.

QC check standards or CVS — The QC check standards are standard solutions originated from different source other than the calibration standards. QC check standards are analyzed after the calibration of the analytical system and before to run the samples. The QC check standards are used to verify the accuracy of the calibration (also called as CVS), as was stated in Chapter 7. Run one in an analytical set or in 5% frequencies.

Duplicate samples — Samples that have been collected at the same time from the same source are called field duplicates and the analyst knows that the samples are duplicates.

Blind field duplicates — Duplicate samples, collected at the same time and from the same source, but submitted and analyzed as separate samples. The analyst does not know that they are duplicates.

Laboratory duplicates — Aliquots of the same sample that are prepared and analyzed at the same time. Duplicate samples are used to determine the precision of an analytical performance.

Duplicate matrix spike samples — Aliquots of the same sample spiked with the same concentration of the analyte. May be used to check the precision of the analytical system instead of duplicate samples.

Calibration standards, CCS, CVS, preparation blanks, and LCS — These standards are also part of the laboratory QC check. Detailed discussion of initial and continuing calibration and the functions of these standards is briefly discussed in Chapter 7.

Cleanups — All batches of adsorbents (Florisil, Alumina, Silica gel, etc.) prepared for GC shall be checked for analyte recovery by running the elution pattern with standards as a column check. The elution pattern shall be optimized for maximum recovery of analytes and maximum rejection for contaminants.

Column check samples — The elution pattern shall be confirmed with a column check of standard compounds, after activating or deactivating a batch of adsorbent. These compounds shall be representative of each elution fraction. Recovery as specified in the methods is considered an acceptable column check. A result lower than specified indicates that the procedure is not acceptable or has been misapplied.

Column check sample blank — The check blank must be run after activating or deactivating a batch of adsorbent.

Internal standard — Internal standard is a compound having similar chemical characteristics to the compounds of interest, but which is not normally found in the environment or does not interfere with the compounds of interest. The concentration in the sample is based on the response of the internal standard relative to that of the calibration standard and the compound in the standard. This is used in chromatographic analysis.

The internal standard calibration procedure is as follows. Prepare calibration standards at a minimum of five concentration levels for each analyte of interest. To each calibration standards, add a known constant amount of one or more internal standards and dilute to volume with an appropriate solvent. Prepare calibration curve by plotting the response ratios vs. response factors. Calculate the response factor (RF) for each analyte:

$$RF = (A_s\ C_{is})/(A_{is}\ C_s)$$

A_s = response of the standard
A_i = response of the internal standard
C_s = concentration of the standard
C_{is} = concentration of the internal standard

The working calibration curve or RF must be verified on each working day by the measurement of one or more calibration standards. If the response for any analyte varies from the predicated response by more than ±15%, a new calibration curve must be prepared.

Performance evaluation samples — Performance evaluation samples are reference materials obtained from an independent source for which the level(s) of analytes have been validated. Performance testing is to evaluate the ability of the laboratory to produce a specified quality of data. It measures the capability of the laboratory for certification and also serves as a tool to evaluate the qualification of the laboratory for contractual purposes. Usually the laboratory participates in performance evaluation programs because they want to offer their services for evaluation. Along with the analytical data, laboratories should have available documented evidence of their accuracy and precision, a complete Standard Operation Procedure (SOP), and an operational Quality Assurance (CompQA/QC) program. Evaluation of the results is an essential part of performance testing. Each participant should be informed of their performance in relation to an acceptable standard. (Coding the results for discretion is advised). Information for each analyte should include the true value, acceptance range, reported value, and the evaluation as "acceptable", "not acceptable", or "check for error".

Laboratories overview their own results and conduct corrective actions when needed and document all steps of the remedial actions. If no actions are deemed necessary, it improves the special performance of the laboratory.

Method of Standard Addition

In the presence of matrix interference, the application of the method of standard addition technique is useful. In this technique, the accurate concentration of the analyte is obtained without the elimination of the interfering substance. Aliquot of standards are added to portions of the sample allowing the interfering substance in the sample to also affect the standard. The work outline for the standard addition technique consists of the following steps:

- Take an equal volume of aliquot from the sample.
- Add nothing for the first aliquot.
- Add a measured amount of standard to the second aliquot. The volume of the standard is selected to give the approximate concentration of the analyte in the sample.
- Add twice the volume of the same standard to the third aliquot of the sample.
- Add three times the first addition standard volume to the next aliquot of the sample.

The measured absorbance for all of the solutions must fall within the linear portion of the working curve. All portions contain the same amount of sample volume so that the final concentrations of the original sample constituents are the same in each case. Only the added analyte is different by a known amount. If there is no interference in the sample, a plot of the measured absorbance vs. the concentration of the added standard would be parallel to the aqueous standard calibration. If an interfering substance is present, the absorbance also increases the same as the added standards and will be proportional to the analyte in the sample. Therefore, the result is also a straight line, but because of the interference substance, its slope will be different than the aqueous standards. In this situation, negative errors result. By continuing the concentration calibration on the abscissa backward from zero and extrapolating the calibration line backward until it intercepts the concentration axis, the concentration responsible for the absorbance of the unspiked sample is indicated. The graph of a standard addition technique is shown in Figure 9-1.

9.3. Detection Limits

Detection limits are the smallest concentration of an analyte of interest that can be measured with a stated probability of significance.

Method Detection Limit (MDL)

The method detection limit (MDL) is the smallest concentration of the analyte of interest that can be measured and reported with 98% confidence that the concentration is greater than zero. To determine the MDL for a certain analyte, first estimate the MDL by using the minimum detection concentration described by the method, or based on previously practice. Prepare 1 L of standard by adding the analyte to analyte-free water to make the concentration near to the estimated MDL. Analyze 7 portions

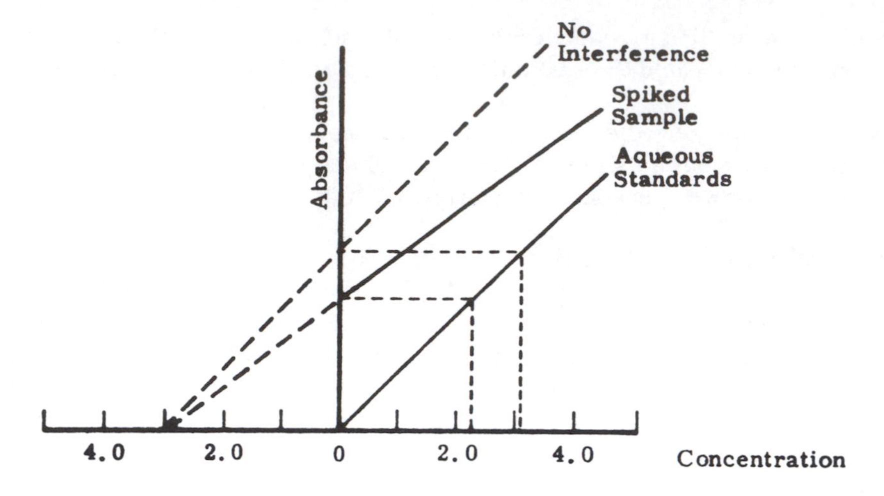

FIGURE 9-1. A set of aqueous standards represents a typical calibration line, passing through the origin. Zero absorbance is defined with analyte-free water blank. If there is no interference in the sample, a plot of the measured absorbances vs. the concentration of the added standards would be parallel to the aqueous standard calibration. If interfering substance is present, a plot of the measured absorbances vs. the concentration of the spiked sample will also result in a straight line, but because of the interference substance, its slope will be different from the aqueous standard. The result is a negative error. The concentration of the unspiked sample will be at the point where the calibration line of the spiked sample intersects the concentration axis (backward from zero). For example of standard addition technique:
(1) 10 ml sample + 10 ml analyte-free water 0 ppm added
(2) 10 ml sample + 5 ml 2 ppm standard 1.0 ppm added
(3) 10 ml sample + 10 ml 2 ppm standard 2.0 ppm added
(4) 10 ml sample + 15 ml 2 ppm standard 3.0 ppm added
(5) 10 ml sample + 20 ml 2ppm standard 4.0 ppm added
The measured absorbances are plotted against the concentration. The standard addition curve and the concentration of the sample (3.0 ppm) are represented here.

of this solution and calculate the standard deviation (s) for the results. From the "Student's t-variant table" (Table 9-1), select the value of "t" for df of 6. (df = degree of freedom, which is the number of values minus one, so 7 - 1 is 6 in this case). The value of "t" at the 98% level is 3.14. The calculation of the MDL is as follows:

$$\text{MDL} = s \times 3.14$$

In this formula, "s" is the standard deviation calculated from the analytical values of the prepared standard solution containing the analyte with the concentration of the estimated detection limit.

Practical Quantitation Limit (PQL)

The practical quantitation limit (PQL) is the smallest concentration of the analyte of interest that can be reported with a specific degree of confidence. PQL has been proposed as the lowest level achievable among laboratories within specified limits during routine laboratory operations. PQL is significant because different laboratories will produce different MDLs even though they are using the same analytical procedures, instruments, and sample matrices. The concentration related to PQL is about five times the MDL, and represents a practical and routinely achievable detection limit with a relatively good certainty that any reported value is reliable. To

Table 9.1. Student "t" Variate Table

*	80%	90%	95%	98%	99%	99.73%
df	$t_{.90}$	$t_{.95}$	$t_{.975}$	$t_{.99}$	$t_{.995}$	$t_{.9985}$
1	3.078	6.314	12.706	31.821	63.657	235.80
2	1.886	2.920	4.303	6.965	9.925	19.207
3	1.638	2.353	3.182	4.541	5.841	9.219
4	1.533	2.132	2.776	3.747	4.604	6.620
5	1.476	2.015	5.571	3.365	4.032	5.507
6	1.440	1.943	2.447	3.143	3.707	4.904
7	1.415	1.895	2.365	2.998	3.499	4.530
8	1.397	1.860	2.306	2.896	3.355	4.277
9	1.383	1.833	2.262	2.821	3.250	4.094
10	1.372	1.812	2.228	2.764	3.169	3.975
11	1.363	1.796	2.201	2.718	3.106	3.850
12	1.356	1.782	2.179	2.681	3.055	3.764
13	1.350	1.771	2.160	2.650	3.012	3.694
14	1.345	1.761	2.145	2.624	2.977	3.636
15	1.341	1.753	2.131	2.602	2.947	3.586
16	1.337	1.746	2.120	2.583	2.921	3.544
17	1.333	1.740	2.110	2.567	2.898	3.507
18	1.330	1.734	2.101	2.552	2.878	3.475
19	1.328	1.729	2.093	2.539	2.861	3.447
20	1.325	1.725	2.086	2.528	2.845	3.422
25	1.316	1.708	2.060	2.485	2.787	3.330
30	1.310	1.697	2.042	2.457	2.750	3.270
40	1.303	1.684	2.021	2.423	2.704	3.199
60	1.296	1.671	2.000	2.390	2.660	3.130
	1.282	1.645	1.960	2.326	2.576	3.000

** Columns to be used in calculating corresponding two-sided confidence interval*
df = degree of freedom = n - 1 (n = number of measurements)
After John Keenan Taylor "Quality Assurance of Chemical Measurements" (1988. p.267.)

determine PQL, take the standard deviation (s) from the determination of MDL, and multiply by 10.

$$PQL = s \times 10$$

s = standard deviation determined by establishing MDLs, as described above.

The 10 standard deviation corresponds to an uncertainty of ±30% in the measured value at the 98% confidence limit.

Instrument Detection Limit (IDL)

The instrument detection limit (IDL) is the concentration of the analyte that produces a signal greater than five times the signal/noise ratio of the instrument. An operating analytical instrument usually produces a signal greater than three standard deviations of the mean noise. To determine the IDL, analyze the analyte free water seven times and record the signal of the instrument. Calculate the standard deviation of the instrument responses and multiply by three.

$$IDL = s \times 3$$

s = standard deviation of the average signal of the instrument by analyzing analyte-free water

9.4. Precision, Accuracy, and Bias

The satisfying quality of analytical data may be characterized by accuracy, precision, and bias.

Accuracy: The degree of agreement of a measured value with the true or expected value of the quantity of concern.

Precision: The degree of mutual agreement among individual measurements as the result of repeated application under the same condition.

Bias: A systematic error in the method, for example, temperature effect or extraction inefficiencies, contamination, calibration errors, etc. A measurement is biased when the error of a limiting mean is not zero caused by systematic error.

Accuracy

Accuracy is measured and expressed as *% Recovery* and calculated according to the following formula:

$$\% \text{ Recovery } (\% \text{ R}) = \text{Analytical value} \times 100 \;/\; \text{True value}$$

$$\% \text{ Recovery on spike } (\% \text{ R}_{sp}) = (\text{Spiked sample value - Sample value}) \times 100 \;/\; \text{Spike value}$$

Precision

Precision measures the variation among measurements and may be expressed in different terms.

Standard Deviation(s):

$$s = \sqrt{E(x - \overline{X})^2 / n - 1}$$

E = sum
x = measurements
$\overline{x}$ = mean
n = number of measurements

Relative Standard Deviation (RSD):

- Calculate the mean $(\overline{x})$.
- Calculate the standard deviation (s).
- Calculate the coefficient of variant (CV).
  CV = standard deviation(s)/mean $(\overline{x})$.
- Calculate the Relative Standard Deviation (RSD).

$$\text{RSD} = \text{CV} \times 100$$

Relative Percent Difference (RPD): RPD is the difference between the duplicate values divided by the average of the duplicate values and multiplied by 100.

$$\text{RPD} = [(A - B)/(A + B/2)] \times 100$$

or shortly

$$\text{RPD} = [(A - B)/(A + B)] \times 200$$

9.5 Quality Control Limit Delineation for Precision and Accuracy

The calculated accuracy (% recovery) and precision (s, RSD, or RPD) values are used for the determination of QC limits per parameters.

Representation of Control Limits for Accuracy

- Collect twenty % Recovery data for the interested parameter.
- Calculate the mean $(\bar{x})$ and the standard deviation (s).
- Calculate the accuracy QC limits as % Recovery.

$$\text{WARNING LIMITS} = \bar{x} \pm 2s$$

$$\text{Upper Warning Limit (UWL)} = \bar{x} + 2s$$

$$\text{Lower Warning Limit (LWL)} = \bar{x} - 2s$$

$$\text{CONTROL LIMITS} = \bar{x} \pm 3s$$

$$\text{Upper Control Limit (UCL)} = \bar{x} + 3s$$

$$\text{Lower Warning Limit (LWL)} = \bar{x} - 3s$$

% Recovery data are outside warning limits indicating that the system is approaching an out of control situation and may require corrective action. Any data falling outside control limits signifies an out of control system. Analysis must be stopped and corrective action taken before further analysis. Samples analyzed with the failed QC check sample should be analyzed again.

Representation of Control Limits for Precision

- Collect 20 data, as precision expressed (for example Relative Percent Difference, RPD) for the interested parameter.
- Calculate the mean $(\bar{x})$.
- Calculate standard deviation (s).
- Calculate the QC limits for precision using RPD values. Because the minimum value for precision is zero, the lower limits should always be zero (cannot be less than zero!).

$$\text{WARNING LIMITS} = 0 - (\bar{x} + 2s)$$

$$\text{Lower Warning limit (LWL)} = 0$$

$$\text{Upper Warning Limit (UWL)} = \bar{x} + 2s$$

$$\text{CONTROL LIMITS} = 0 - (\bar{x} + 3s)$$

$$\text{Lower Control Limit (LCL)} = 0$$

$$\text{Upper Control Limit (UCL)} = \bar{x} + 3s$$

Any RPD value falling above the warning limit should act as a signal that the system is approaching an out of control situation and may indicate the need for corrective action. Data falling out of the control limits indicate that the system is out of control.

Quality Assurance Confidence Limits for Precision and Accuracy

By the completion of any analysis, the calculated precision and accuracy values are documented on the related QC charts, and recorded in summary log forms, as on

Tables 9-2 , 9-3, and 9-4. Based on these collected statistical results, QA "target limits" should be established for each analyte per matrices. These limits serve as a confidence range for precision and accuracy values.

Calculation of Confidence Limit or Target Limit for Precision and Accuracy

- Collect minimum 20 RPD data (for precision) or 20 % Recovery data (for accuracy).
- Calculate the mean ($\bar{x}$) of these data.
- Calculate the standard deviation (s).
- Calculate the Confidence Interval (CI) according to the following formula:

$$CI = \bar{x} \pm ts/\sqrt{n}$$

The value for t depends on the number of degrees of freedom (n - 1) for a 98% level of confidence, and taken from the student's "t" table (Table 9-1). For example:

$\bar{x} = 98$ $n = 20$ $n - 1 = 19$

$s = 5.26$ $t = 2.539$

Table 9-2. Monitoring Form for Precision (RPD) Values

ANALYTE __________ *METHOD* __________ *MATRIX* __________

Unit, values are expressed : __________

Date	Sample value (A)	Duplicate value (B)	RPD %	Control Limit	Remark	Sign.

RPD = Relative Percent Difference

RPD = [(A-B)/(A+B)/2)] x 100 or shortly

(A-B / A+B) x 200

$$98 + (2.539)(5.26)/4.47 = 98 + 2.99 = 101$$

$$98 - (2.539)(5.26)/4.47 = 98 - 2.99 = 95$$

The confidence limit for accuracy is 95 to 101%.

9.6. Quality Control Charts

Control charts are statistical tools for monitoring the performance of a particular task on a continuing basis. The concept of control charts was developed in 1934 by Walter Shewhart to monitor industrial processes. Later these charts were adapted to monitor chemical measurements and became very useful in demonstrating statistical control.

Accuracy Control Chart

Accuracy Control Chart monitors the % Recovery with standard deviation being the limiting control. The control chart is prepared for each test parameter after 20 determinations have been performed. The mean is plotted with the warning control limit being ± 2 SD and the upper and lower control limits, respectively, being ± 3 SD. In other words, the upper and the lower warning limits (UWL, LWL) are defined as the limits that would encompass 95% of the measured values for % Recovery, and the upper and lower control limits (UCL, LCL) are defined as the limits that would encompass 99% of the measured values of the % Recovery.

Table 9-3. Monitoring Form for Accuracy (% Recovery) Values

ANALYTE ______________ ***METHOD*** ______________ ***MATRIX*** ______________

Unit, values are expressed: ____________________

Date	QC check sample true value	QC check sample measured value	Recovery %	Control limit	Remark	Sign.

% Recovery (%R) = measured value x 100 / true value

Table 9-4. Monitoring Form for Spike Recovery ($\%R_{spike}$) Values

ANALYTE ________________ *METHOD* ________________ *MATRIX* ____________

Unit, values are expressed: ______________________

Date	Sample value (SV)	Spike added (SA)	Spiked sample value (SSV)	Recovery %	Control limit	Remark	Sign.

$\% R_{spike} = [(SSV - SV) / SA] \times 100$

______________________________ UCL ($\bar{x} + 3s$)
______________________________ UWL ($\bar{x} + 2s$)
______________________________ $\bar{x}$ (mean)
______________________________ LWL ($\bar{x} - 2s$)
______________________________ LCL ($\bar{x} - 3s$)

Precision Control Chart

The precision control chart monitors the repeatability of a measurement system disregarding accuracy. It is based on the term that was selected to express precision values. These charts are prepared from the duplicate sample results and are prepared after accumulating 20 data points. Precision control charts have only upper warning and upper control limits. The lower limit for both warning and control limit is 0.

______________________________ CL ($\bar{x} + 3s$)
______________________________ WL ($\bar{x} + 2s$)
______________________________ $\bar{x}$ (mean)
______________________________ 0

Both the accuracy and the precision control charts should contain the name of the laboratory, the test parameter, analytical method (method number with reference), range, and calculated standard deviation. All the initial data used to prepare the control limits is established by the methodology and it should be less than 5%

exceeding the upper and lower warning limits. QC charts are used on a daily basis by the analyst.

9.7. Interpretation of the Control Charts

Valuation of the Conditions of the Accuracy Control Charts

The accuracy control chart is based on % Recovery data that may exhibit results exceeding upper and lower limits, as stated above. Possible conditions appear on the Accuracy Control Chart and are represented on Figure 9-2.

Condition is satisfactory — Data is variable showing no trends and remaining within the warning limits.

Condition is critical — Any point outside the UWL and LWL. Seven successive points in the same direction causing either an upward or downward trend. Ten successive points on the same side of the average value (x) of the chart

Condition is out-of-control — Any points outside the UCL and LCL.

Valuation of the Conditions of the Precision Control Chart

The precision control chart is based on Relative Percent Differences (RPD), which are absolute numbers. Therefore, RPD values can never be less than zero, as reflected in the control chart. Possible conditions appear on the chart is represented on Figure 9-2.

Condition is satisfactory — Data is variable showing no trends and remaining below the warning limits.

Condition is critical — Any point above warning limit (WL); seven successive points in the same direction causing an upward trend

Condition is out-of-control — Any point outside the control limit (CL).

Each analytical method has established Accuracy and Precision Control Limits according to the control charts that are used to determine the acceptability of data on a continuous basis. The response to an out-of-control event must be immediate. The conditions need to be documented. The report includes the out-of-control test parameter, date, description of the QC problem, and the necessary corrective action. Once the out-of-control condition has been corrected, the QC requirements will be doubled until ten satisfactory data points have been plotted on the control chart.

9.8. Quality Control Checks in Microbiology Testing

Microbiological examinations determine the sanitary quality of environmental samples. The significance of coliform group density is a criteria of the degree of pollution and thus of sanitary quality. Coliform bacteria group, including *Escherichia coli* (*E. coli* or *Fecal coli*) bacteria, is a natural inhabitant of the human digestive tract, and its presence in water indicates that the water is contaminated with fecal material. Therefore, they are called "indicator bacteria". Routine microbiological tests are included in the detection and enumeration of total coliform, and *E. coli* or Fecal coliform bacteria. Heterotrophic plate count is also routinely used to measure water treatment plant efficiency, and general bacterial composition of source water. Non-routine microbiological examination includes the differentiation of total coliform group, detection of pathogenic organisms, pathogenic protozoa, and viruses. A laboratory quality assurance program is necessary to remove and reduce errors that may occur in any laboratory operation, including microbiological testing. Minimal Quality Assurance criteria are discussed in the following areas:

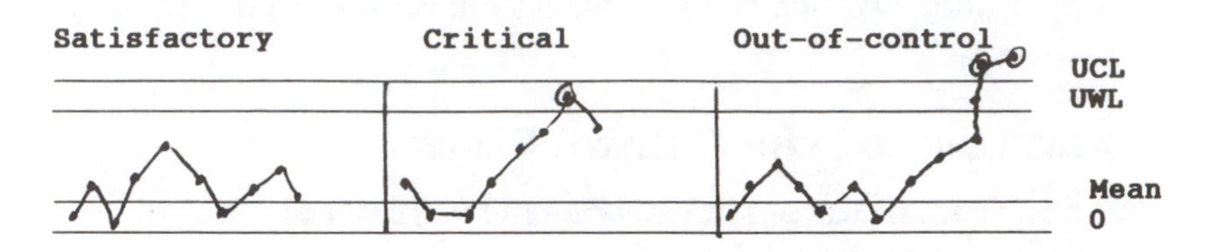

Satisfactory - Data is variable, showing no trends and remaining below the warning limit

Critical - Any point above Upper Warning Limit (UWL)
Seven (7) successive pints in the same direction causing upward trend

Out-of-Control - Any point outside the Upper Control Limit (UCL)

ACCURACY CONTROL CHARTS

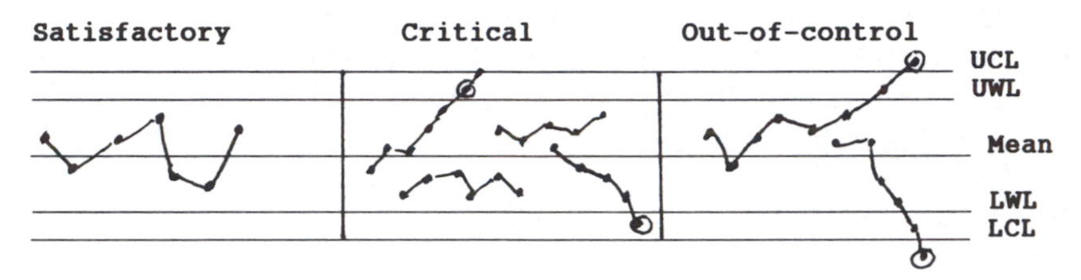

Satisfactory - Data is variable showing no trends and remaining within the warning limits

Critical - Any point outside the Upper and Lower Warning Limits (UWL and LWL)
Seven (7) successive points in the same direction causing either an upward or downward trends
Ten (10) successive points on the same side of the average value of the chart

Out-of-Control - Any points outside the Upper and Lower Control Limits (UCL, LCL)

FIGURE 9-2. Interpretation of QC Charts

- Facilities and personnel
- Laboratory equipments and instrumentation
- Laboratory supplies
- Analytical QC procedures
- Records and data reporting

Facilities and Personnel

Laboratory Design

- Well-ventilated laboratory area recommended, free from dust, drafts, and extreme temperature changes. Central air-conditioning is preferred.
- The laboratory design and operation should avoid or minimize heavy traffic and visitors in the area. It is desirable to provide separate area for sterile work.
- Bench spaces for stand-up work and for sit-down activities should be available.

Bench tops should be stainless steel, epoxy plastic, or other smooth surface that is inert and corrosion-resistant.

- Walls and floors must cover with smooth finish that is easily cleaned and disinfected.
- Maintain high standard of cleanliness in the laboratory. Regularly clean the laboratory, wash benches, shelves, floors, and windows. Wet-mop floors and treat with a disinfectant solution. Do not sweep or dry-mop. Wipe bench-tops and wash with disinfectant before and after work.
- Check and monitor air quality of the laboratory by air density plates, and the bench tops by swab test.
- Microbiological testing should be done by professional microbiologist, or by a person who is specially trained and regularly supervised by a microbiologist.

QC of Equipments and Instruments

Thermometers and Temperature Recording Devices

Check the accuracy of thermometers and temperature recording devices at least semiannually against a certified National Bureau of Standards (NBS) thermometer (Table 7-14).

Balances

Wipe balance before and after each use, and wipe spills immediately with a damp towel. Check weights monthly against certified weights and document as on Table 7-7. For weighing 2 g or less, use an analytical balance, and for larger quantities, use a toploading balance with a sensitivity of 0.1 g at a 150 g load. Protect balance and weights from laboratory atmosphere, corrosion, and moisture. Maintenance of the balance should be done by qualified experts on a base of annual contract. Maintenance activities are also kept on record.

pH Meter

Calibrate the meter and electrode each time before use with pH 7.00, 4.00, and 10.00 pH buffers, and adjust temperature correction. Record the source, date of arrival, date of opening, and expiration date of the buffers as on Figure 7-1. Follow calibration procedure as described by the manufacturer. Detailed calibration procedure and regular maintenance of the meter and electrode are discussed in Chapter 2.

Water Deionization Unit

Commercial systems are available that contain prefiltration, mixed-bed resins, activated carbon, and final filtration to produce ultra-pure water. Monitor daily pH and conductivity check, monthly bacterial population check by Heterotrophic Plate Count method, and trace metal content yearly. When bacterial population exceeds 1000/ml, change cartridge, or filter product water through 0.22 um pore size membrane filter to remove bacterial contamination.

Laboratory Pure Water Quality

Check and monitoring criteria for the laboratory pure water used in the microbiology laboratory is similar to that discussed in Chapter 8, as on Figure 8-1. Test for water bacteriological quality (suitability test) is based on the growth of Enterobacter aerogenes in a chemically defined minimal growth media. The presence of a toxic agent or a growth promoting substance will alter the 24 hour population by an increase or decrease of 20% or more when compared with a control. Detailed procedure should be in the Microbiology Laboratory Standard Operation Procedures (SOP).

UV Sterilizer

- Remove plug from outlet and clean UV lamp monthly by wiping with a soft cloth moistened with ethanol.
- Test UV lamp by light meter quarterly. If it emits less than 80% of initial rated output, replace lamp.
- Perform spread plate irradiation test quarterly, as follows:
  - (a) Prepare 100 ml Heterotrophic plate count agar.
  - (b) Pour 10 to 15 ml of the melted agar into Petri-dishes. Keep covers opened slightly until agar has hardened and moisture and condensation is evaporated. Close dishes and keep in refrigerator until needed.
  - (c) Prepare a Coliform culture and dilute to give about 200 to 250 count per 0.5 ml.
  - (d) Pipet 0.5 ml from the culture for selected dishes. Spread inoculum with a sterile glass rod or sterile spreader, or rotate plates several revolutions so the inoculum will spread uniformly over the surface of the agar.
  - (e) With cover removed, place agar spread plates under the UV lamp.
  - (f) Place one inoculated plate under ordinary laboratory lighting as a control.
  - (g) Expose plates for 2 minutes.
  - (h) Close plates and incubate at 35°C for 24 hours. Count the colonies. Control plate should contain 200 to 250 bacterial colonies.

UV treated plates should show 99% reduction in counts. If the reduction is less than 80%, replace lamp.

Membrane Filter Apparatus

- Check funnel support for leaks on a daily basis.
- Check funnel and funnel support to make certain they are smooth.
- Discard funnel if inside surfaces are scratched.
- Clean thoroughly after each time used, wrap in aluminum foil, and autoclave.

Centrifuge

Check brushes and bearings semiannually, and check the rheostat control against a tachometer at various loadings every six months to ensure proper gravitational fields.

Microscope

Clean optics and stage after each use; use only lens paper for cleaning. Keep microscopes covered when not in use. Establish yearly maintenance contract.

Water Bath

Check and record temperature daily, by keeping a thermometer completely immersed in the water. Clean bath monthly, or as needed.

Use only stainless steel, rubber, plastic coated, or other corrosion-proof racks in waterbath.

Incubators

Check and monitor temperature daily on top and bottom shelves. If a glass thermometer is used, bulb and stem must be immersed in water to the mark on the stem. For recording temperature use forms on Tables 7-11 and 7-12.

Hot Air Oven

Monitor temperature with a thermometer is accurate in the 160 to 180°C range. Test the performance of the oven with commercially available spore strips or spore suspension quarterly. Use heat-indicating tapes to identify supplies and materials that have been exposed to sterilization temperature. Daily monitoring form of the temperature is on Table 7-9.

Refrigerators and Freezers (4°C)

Check and record temperature daily as on Table 7-13. Monthly cleaning and rearrangement of the interior of the refrigerator is recommended. All material should be dated and identified clearly in the refrigerator. Defrost and discard updated material in refrigerator and freezer compartment every three months and record the activities.

Autoclave

Record temperature and pressure each time autoclave is used. Check operating temperature weekly. Test performance with spore strip or suspension weekly, and record. If evidence of contamination occurs, check until the cause is identified and eliminated. Use heat- sensitive tape to identify supplies and materials that have been sterilized.

Quality Control of Supplies

Utensils and Containers for Media Preparation

Use containers of non-corrosive and non-contaminating materials such as Pyrex glass, stainless steel, or aluminum.

Glassware

With each use, examine glassware, especially screw-cap dilution bottles and flasks, for chipped or broken edges and etched surfaces. Discard chipped or badly etched glassware. Inspect glassware after washing: if water beads are excessive on the cleaned surface, wash again.

Test for Alkaline or Acid Residues

Adding bromthymol blue is particularly advantageous for this check because it shows color changes of yellow (acid) to blue-green (neutral) to blue (alkaline) in the pH range of 6.5 to 7.3. Preparation of bromthymol blue indicator (0.04 %) is as follows: 18 ml 0.01 M NaOH, 0.1 g bromthymol blue and dilute to 250 ml with DI water.

Test for Residual Detergent

- Wash and rinse six petri-dishes in the usual manner (group A).
- After normal washing, rinse a second group of six Petri-dishes one time with DI water (group B).
- Wash six petri-dishes with the detergent and dry without rinsing (group C).
- Sterilize all three groups of dishes.
- Proceed heterotrophic count determination by using Coliform inoculates, 0.5 ml.

The results are interpreted as: differences in bacterial counts of 15% or more among all groups indicate the detergent has no toxicity or inhibitory effect. Differences in bacterial growth of 15% or more between group A and B demonstrate that inhibitory residues are left on glassware after normal washing procedures are used. Disagreement in averages of less than 15% between group A and group B, and greater than 15% between group A and group C indicate that the detergent used has inhibitory properties that are eliminated during routine wash.

Sterility Check on Glassware

After sterilization of bottles, flasks, or tubes, and after one of each bench has been sterilized, add aerobic and anaerobic broth, incubate, and check for bacterial growth. Lauryl triptose broth is excellent for this purpose.

Reagents Used in Microbiological Testing

Use only chemicals with ACS or equivalent grade, because impurities can inhibit bacterial growth, provide nutrients, or fail to produce the desired reaction. Date of receiving and first opening of the chemicals should be written on the container. Prepared reagents are properly labeled with name and concentration, date of preparation, and with the initial or signature of the preparer. Care should be taken for the storage of the reagents (refrigerator, separate from other reagents, etc.). Refrigerator for reagents used in microbiological testing should be separate for the purpose. Dyes and stains are commercially available. Use only dyes certified for biological use by the biological stain commission.

Quality Control on Rinse and Dilution Water

Record should be kept on:

- Preparation of stock buffer solution and its pH (pH should be adjusted to pH 7.2).
- Stock buffer autoclaved at 121°C, and stored at 4°C. Label correctly, by name, date, signature of preparer, and exact pH.
- Working buffer solution, diluted from stock buffer solution. Label correctly, with the date of preparation and the exact pH.

Quality Control Check on Media

Order media in quantities to last no longer than one year. When practical, order media in the smallest bottles available and multiply to the desired quantity, to keep sealed longer. When the media is prepared, record manufacturer, lot number, amount and appearance of media, date received, date opened, and expiration date. Check inventory regularly for reordering. Discard media that have changed from the original appearance, for example, discolored or hardened.

- The shelflife for unopened bottles is two years at room temperature.
- Use open bottles of media within six months after opening.
- Date of preparation, pH of the media, and sterilization should be monitored.
- Ready media stored in refrigerator, with name of the media and date of preparation. Hold no longer than three months.
- Media stored in refrigerator are incubated overnight prior to use and tubes with air bubbles are discarded.
- Broth media are used within 96 hours.
- Agar media are used within two weeks. Seal prepared agar plates in plastic bags and refrigerate to retain moisture.
- Ampouled media stored in refrigerator, expiration date designated by the manufacturer.

Quality Control on Membrane Filters and Pads

Membrane filters should be 47 mm diameters, with 0.45 um pore size. Absorbent pads diameter is 47 mm, thickness is 0.8 mm, capable of absorbing 2.0 ml Endo broth. Manufacturers provide information and certify that their membrane filters are satisfactory for water analysis. To maintain quality control for membranes, inspect each lot of membranes before use and during testing to insure that they are round and pliable, with undistorted gridlines.

Quality Control in Routine Analysis

Check precision of *duplicate analyses* for each different type of sample examined. Run duplicate analysis in 10% frequency.

Negative (Sterile) Controls

Each series of samples should include control samples, to check sterility of the media, and the dilution water (buffered water). Empty media and dilution water sample is tested at the beginning of the test, following every tenth samples, and at the end of the run. When sterile controls indicate contamination, samples affected should be rejected and a request made for immediate resampling of those samples involved.

Colony Verification

Colonies from positive test plates are verified by biochemical reaction according to the methodology.

Positive Control Sample

At the end of the analysis, one positive control culture must filter and incubate, to check the media. Always make positive control at the end to avoid contamination of the samples.

Air Density Plates

Prepare heterotrophic plate count agar. Remove petri-dish covers and place dishes top side up. Expose dishes with the agar in selected work sites for 15 minutes. Replace covers and incubate at 35°C for 48 hours. The number of organisms that settle after 15 minutes exposure on one Petri-dish is equivalent to that for 1 ft^2/min, because the area of the Petri-dish is approximately 1/15 ft^2. The microbial density normally should not exceed 15 colonies per square foot. Air density plates should be taken weekly.

Summary List of the Quality Control Documentation in the Microbiology Laboratory

pH Meter

Manufacturer ________________ Model ______________________

Clean, calibrated to 0.1 pH unity each time used, record must be maintained.

Balance Toploading

Manufacturer ________________ Model ________________________

Clean, detect a 50 mg weight accurately (for a general media preparation of >2 g quantities).

Temperature Monitoring Devices
Accuracy checked against a certified thermometer. Thermometer legible graduations in 0.5°C increments.

Incubator
Manufacturer ________________ Model ______________________

Sufficient size for work load. Uniform temperature maintained on shelves (35°C). Temperature recorded daily; temperature measured by calibrated thermometer with bulb immersed in liquid and located on shelves in use.

Autoclave
Manufacturer ______________ Model ______________________

Reaches sterilization temperature 121°C and maintains temperature during sterilization cycle, requires no more than 45 minutes for a complete cycle. Pressure and temperature gauges on exhaust side and an operating safety valve. No air bubbles produced in fermentation vials during depressurization. Record maintained on time and temperature for each sterilization cycle. Thermometer bulb in sand.

Refrigerator
Temperature maintained at 1°C to 4.4°C (34 to 40°F).

Membrane Filtration Equipment
Manufacturer________________ Model______________________

Made of stainless steel, glass or autoclavable plastic that is non-leaking and uncorroded.

Membrane Filters and Pads
Manufacturer ______________ Model ____________________

Filters must be recommended for water analysis by the manufacturer. Filters and pads are presterilized or autoclavable.

Glass, Plastic, and Metal Utensils for Media Preparation
Washing process provides glassware free of toxic residue as demonstrated by the inhibitory residue test. Glass items are free of chips and cracks. Utensils are clean and free from foreign residues or dried media. Plastic items are clear with visible graduations.

Sample Bottles
Wide mouth, hard glass bottles; stoppered or plastic screw capped, capacity at least 120 mls. Glass stoppered bottles with tops covered with aluminum foil or kraft paper. Screw caps have leak proof nontoxic liners that can withstand repeated sterilization. Sterile sample bottles contain 10 ml $Na_2S_2O_3$ as dechlorinating agent per 100 ml sample. Sterile plastic cups or sterile plastic bags are commercially available.

Hot Air Oven
Manufacturer________________ Model ____________________

Operates at minimum 170°C. Thermometer inserted or oven equipped with temperature recording thermometer device is available.

Pipets

Brand ______________________ Type ____________________________

Sterile glass or plastic, with a 2.5% tolerance. Tips unbroken, graduation distinctly marked.

Pipet Containers

Aluminum stainless steel or pipets wrapped in quality kraft paper. Open packs of disposable sterile pipets resealed between uses.

Culture Dishes

Brand ________________________ Type _________________________

Sterile plastic or glass. Open packs of disposable sterile plastic dishes resealed between uses. Dishes are in containers of aluminum or stainless steel with covers or are wrapped with heavy aluminum foil or char-resistant paper.

Culture Tubes and Closures

Sufficient size to contain sterile medium and sample without danger or spillage. Metal or plastic caps. The material is borosilicate glass or other corrosion-resistant glass.

Part 3

Data Quality and Reporting

Chapter 10
Raw Data Conversion Into Reportable Results

Raw data is data generated by analytical performances, including Quality Control checks. *Reportable data* or reportable results are generated from raw data by mathematical or statistical calculations. Final results may be produced by direct reading from the instrument or calculating from readings or instrument outputs by using formulas. Before starting calculations, assure that all readings or outputs are correct and the selected formula is appropriate. The formula and calculations used should be recorded in the laboratory notebook or on the worksheet. As for all records, calculations are entered in ink and mistakes are never erased, just cross through the error with a single line, date and sign. Generation of raw data and all of the related calculations and complete recordkeeping are the responsibility of the analyst.

10.1. Responsibilities of the Analyst

The following list contains the duties of the analyst from the choosing of proper samples for the particular analysis through the actual analytical performance, QC checks, documentation, calculations, recognition and correction of problems, and giving the accurate, defendable result.

- Analytical work should be planned and organized so the laboratory time will be used efficiently.
- The analyst should be familiar with the test method described in the laboratory SOP.
- If the samples need pretreatment (drying, distillation, digestion, extraction, etc.), start as soon as possible because they are usually time consuming procedures. If sample preparation is the duty of another laboratory section, the analyst has to be sure that the preparation of the sample was correct and collect all of the information needed to calculate the final result (volume or weight of the sample used for treatment, final volume after treatment, moisture of solid, any dilution, or concentration etc.) was collected. Carefully select the accompanying pretreated QC check samples (duplicates, spiked samples, laboratory control samples, preparation blank, equipment blanks, trip blanks, etc.).
- When samples are stored in a refrigerator, remove the bottles prior to starting the analysis to allow samples to warm up to room temperature.
- Carefully check sample label for identification and scrupulously select samples according to the proper bottle-type and preservation.
- Collect and check stocks, standards, reagents, QC check standards and preparation date. When the expiration date indicates, discard these solutions,

prepare new ones by following the SOP and be sure that all of the related preparation log forms are completed. When reagents are stored in a refrigerator, leave time to warm up to room temperature before using them!

- Standardize solutions, when applicable, and document on the designated form.
- Take the suitable workpaper form, according to the test and matrix, and prepare for starting.
- Collect the necessary glassware, check the cleanliness and the appropriate washing procedure. Improperly cleaned glassware will easily ruin the test! Mark glassware with sample ID number and organize.
- Switch instrument "on" (if applicable) to give time to "warm up".
- Prepare calibration standards, when the method indicates, by choosing the proper range, and document each step properly as stated in Chapter 7.
- Prepare calibration curve and verify its acceptance according to the initial and continuing calibration criteria as detailed in Chapter 7. Be sure that calibration curve and linear regression calculations are documented, and keep the records by manual forms, strip charts, or tabular printouts.
- When a previously approved calibration curve is available, perform continuing calibration check, and identify the acceptance or rejection of the calibration curve. In the case of adoption of the existing curve, analysis of the samples may begin. If the acceptance criteria fails, discard the former calibration and prepare fresh one.
- Measure the precision and accuracy of the analytical performance by the required QC checks (duplicate, QC check standard, QC check sample, matrix spike, etc.) as described in the methodology and in the SOP. When the QC data are not acceptable, look for the corrective action and solve the problem. Never report any results with a doubtful QC check!
- If the analyte of interest shows a higher value than the concentration of the highest calibration standard, the analyst should dilute the sample for an accurate reading. Dilution technique and proper calculation for the final value should be documented on the working paper or in the laboratory notebook.
- Turbidity, color, or other interferences should be corrected according to the methodology.
- The analyst should run the test by following the approved method and incorporate all QC checks required by the laboratory Comprehensive Quality Assurance/Quality Control (CompQA/QC) Program or by the QA/QC Project Plan.
- The analyst should be knowledgeable to recognize problems, initiate and conduct corrective actions, and keep all documentation related to the analysis clean and in order to be ready for inspection at any time.
- The analyst should be able to protect and defend all the raw data as well as the reported results.
- By the end of the analysis, switch "off" the instrument, collect analytical wastes in designated containers (if applicable), collect dirty glassware and transfer to the washing area, clean the used working area, and transfer stocks, standards, reagents, and samples to the appropriate storage area or to the designated refrigerators.
- Collect all documentation, calculate results and QC checks, plot the accuracy and precision data on QC charts, and save on the tabulated summary log form. If there is any deviation from the confidence limits, the analyst should find the

cause of the failure and correct it, or report to the laboratory supervisor and QA Officer for further assistance. All the sample data determined at the time the "out of control" condition occurred must be labeled as "suspect data" and reanalyzed after the problem is solved.

- If all the QC data agree with the analytical values, the analyst transfers the result to the analytical report summary log paper or enters it into the computer. Collect all documentation to be ready for further questions or checking.

10.2. Calculations for Final Values

Raw data produced by analytical performances should be converted to reportable, final results in the appropriate unit. By using the proper formula, the analyst calculates the final values and registers them on the laboratory "result" sheet or enters them into the laboratory computer system, to be ready for further supervision and checks. The most commonly used calculations are as follows.

Temperature Correction for Correct pH Values

pH meters are usually automatically adjusted to the desired temperature. If this correction is not available, the temperature of the sample should be associated with the pH value. When the temperature is not specified in the report, it should be 25°C.

Temperature Compensation for Conductivity

Electrolytic conductance increases with temperature at a rate of approximately 2% deviation per degree Celsius. If the temperature of the sample is below 25°C, add 2% of the reading per degree difference. If the temperature of the sample is above 25°C, subtract 2% of the reading per degree.

The following formula for temperature correction may be used as an alternative:

$$\text{umhos/cm at } 25°\text{C} = \text{measured conductivity}/1 + 0.019(t - 25)$$

Always report conductivity at 25°C.

Solids, Moisture, and Ash Calculations

The given unit with the complete and clear description of the reported analyte is an important and critical part for a correct analytical result. Determination of solids, moisture, or ash may be expressed as milligram per liter or as percent. Solids as Total Solids (TS), Suspended Solids (SS), and Total Dissolved Solids (TDS) are reported in milligrams per liter, moisture and ash are reported in percent, according to the following calculations.

Solids

$$\text{mg/L solids (TS,TDS,SS)} = [(A - B)1000/\text{ml sample}] \times 1000$$

A = TS: weight of dish and residue dried at 105°C in grams
TDS: weight of dish and residue dried at 180°C in grams
SS: weight of dish and residue dried at 105°C in grams
B = TS, TDS, SS: weight of dish in grams

$$\%\ \text{Solids (TS, TDS, SS)} = (A - B)\ 100/\text{ml sample}$$

% Solids may also be calculated from the determined milligram per liter values, by dividing the milligram per liter value by 10,000.

$$\% \text{ Solids} = (\text{mg/L})/10{,}000$$

Moisture

Moisture of any solid is determined by drying a known quantity aliquot of the well-mixed sample at 105°C in a laboratory oven. Determine the weight of the dried solid and calculate the percentage of moisture by using the following formula:

$$\% \text{ moisture} = [(\text{g of solid} - \text{g of dried solid}) \times 100]/\text{g of solid}$$

Ash

The ash content of any solid is determined by igniting a known quantity aliquot of the well-mixed sample at 1000°C in a muffle furnace. By knowing the weight of the original sample and the weight of the remaining ash, and using the formula below, the percentage of ash content of the sample is determined and calculated.

$$\% \text{ ash} = \text{g of ash} \times 100/\text{g of original sample}$$

Composition Percentage of a Solid Sample

- Determine the moisture at 105°C as described above and calculate as percentage of moisture as described above.
- The remaining residue from the above determination will be ignited at 1000°C in a muffle furnace. The difference between the weight of the ignited residue (1000°C) and the dried solid will give the volatile and organic compounds in the sample. It is also calculated in percent.
- The remaining residue is considered as ash or inorganics. It is calculated and expressed as percent, as mentioned above.
- The sum of the percent moisture, percent organics, and percent ash should be 100%. If the sum of the calculated fractions is not 100% and the checked calculations are correct, the whole process should be repeated.

Calculations Related to Titrimetric Methods

The general formula for calculating results for titrimetric determination expressed in milligrams per liter is as follows:

$$\text{mg/L} = (\text{ml} \times \text{N} \times \text{Eqw} \times 1000)/\text{ml sample}$$

ml = volume of the used titrant solution
N = normality of the titrant solution
Eqw = equivalent weight of the determined substance

Results for titrimetric determination expressed as percent:

$$\% = (\text{ml} \times \text{N} \times \text{mEqw} \times 100)/\text{ml sample}$$

ml = volume of the used titrant solution
N = normality of the titrant solution
mEqw = milliequivalent weight of the determined substance (Eqw/1000)

When any *dilution* has been used during the test, the calculated result of the titration should be multiplied by the dilution factor for the final report.

Calculations Related to Calibration Curve

Linear Regression Calculation

In chemical analysis, it is often necessary to prepare calibration plots for use in calculating the results of analytical measurements. It is prepared by taking standards of known concentrations, measuring the desired property of each standard, and plotting the property against the concentration. When such a plot is prepared, the sample concentration is obtained by measuring the instrument's response (absorbance, transmittance, peak height, etc.) under the same conditions that were used for the standards and the concentration of the sample can be read from the horizontal axis of the plot.

Although unknown concentrations can be read directly from the graphical plot, better accuracy is possible by using the *Linear regression calculation*, also called the *Least of square calculation*. In this calculation, the possible straight lines that can be drawn through or near the data points, the one chosen minimizes the sum of the squared deviations. The deviation for each point is the difference between the actual data points with the same "x" axis value that lies exactly on the straight line. It gives information about the best straight line through the points entered, including its correlation coefficient, intercept, slope, and its predicted "x" and "y" values.

Correlation Coefficient

It gives the correlation between the "x" and "y" values in a set of data points. A result near one indicates that the values have a strong linear relationship. A result near zero indicates that the values are only slightly related. A value near minus one indicates that the values are very closely related, but in a negative way; that is, an increase in one is related to a decrease in the other. The value of <0.9998 is accepted.

Intercept (b)

The intercept tells whether there is a significant blank measurement even when the concentration of the blank is zero. The intercept value gives the "y" intercept of the best straight line through the points.

Slope (m)

The calculated slope value gives the slope of the best straight line. In the routine work, the calculation of these values is by calculator or by computer software. The outline of the calculation is as follows:

$$m = n\ Exy - ExEy/n\ Ey^2 - (Ey)^2$$

$$b = Ey^2Ex - EyExy/n\ Ey^2 - (Ey)^2$$

n = number of measurements
E = sum
x = concentration
y = absorbance (or other response)
m = slope
b = intercept

Predicting "x" and "y" Values

Using the formula $x = my + b$

x = concentration
y = absorbance (or other response)
m = the calculated slope
b = the calculated intercept

For example, for calculating the concentration of an unknown sample by using the linear regression calculation:

n = 5

x = 0.2, 0.4, 0.6, 0.8, 1.0 mg/L
y = 0.196, 0.408, 0.596, 0.810. 0.992 absorbance

	x	y	y^2	xy
	0.2	0.196	0.038	0.039
	0.4	0.408	0.166	0.163
	0.6	0.596	0.355	0.357
	0.8	0.810	0.656	0.648
	1.0	0.992	0.984	0.992
E =	3.0	3.002	2.199	2.199

$m = (5 \times 2.199) - (3 \times 3.002)/(5 \times 2.199) - (3.002)^2 = 1.989/1.983 = 1.003$

$b = (2.199 \times 3) - (3.002 \times 2.199)/(5 \times 2.199) - (3.002)^2 = -0.003/1.983 = -0.0015$

After the calibration of the instrument, one unknown sample was measured with an absorbance of 0.625. Calculate the concentration of the sample for the particular parameter by using the calculated "m" and "b" values:

$x = my + b$ $\qquad$ $x = (1.003)(0.625) + (-0.0015)$

$x = 0.625$

So, the reported result of the sample will be 0.625 mg/L.

Logarithmic Regression Calculation

When ions are measured by Ion Selective Electrode (ISE) methods, an "Ion analyzer" may be used, which expresses the results in direct concentration. If the ion analyzer is not available, a pH meter may be used. By using the pH meter, the pH and its reference electrode change to one of the ion selective electrodes and the accompanied reference electrode, and the reading position of pH turns to a millivolt (MV) reading. Different concentrations of standards will produce different millivolt (MV) readings on the instrument. With the increase in the concentration, MV readings will decrease. The curve is prepared by plotting the concentration against the MV reading. Therefore, the plot will be different from the usual calibration curves where the increased concentration increases the absorbance reading. For plotting the curve, use *semilog graphpaper* by using the "x" axis for MV and the "y" axis for concentration. Instead of using linear regression calculations, use *logarithmic regression calculation.* This is similar to the linear regression calculation, except the concentration values should be logarithmic values. The final prediction of the "x" and "y" values are calculated the same way as the linear regression, except the final result will be converted to the reporting form, calculated as "antilog". For example, concentrations "x": 0.2, 0.5, 0.8, 1.0, 2.0 mg/L; MV readings "y": 262, 248, 234, 224, 210.

y	x(log)	x^2	xy
262	– 0.699	0.488	–183.138
248	– 0.301	0.090	– 74.648
234	– 0.097	0.009	– 22.698
224	0.000	0.000	0.000
210	0.301	0.090	63.210

E = 1,178 - 0.795 0.677 - 217.274

m = 5 (-217.274) - (-0.795)(1,178)/5 (0.677) - $(0.795)^2$ = (-1086.37) - (-936.51)/ 3.385 - 0.632 = -149.86 / 2.753 = -54.4

b = (0.677)(1,178) - (-0.795)(-217.274)/2.753 = 797.606 - 172.73/2.753 = 228.85 = 229

If the unknown sample gives an MV reading of 250, the calculated milligrams per liter value will be:

$y = mx + b \quad 250 = -54.4\,x + 229$

$x = -0.386$

$x = \text{antilog}(-0.386) = 0.411$ mg/L

The reported value for this particular sample will be 0.411 mg/L.

Calculating the Results for Solid Matrices

For solid matrices, the final report is expressed as ppm, milligram per kilogram (mg/Kg) or as ppb, microgram per kilogram (ug/Kg). For the calculation, use the following formula:

$$\text{mg/Kg on wet base} = \text{mg/L} \times \text{final volume of the sample after treatment/ g sample} \quad (1)$$

$$\text{mg/Kg on dry base} = \text{mg/Kg on wet base/decimal fraction of dry solid} \quad (2)$$

For example, 5 g soil sample was weighed and digested for lead (Pb) analysis. After the preparatory process, the final volume of the "ready for analysis sample" was 100 ml. The Pb content of the digestate was found to be 0.56 mg/L or 0.56 mg in 1000 ml, what is 0.056 mg in 100 ml. The 100 ml soil digestate corresponds to the 5 g soil sample; therefore, the 5 g soil sample contains the 0.056 mg Pb. 1000 g (1Kg) will have 200 times more Pb, so the final report will be 0.056 × 200 = 11.2 mg/Kg Pb. To save time, the easiest way to follow Equation 1 is

$$\text{mg/Kg Pb on wet base} = 0.56 \times 100/5 = 11.2 \text{ mg/Kg}$$

"Wet base" means that the original soil or sediment sample contains various quantities of moisture in it; therefore, the weight of the soil is incorporated in the weight of the moisture. Consequently, the weight of the soil is incorrect. With the knowledge of the moisture content of the soil sample, the analyst can make the necessary corrections. The previously determined Pb content of the soil was 11.2 mg/Kg and corresponds to the original, wet sample. The soil has 12% moisture and 88% dry soil (100% – 12% moisture = 88% dry soil). Using the Equation 2,

$$\text{mg/Kg Pb on dry base} = 11.2 / 0.88 = 12.72 \text{ mg/Kg}$$

Calculation of Total Hardness From Calcium and Magnesium Values

With available calcium (Ca) and magnesium (Mg) values, the Total Hardness, expressed in $CaCO_3$ (Calcium carbonate) may be calculated. Total Hardness of a water is principally caused by Ca and Mg salts. Therefore, hardness in the term of $CaCO_3$ refers to the sum of the calcium as $CaCO_3$ and magnesium as $CaCO_3$.

- Convert Ca into $CaCO_3$ by dividing the Ca mg/L value with a factor of 0.4. How did we calculate the factor of 0.4?
  - (a) The equivalent weight (Eqw) of calcium is the atomic weight divided by the valency, so 40/2 = 20. (Equivalent weight of the elements: Atomic weight divided by the valency of the element.)
  - (b) The equivalent weight of Calcium carbonate ($CaCO_3$) is the molecular weight divided by the valency of the calcium 100/2 = 50. (Equivalent weight of salts: molecule weight of the salt divided by the valency of the metal ion in the compound.)
  - (c) The factor derived from the ratio of the equivalent weight of the calcium and the equivalent weight of the calcium carbonate, 20/50 = 0.4.
- Convert Mg into $CaCO_3$ in the same way except the ratio is 0.24, respectively to the Eqw of Mg, 24/2 = 12. The factor will be 12/50 = 0.24.

For example:

Calcium, as Ca	= 24 mg/L
Calcium, as $CaCO_3$	= 24/0.4 = 60 mg/L
Magnesium, as Mg	= 8 mg/L
Magnesium, as $CaCO_3$	= 8/0.24 = 33 mg/L
Total Hardness, as $CaCO_3$	= 60 + 33 = 93 mg/L

Calculation of Magnesium Values from Total Hardness, as $CaCO_3$ and Calcium, as $CaC0_3$

With the knowledge of the Total Hardness as $CaCO_3$ and calcium as $CaCO_3$, magnesium can be calculated.

$$\text{Total Hardness as } CaCO_3 = \text{Calcium as } CaCO_3 + \text{Magnesium as } CaCO_3$$

$$\text{Magnesium, as } CaCO_3 = \text{Total Hardness as } CaCO_3 - \text{Calcium as } CaCO_3$$

Ca and Mg may be calculated from their $CaCO_3$ form by multiplying it with the factors of 0.4 and 0.24, respectively. (The calculation of these factors discussed above.)

For example:

Total Hardness, as $CaCO_3$		= 120 mg/L
Calcium, as $CaCO_3$		= 70 mg/L
Magnesium, as $CaCO_3$		= 50 mg/L
Calcium, as Ca	= 70 × 0.4	= 28 mg/L
Magnesium, as Mg	= 50 × 0.24	= 12 mg/L

Calculation of Carbonate, Bicarbonate, and Hydroxide From Alkalinity

There are three kinds of alkalinity: hydroxide OH^-, carbonate CO_3^-, and bicarbonate HCO_3^-. In order to distinguish between the kinds of alkalinity present in a sample and to determine the quantities of each, a titration is made with a standard acid using two indicators successively. Carbonate alkalinity may be present with either hydroxide or bicarbonate alkalinity, but hydroxide and bicarbonate cannot be present together in the same sample. Alkalinity is expressed in the term of $CaCO_3$ and titrated in the presence of phenolphthalein and methylorange indicators. “P” (or phenolphthalein) alkalinity is a result of titration of the sample with a standard acid solution in the presence of phenolphthalein indicator. “M” (or methylorange) alkalinity is obtained by the titration in the presence of a methylorange indicator. “T” (or total) alkalinity is the sum of the results of the previously mentioned titration. The updated method for the determination of "M" Alkalinity used the "mixed bromcresol green-methyl red indicator" instead of "methylorange" indicator. Alternatively, a potentiometric titration may be applied for alkalinity analysis by using preselected pH-end points.

There are five alkalinity conditions possible in the sample:

(1) Hydroxide alone
(2) Hydroxide and carbonate
(3) Carbonate alone
(4) Carbonate and bicarbonate
(5) Bicarbonate alone

The results obtained from the phenolphthalein and total alkalinity determinations offer the calculation of these alkalinity conditions.

1.	$P = T$	Hydroxide Alkalinity, as $CaCO_3 = P$
2.	$P > 1/2\,T$	Hydroxide Alkalinity, as $CaCO_3 = 2P - T$
		Carbonate Alkalinity, as $CaCO_3 = 2(T - P)$
3.	$P = 1/2\,T$	Carbonate Alkalinity, as $CaCO_3 = T$
4.	$P < 1/2\,T$	Carbonate Alkalinity, as $CaCO_3 = 2P$
		Bicarbonate Alkalinity, as $CaCO_3 = T - 2P$
5.	$P = 0$	Bicarbonate Alkalinity, as $CaCO_3 = T$

P = Phenolphthalein Alkalinity as $CaCO_3$ mg/L

T = Total Alkalinity as $CaCO_3$ mg/L

Hydroxide, Carbonate, and Bicarbonate as $CaCO_3$ may be converted to the corresponding anions by using conversion factors. Conversion factors are obtained from the ratio of the equivalent weight of the anion and the equivalent weight of $CaCO_3$.

Equivalent weight of hydroxide ion (OH^-) is 17/1 = 17
Equivalent weight of carbonate ion (CO_3^{2-}) is 60/2 = 30
Equivalent weight of bicarbonate ion (HCO_3^-) is 61/1 = 61
Equivalent weight of calcium carbonate ($CaCO_3$) is 100/2 = 50

To calculate the factors, the equivalent weight of the polyatomic ion should be divided by the equivalent weight of the $CaCO_3$.

Anion		Factor
OH^-	= 17/50	= 0.34
CO_3^{2-}	= 30/50	= 0.6
HCO_3^-	= 61/50	= 1.22

For example, the results of one alkalinity determination are: "P" Alkalinity, as $CaCO_3$ = 12 mg/L; "T" Alkalinity, as $CaCO_3$ = 126 mg/L. The situation corresponds to Condition 4; therefore, Carbonate Alkalinity, as $CaCO_3 = 2 \times 12 = 24$ mg/L; Bicarbonate Alkalinity, as $CaCO_3 = 126 - 24 = 102$ mg/L.

By using the conversion factors calculated above, the values for the participating anions will be: Carbonate, as $CO_3^{2-} = 24 \times 0.6 = 14.4$ mg/L; Bicarbonate, as $HCO_3^- = 102 \times 1.22 = 124$ mg/L.

Dilutions and Concentrations

Any dilution applied during the test should be recorded and the final result will obtained by multiplying the measured value by the dilution factor. Any concentration technique used during sample preparation should be documented and the analytical value divided by the concentration factor will produce the result.

For example, a water sample analyzed for phosphorus and for the appropriate reading should be diluted five times. If the result of the diluted sample was 0.8 mg/L phosphorus, the final result will be $0.8 \times 5 = 4.0$ mg/L.

During the digestion technique, a water sample analyzed for chromium was concentrated to 10 times (original 100 ml sample cook down to 10 ml final volume) because of the low concentration of the metal in the sample. The reading of this concentrated sample was 0.06 mg/L, so the final result will be $0.06/10 = 0.006$ mg/L or 6 ug/L.

10.3. Significant Figures

Number of significant figures refers to the number of digits reported for the value of a measured or calculated quantity indicating the accuracy or precision of the value. The number of significant figures is said to be the number of digits remaining when the data is rounded. For example, when a measured value is 10.6 mg/L, the analyst should be quite certain of the 10, but may be uncertain as to whether the 0.6 should be 0.5 or 0.7, or even 0.4 or 0.8, because of unavoidable uncertainty in the analytical procedure. The 10.6 mg/L reported value has three significant figures. To count the number of significant figures in a given measured quantity, the following rules exist.

Nonzero integers — They always count as significant figures.

Leading zeros — They are zeros that precede all of the zero digits. They do not count as significant figures. In the number of 0.0025, the three zeros simply indicate the position of the decimal point. This number has only two significant figures.

Captive zeros — They are zeros between nonzero digits, always count as significant figures. The number 1.008 has four significant figures.

Trailing zeros — They are zeros at the right end of the number. They are significant only if the number contains a decimal point. The number 100 has only one significant figure and 100. has three significant figures. It can cause confusion. For example, when reporting 4600 ug/L, it is better to report it as 4.6 mg/L, if the two significant figures are trusted. The concept of significant figures is especially important in mathematical calculations.

Significant figures in multiplication and division — The number of significant figures in the result is the same as the number in the least precise measurement used in the calculation. For example, consider this calculation: $4.56 \times 1.4 = 6.38$, by rounding 6.4 (two significant figures). The correct result has only two significant figures, since 1.4 has two significant figures.

Significant figures in addition and subtraction — The result has the same number of decimal places as the least precise measurement used in the calculation. For example, consider the following sum:

$$12.11 + 18.0 + 1.013 = 31.123, \text{ rounded to } 31.1$$

The correct result is 31.1, since 18.0 has only one decimal place.

The following rules should be used in rounding data consistent with its significance:

- If the digit to be removed is less than 5, the preceding digit stays the same. For example, 1.33 is rounded to 1.3.
- If the digit to be removed is greater than 5, the preceding digit is increased by 1. For example, 1.36 is rounded to 1.4.

- If the digit to be removed is 5, round off the preceding digit to the nearest even number: thus 2.15 becomes 2.2 and 2.35 becomes 2.4.

Exponential notation — Exponential notation, for example 1.15×10^3 or 11.5×10^2, is an acceptable way to express both the number and the significant figures. This form is not generally used because it would not be consistent with the normal expression of results and might be confusing. The general rule is to report only such figures that are justified by the accuracy of the method.

10.4. Records for Raw and Calculated Data

All documentation related to the raw data and the reported results should be in a bound form, easy to identify, and ready for inspection at anytime.

Field and Laboratory Notebook

A bound notebook is preferred to a loose-leaf one. The size of the notebook should be good and comfortable to work with. Preferably, choose one to fit easily around the balance and working table leaving space for glassware, reagents, etc. necessary for the analysis. Larger notebooks, especially when opened, tend to get in the way and can cause spills and other problems. The first two pages should be reserved for an index, in which you can list the page numbers and the title of the subject. If the pages of the notebook are not numbered, the analyst must number each page in the upper corner. Use the title heading to identify data on each page and date it. Write all data in ink to avoid smearing and erasures. All raw data and any observations should be written in the notebook. Nothing should be erased; anything not needed should be crossed out with a single line, followed by an explanatory notation, initialed, and dated.

Worksheets

Each analysis has specified worksheets. The analyst must complete all parts of the worksheet that contain the information relating to the sample (I.D. number, sample type, matrix, source, etc.), analytical method or method number with reference, method detection limit (MDL), instrumentation, analytical and QC raw data, calculated values, and the proper unit. The worksheet should clearly and completely contain all information necessary for validation of the analytical process. Worksheets for field tests, as pH, conductivity, dissolved oxygen, and residual chlorine measurements are on Tables 2-1, 2-2, 2-3, and 2-4, respectively. The same form of documentations are also used in laboratory testing for these parameters. Tables 10-1 through 10-6 give examples for recording different analytical performances on different matrices. The design of these documentation forms are just examples, each institution and laboratory has their own patterns as described and approved in their QA/QC programs.

Other Documentation

Documentation should stored in bounded forms, with the date started, date ended, identification number (I.D.) started and ended, document title, analytical group, and parameter. The storage area should be large enough, to keep records on shelves or storage cabinets and easy to find. Strip charts, documented AA calibration curves, and raw data collected should be stored in file boxes and identified as mentioned above.

Documents to be Saved

- Chromatograms, charts, and other instrument response readout records
- Calibration curves
- QC charts
- Target Limits
- Method Detection Limits (MDLs)
- Summary log forms for precision and accuracy data
- Working sheets
- Records concerning receipt of stock solutions
- Preparation of calibration standards
- All notebooks, data forms, and log forms that belong to laboratory operations
- Laboratory custody reports (holding times, sample transmittal forms, sample storage log, sample disposal log)
- Sample preparation log
- All calculations related to sample results and statistical calculations for QC limits
- Instruments maintenance log and instrument performance check logs
- Copies of final reports
- Field records such as field notebooks, field tests data, and field custody record

Records must be retained for a period of at least three years. Drinking water reports require a retention time of up to ten years.

Table 10-1. Working Paper for Titrimetric Procedures (Water Samples)

Parameter Method No. Reference ____________
Method Detection Limit (MDL)____________________________
Titrant and Normality (standadized) ____________________
Analyzed by _________________________ Date ____________

ID No.	*Sample Identification*	*Sample type*	*Sample ml*	*Titrant ml*	*Result mg/L*	*Remark*
	Blank					
	QC sample					
	Duplicate					
	Spiked sample					

INFORMATIONS RELATED TO QC CHECKS

True value of QC sample ________ ID of duplicate sample ________
ID No. of spiked sample ____________ Sample ml, spiked__________
Concentration of spike stock solution __________________________
ml or ul spike stock added _______ Added spike value __________

Precision as Relative Percent Deviation (RPD) ___________
Accuracy as % Recovery (% R) ____________
% Recovery of Spike (% R_{spike}) _________________________

Approval of Supervisor________________ Date _______________

Table 10-2. Working Paper for Titrimetric Procedures (Solid Matrices)

Parameter ___________ Method No.__________Reference______________
Method Detection Limit (MDL) ________________________________
Titrant and Normality (standardized) __________________________
Analyzed by _______________________ Date ______________________

ID	Sample identification	g	Final ml	Test ml	Titr ml	Test res. mg/L	Moisture %	Rep. wet mg/Kg	Rep. dry mg/Kg
	Blank								
	QC sample								
	Duplicate								
	Spike								

INFORMATIONS RELATED TO QC CHECKS

True value of QC sample___________ID No. of Duplicate __________
ID No. of spiked sample __________ Sample ml spiked __________
Concentration of spike stock solution __________________________
ml or ul spike stock added ___________ Added spike value ________
Precision as Relative Percent Deviation (RPD) ________________
Accuracy as % Recovery (% R) _______________
% Recovery on spike (% R_{spike}) _______________

Approval of supervisor _______________________ Date _______________

Table 10-3. Working Paper for Spectrophotometric Analysis (Water Samples)

Parameter ____________ **Method No.** _______ **Reference** ________________
Method Detection Limit (MDL) ______________________________________
Model of spectrophotometer ______________ **Wavelength** _________
Correlation Coefficient of the calibration curve ____________
Analyzed by ______________________ **Date** _________________

ID No	*Sample Identification*	*Sample type*	*Sample ml*	*Absorbance*	*Dilution*	*Result mg/L*	*Remarks*
	Blank						
	Standard 1						
	2						
	3						
	4						
	CCS						
	CVS						
	Duplicate						
	Spike						
	Blank						

INFORMATIONS RELATED TO QC CHECKS

% Recovery of Continuing Calibration Standard (CCS) _____________
True Value of Calibration Verification Standard (CVS) ____________
% Recovery of CVS ______________ **ID of Duplicate Sample** ___________
ID No. of Spiked Sample _________ **Sample ml spiked** ____________
Conc. of Spike Stock solution ________ **Added spike value** __________
Precision as Relative Percent Deviation (RPD) ____________
Accuracy as % Recovery (%R) ________
% Recovery of spike ($\%R_{spike}$) __________

Approval of supervisor ________________ Date ______________

Table 10-4. Working Paper for Spectrophotometric Analysis (Solid Samples)

Parameter ____________ **Method No.**_____ **Reference**______________
Method detection Limit (MDL) ______________________________
Model of Spectrophotometer ______________ **Wavelength** _________
Correlation Coefficient of the calibration curve ___________
Analyzed by __________________ **Date** _______________

ID No	*Sample identification*	*g*	*Final ml*	*Tested ml*	*Abs*	*Dil*	*Res mg/L*	*Mois %*	*Rep. wet mg/Kg*	*Rep. dry mg/Kg*
	Blank									
	Stand.1									
	2									
	3									
	4									
	CCS									
	CVS									
	Duplic.									
	Spike									
	Blank									

INFORMATIONS RELATED TO QC CHECKS

% Recovery of Continuing Calibration Standard (CCS)______________
True value of Calibration Verification Standard (CVS)___________
% Recovery of CVS __________ ID of Duplicate Sample _____________
ID No. of Spiked sample __________ Sample ml spiked _____________
Concentration of spike stock solution ____________________________
ml or ul spike stock added _________ Added spike value ___________
Precision as Relative Percent Deviation (RPD) ____________________
% Recovery of Spike (% R_{spike}) __________________

Approval of supervisor ________________ Date ______________

Table 10-5. Working Paper for Ion Selective Electrodes (Water Samples)

Ion Analyzer model ________________ *Parameter* ________________
Method ________ *Reference* ____________ *Detection Limit (MDL)* ________
Reference electrode ________ *single junction*________ *double junction*
Reference electrode filling solution ________________
Slope ____________ *Analyzed by* ____________ *Date* ____________

ID No	*Sample Identific ation*	*Sample type*	*Sample ml*	*Reading mg/L*	*Dilut ion*	*Result mg/L*	*Remark*
	Stand. 1 ppm						
	Stand. 10 ppm						
	QC sample						
	CCS						
	CCS						
	QC sample						
	Duplicate						
	Spike						

INFORMATIONS RELATED TO QC CHECKS

% Recovery on Continuing Calibration Standard (CCS) ____________
True Value of QC sample ________ % Recovery on QC sample ________
ID of Duplicate sample ________ Precision as RPD ____________
ID of Spiked sample ____________ Sample ml spiked ____________
Spike Stock solution conc ________ ml or ul stock added ____________
Added Spike Value ________ % Recovery on spike (% R_{spike}) ________

Approval of Supervisor ______________ Date ____________

Table 10-6. Working Paper for Ion Selective Electrodes (Solid Samples)

Ion Analyzer model ____________________ Parameter ________________
Method ________Reference ____________ Detection Limit (MDL)________
Reference electrode ______ single junction _______ double junction
Reference electrode filling solution ________________________________
Slope ___________ Analyzed by __________________ Date ______________

ID No	*Sampl Ident ific.*	*g*	*Fin al ml*	*Tes ted ml*	*Read mg/L*	*Dil*	*Res ult mg/L*	*Mois ture %*	*Rep. wet mg/Kg*	*Rep. dry mg/Kg*
	Stand. 1 ppm									
	Stand. 10 ppm									
	QC									
	CCS									
	CCS									
	QC									
	Dupl.									
	Spike									

INFORMATIONS RELATED TO QC CHECKS

% Recovery on Continuing Calibration Standard (CCS) ______________
True value of QC sample _________ % Recovery on QC sample __________
ID of duplicate sample __________ Precision as RPD _________________
ID of spiked sample _____________ Sample ml spiked _________________
Spike stock conc ________________ ml or ul stock added _____________
Added Spike value _____________ % Recovery on spike ($\%R_{spike}$)________

Approval of Supervisor________________ Date ________________

Chapter 11
Evaluation and Approval of Analytical Data

The responsibilities of the analyst have been discussed in previous chapters. Those responsibilities are to prepare reagents and standards, make the calibrations and calibration checks, run the test as described in the SOP, convert raw data to reportable data by using correct calculations, apply all required QC checks according to the approved QA/QC program of the organization to defend the reportable data, and recognize errors and conduct corrective actions. The first step in the evaluation of both the analytical and QC data is the job of the analyst. This chapter presents the responsibilities of the laboratory supervisor and QA officer relating to these checks, and the final validation and approval of the reporting values. To evaluate and approve reportable results, checks should be performed on sampling, sample handling and storage, methodology, calibration, and QC. If any problems arise through doubtful data, initiation of corrective action is necessary. Leading and advising the corrective actions, giving the final conclusions, and finally giving the approved results are the responsibilities of the above mentioned parties.

11.1. Validation of Analytical QC Checks

QC is the check of the result of an analytical measurement. It helps to indicate that the reported value is correct and acceptable. When any error occurs, it should be identified in detail, and should be corrected into valid, reportable value. When a QC error has been recognized, testing cannot continue until the QC check indicates the correction. The reviewer should be concerned about the acceptance criteria of the QC check, the probable source of the out-of-control item, and the steps taken in the specified corrective action. The QC check must include QC measures such as blanks, calibration processes, QC check standards, QC check samples, certified reference materials, matrix and reagent water spikes, duplicates, surrogate standards, internal standards, standardization of titrant, instrument's performances, split samples, quality of laboratory pure water, positive and negative control samples, and confirmation of bacterial colonies in microbiological testing, as discussed in Chapter 8.

Blanks

Analytical blanks measure the degree of contamination and seriously affect the accuracy of low-level trace determinations. Blanks are known as Method blank, Reagent blank, Calibration blank, Equipment blank, and Trip blank as described in Chapter 9. **Blank values should be less than the Method Detection Limit (MDL).**

Possible Causes of Unaccepted Blank Values and Their Correction

Contaminated reagents — Reagent blank contamination may be produced by chemical contact from reagents used for the test. Review the source and the expiration date of the chemicals and reagents and check preparation and expiration date. To avoid further problems, discard old reagents. Deionized water used for rinsing, for solvent, for dilution, or other laboratory operations can be a great source of contaminated blanks. To find this problem, overview the routine checks of the used laboratory pure water, as outlined in Chapter 8.

Contaminated glassware or sample containers — Contaminated apparatus and glassware may produce high blank readings. To eliminate this source of the contamination, check cleaning procedures.

Contamination from environmental conditions — Blanks may be produced by contamination from analytical situations, from sample collection, or from unclean sample handling. Contamination from the garments of the analyst from the bench areas may also effect blanks. Environmental control by good housekeeping and by clean, contamination-free laboratory operation is the answer for this problem.

Contaminated analyte-free water — Reagent grade water should be free of substances that interfere with analytical methods. Requirements for water quality may differ for organic, inorganic, and microbiological constituents, depending on the uses for which the water is intended. (Reagent water quality is in Chapter 8.) Check the source and storage conditions to verify contamination.

Calibration

Calibration in chemical measurements refers to the process by which the response of a measurement system is related to the different concentrations of the analyte of interest. Generally, a measurement is a comparison process in which an unknown is compared with a known standard. (Detailed calibration procedures are in Chapter 7.) The first criteria for correct calibration is the purity and accuracy of the standards. Standards may be prepared by the analyst or suppliers, and should have an assigned expiration date indicating the stable life expectancy. Standards should never be used beyond such date. Storage and life expectancy for standard solutions are on Table 8-3.

Acceptance Criteria for Calibration

Initial calibration is accepted, when the calculated correlation coefficient for the calibration curve is >0.9998 or specified by the method. The deviation of the continuing calibration standard (CCS) from the original calibration standard should be ±5% in inorganic analysis, and ±10% in organic analysis. The accepted criteria for the Calibration Verification Standard (CVS) or QC Check Standard is ±5% of true value for inorganic analysis, and ±10% of true value for organic analysis.

Probable Sources of Unaccepted Calibration Criteria and Their Corrections

Check calibration curves and calculations — Recalculate by using the same method as designed by the approved method and by the laboratory SOP.

Incorrectly prepared stocks, intermediate and working standard solutions — Check preparation logs for the stocks and standards, look for errors in calculations, dilution techniques, check chemicals used in preparation. When stocks and standards are from an outside source, look for proof of one reliability of the supplier. Prepare or supply fresh standards and repeat calibration.

Outdated stocks and standards — Every stock and standard should have an assigned expiration date. Discard expired solutions immediately because they may cause further problems! Storage area should be routinely checked for expired chemicals, stocks, and standards.

Faulty or expired QC check standard (CVS) — Check for source, preparation, and expiration date. If expiration date is indicated, discard the standard immediately. If the date is current, control the accuracy of the standard by analyzing a new standard, and identify the unaccepted criteria caused by faulty CVS standard or by an incorrect calibration, and make actions according to this verification.

Improperly stored stocks, standards, and reagents — Storage of stocks, standards, and reagents must follow the method outlined here, and as described in the approved laboratory QA/QC program. Improperly stored stocks and standards can be a good source of unaccepted calibration.

Contamination from incorrectly cleaned glassware, containers, or from a dirty environment — Check cleaning processes, storage of the clean glassware, and the possibility of contamination from unclean laboratory facilities. Make sure that cleaning procedures are followed exactly as outlined in the laboratory SOP and in the updated QA/QC program. Check spectrophotometric cells or cuvette for sparkling cleanness; any scratch or chipping in this glassware will indicate their elimination. Make an effort to obtain clean contamination-free atmosphere of the working area.

Using improper volumetric glassware — Analyst should be familiar with basic laboratory techniques. Unprofessionally using or selecting volumetric glassware can ruin analytical tests. If the above facts have been confirmed as the source of the incorrect analytical data, the analytical activities of the analyst should be stopped until he or she is completely trained and his or her ability for correct analytical work has been approved.

Incorrect response of the instrument — Conduct instrument performance check, and perform routine maintenance to recognize if any instrument failure is responsible for the calibration problem. Be sure that the problem has been solved, and the instrument is in good working condition.

Note: Reanalyze all samples measured between the acceptable and non-acceptable calibration check with the corrected calibration.

Duplicates and Split Samples

Duplicates and split samples are used to measure the precision of the analytical system. Duplicates may be field duplicates, laboratory duplicates, or matrix spike duplicates, as stated in Chapter 9. The accepted limit (Target Limit) for the calculated precision is different by laboratories, parameters, and methods. When the measured and calculated precision values are out of the established and targeted acceptance limit, corrective actions are needed to find the cause(s).

Probable Sources of the Unaccepted Precision Values and Their Correction

Sampling error for field duplicates — Review sample collection and preservation protocol. Review sample preparation techniques and, if necessary, repeat the process for both duplicates and then reanalyze. If these activities do not give an answer for the questionable error, sample(s) and duplicate(s) should be resampled.

Preparation error for laboratory duplicates — A crowded working schedule in the laboratory may cause the use of unidentical samples for duplication. Check carefully and identify the sample for duplication prior to reanalyzing. Poor homogenization of the sample before duplication may also produce incompatible values.

Contamination error — Different results for duplicate analyses are possibly caused by contaminations originating from improperly washed or stored glassware and equipment. Check washing procedures and take all precautions to avoid contamination during duplication, sample pretreatment, and analysis.

Calculation error — Mistakes in calculations are a great possibility in contributing to bad reports. An overview of all calculations relating to the data may easily solve the problem.

Note: When these steps do not bring the analysis back within the acceptable precision limits, then the whole analytical procedure must start again. It may be necessary to prepare new standards and calibration. Other sample data generated with the same analytical run are questionable, and have to be reanalyzed after the control limits are justified.

Spikes and Surrogate Spikes

All daily spiking data should agree with the spike accuracy limit established for each parameter, and for each method. When a result falls outside the control limit, steps must be taken to determine the source.

Probable Sources of Unaccepted Spike Recoveries and Their Corrections

Calculation error — Recalculate sample and spiked sample values.

Improperly prepared or stored spike stock solution — Overview all preparation logs and used chemicals. Check preservation or special storage conditions if applicable. If an error is found in this level, discard old spike stock solution and prepare a new one.

Expired spike stock or standard solution — Check expiration date and discard if necessary. Check stock and standard logs for correct documentation.

Error in spiking — Check calculations, and the quantity of added spike stock solution. Error may arise from the misuse of micropipette. If the error is caused by poor knowledge of the spiking technique, schedule additional training programs in this area.

Contamination of the spiked sample during pretreatment or analysis — Remediation of contamination errors is the same as mentioned previously.

Malfunction of the instrument — Check all of the other QC check recoveries for optimum performance. If needed, conduct performance check and preventive maintenance for the instrument, as stated previously.

Note: If all of the above corrective actions do not bring the analysis back within the acceptable control limits, then the spike sample must be prepared again and analyzed. When the system is again under control, all associated samples analyzed with the faulty spiked sample must be reprocessed and analyzed again.

Standard Titrant Solutions

Titrimetric or volumetric analyses are very sensitive for carefully cleaned and properly used glassware. Titrant solutions must be prepared with extreme care and standardized for exact normality each time before use. The first check is to review the preparation and standardization of the titrant solutions if any problem arises in this analytical performance.

Titrant solutions are normal (N) solutions with standardized, exact normality. The normality of the titrant is given by each particular method, or chosen according to the expected concentration of the analyte in the samples. False normality may cause unusable, incorrect results. After the solution has been prepared, the exact normality should be determined as described by the appropriate method, and in the laboratory SOP. Three parallel standardization procedures should be done, and the used normality is given by the average of these three values (see Chapter 7).

The acceptance criteria of the good titrant solution is the production of a ±5% deviation from the true value of the QC check standard, and all QC checks (blank, duplicates, spiked samples) should fall in the acceptable range. If the titrimetric procedure fails this criteria, the source of the problem should be found and corrected.

Probable Sources of Unaccepted Titrant Solutions and Their Corrections

Bad titrant solution — Bad titrant solutions may be produced by different errors.

- Preparation error: check preparation log, and source chemicals.
- Mistake in dilution technique: check preparation log.
- Used overdated stock titrant solution: check source, date of preparation, accepted life-time of the stock, and proper storage. Stock solutions are usually stored in the refrigerator; if this is so, do not give them time to warm up to room temperature before using them. The diluted titrant solution will be inaccurate and, subsequently, the analytical results will be incorrect and unusable. This is a small error that may destroy hours of hard work.
- Standardization error: as mentioned under the acceptance criteria of titrant solutions, exact normality (N) of the titrant solutions is important, and it is chargeable for the correct analytical performance and valuable result. Carefully check the standardization process and the associated calculations for error.

Contaminated or expired reagents, or indicators — Check the dates on the reagent-bottle labels, and check the reagent preparation logs for information. Expired or defective reagents should be discarded, and the analysis should be repeated with new, updated reagents.

Contaminated glassware and laboratory facility — Check for washing procedures, storage for glassware, and for cleanliness of the bench area where titration have been made.

Improperly used volumetric glassware — Volumetric glassware should be handled with knowledge and care. Problems and inaccuracies in the so called "old fashioned wet chemistry" analytical techniques are mostly caused by the lack of proper training in the basic laboratory skills.

Mistakes in endpoint readings — Endpoint reading by indicators needs practice and experience. When a problem arises from this mistake, additional training and practice should be provided to the analyst in this area.

Analytical System Performance Checks

To validate the analytical system or measurement process, laboratories use Reference Standards. *Internal Reference Standards* are prepared by the laboratory. *External Reference Material* is provided by an outside source, and according to the origin of the standard material, can be a certified reference material, originated from a technically accepted and certified organization. Analytical value should be within the certified limits.

Causes of failing this criteria and their corrections are as follows:

- Calculation error — Check calculations
- Improper sample preparation — Review sample preparation techniques and methods, according to the analytical performance and to the matrix. Wrong sample treatment may lead to unusable analytical data, so the preparation and analysis must be repeated.
- Improper analysis — Unaccepted analytical values of reference materials and QC check samples may be caused by improper analysis. In the case of the

second failure of this criteria, all protocols associated with the sample preparation and analysis should be reviewed to find the error that caused the unsuccessful analysis. If samples have to be concentrated or diluted to the proper range of the analysis, carefully check these techniques and the accompanying calculations.

- Contaminations — Check previously listed activities to prevent contamination errors!

Note: After the source of the problem has been eliminated, reprocess and reanalyze all samples measured together with the failed Reference Standard.

Quality of Laboratory Pure Water

The acceptance criteria for the quality of laboratory pure water is discussed in Chapter 8. If the measured values show alteration from this standard, corrective action is needed immediately. Change the cartridges and call the contracted company for regular maintenance of the system.

Microbiological Tests

Positive and Negative Controls

When control samples fail to show the negative or the positive results, the first possible source of the failure is the *improper media.* It may be caused by preparation error, incorrect pH, incorrect incubation temperature, or expired media. Other sources may be contamination of the negative control, or the positive control was not positive control.

Blanks

Blanks contamination can be produced by inadequate sterilization of the filtration unit, and/or the rinse and dilution water. Contamination from pipets, forceps, and other sources during the test is also possible. With contaminated blanks, reject positive sample results and resample.

Inhibitory Residue Test

Detergent residues inhibit bacterial growth. When the residue test is positive, use other detergents or increase the rinsing time until the inhibitory test is negative (see Chapter 7).

11.2. Documentation of Out-of-Control Conditions

All out-of-control conditions must be documented immediately by all levels of responsibility. Documentation by the analyst is done by plotting the out-of-control data on the QC Charts and labeling the sample results pertaining to that test parameter as "suspect data".

The laboratory supervisor will document all "out-of-control" conditions by issuing a non-compliance report to the QA officer. The report must include the out-of-control test parameter, date of occurrence, ID number, copy of the failed QC data, description of the analytical problem, and the corrective action.

The QA officer will issue a written report to the laboratory manager defining any QC non-compliance occurrence along with any suspected sample data. The report must include the out-of-control parameter, date, time, ID number of samples with "suspect data", description of the QC problem, and the corrective action. A typical non-compliance report form is on Table 11-1.

11.3. Checking the Correctness of the Analyses

Analytical data for each parameter, produced by the analyst are checked by QC acceptance criteria, validated, corrected if necessary, and then converted to a reportable value. The approved and checked analytical results for each analyzed parameter will be transferred to the sample report form. The sample report contains the results for each required parameter analyzed for the particular sample. By comparing selected analytical results, there is a procedure to check the correctness of the analyses in one sample.

Calculation of Total Dissolved Solids (TDS)

The correctness of the measured value of the TDS may be verified by using the following formula:

$$TDS = 0.6(\text{Alkalinity}) + Na + K + Ca + Mg + Cl + SO_4 + SiO_3 + (NO_3\text{-}N) + F$$

The concentration of the measured constituents is in mg/L. The measured TDS concentration should be higher than the calculated one because significant contribu-

Table 11-1. Non-Compliance Report Form

Test out-of-control	*Date of problem*	*Name of Analyst*	*Suspect samples ID*	*QC true value*	*QC measured value*	*% R*	*% R limit*

DESCRIPTION OF ANALYTICAL PROBLEM

CORRECTIVE ACTION

Laboratory supervisor __________________ Date _______________

tors may not be included in the calculation. If the measured value is less than the calculated one, the sample should be reanalyzed. If the measured solid concentration is more than 20% higher than the calculated one, the selected constituents for the calculation should be reanalyzed. The accepted ratio of the measured and calculated TDS values is

$$1.0 < \text{measured TDS} / \text{calculated TDS} < 1.2$$

TDS and Electrical Conductance (EC) Ratio

The ratio of TDS, expressed in milligrams per liter and Conductivity, expressed in umhos per centimeter should be between 0.55 and 0.7. If the ratio is outside these limits, the measured TDS and Conductivity are suspect and should be reanalyzed. As the opposite, when conductivity multiplied by a factor between 0.55 and 0.7, the value should give the measured TDS. If the value is outside of these limits, reanalyze both TDS and Conductivity.

Anion - Cation Balance

The sum of the cations, expressed in milliequivalent per liter (mEq/L) and the anions, expressed in milliequivalent per liter must be balanced because all potable waters are electrically neutral. The test is based on the percent difference between the sum of the cations and sum of the anions. The percent difference is calculated as follows:

% difference = [(E cations - E anions)/(E cations + E anions)] × 100
(E = sum)

The acceptance criteria depends on the sum of the anions:

Anion sum, as mEq/L	Acceptance criteria
0 – 3.0	±0.2 mEq/L
3.0–10.0	±2%
20.0 –800	±2–5%

Table 11-2 contains the conversion factors to convert milligram per liter values to milliequivalent per liter and vice versa. The factors are derived from the valency and the atomic weight or the formula weight of the ions.

When the analyzed milligram per liter value of the ion is to be converted to milliequivalent per liter, the milligram per liter value is multiplied with the factor, derived from the ratio of the valency and the atomic or formula weight. For example, Na has a valency of 1, and the atomic weight is 22.9897; the used factor will be 1/22.9897 = 0.04349. For a sulfate (SO_4^{2-}) ion, which has a valency of 2, the formula weight of the substance is 96.0636 (S = 32.066 + 4 O = 63.9976); therefore, the factor will be 2/96.0636 = 0.0282.

When the milliequivalent per liter value is converted to milligram per liter value, the factor is calculated by the atomic weight or formula weight divided by the valency. For example, for Na the factor will be 22.9897 (22,9897/1) and for SO_4^{2-}, the factor becomes 96.0636/2 = 48.03. For example, 20 mg/L of Na is (20 × 0.04349) 0.8698 mEq/L; 0.8698 mEq/L Na is (0.8698 × 23) 20 mg/L

Relation Between Conductivity and Total Cations or Total Anions

The calculated total cations or total anions expressed as milliequivalent per liter multiplied by 100 should be close to the measured Conductivity, as umhos per centimeter. Otherwise, check Conductivity and if it gives the same value, reanalyze cations and anions.

TABLE 11-2. Conversion Factors Milligrams Per Liter -Milliequivalent Per Liter

Cations	Factor mg/L x = me/L	Factor me/L x = mg/L	Anions	Factor mg/L x = me/L	Factor me/L x = mg/L
Al^{3+}	0.1112	8.994	BO_2^-	0.02336	42.81
B^{3+}	0.2775	3.603	Br^-	0.01257	79.90
Ba^{2+}	0.01456	68.67	Cl^-	0.02821	35.45
Ca^{2+}	0.04990	20.04	CO_3^{2-}	0.03333	30.00
Cr^{3+}	0.05770	17.33	CrO_4^{2-}	0.01724	58.00
Cu^{2+}	0.03147	31.77	F^-	0.05264	19.0
Fe^{2+}	0.03581	27.92	HCO_3^-	0.01639	61.02
Fe^{3+}	0.05372	18.62	HPO_4^{3+}	0.02084	47.99
H^+	0.9922	1.008	H_2PO^{4-}	0.01031	96.99
K^+	0.02558	39.10	HS^-	0.03024	33.07
Li^+	0.1441	6.941	HSO_3^-	0.01234	81.07
Mg^{2+}	0.08229	12.15	HSO_4^-	0.01030	97.07
Mn^{2+}	0.03640	27.47	I^-	0.00788	126.9
Mn^{4+}	0.07281	13.73	NO_2^-	0.02174	46.01
Na^+	0.04350	22.29	NO_3^-	0.01613	62.0
NH_4^+	0.05544	18.04	OH^-	0.05880	17.01
Pb_{2+}	0.009653	103.6	PO_4^{3-}	0.03159	31.66
Sr^{2+}	0.02283	43.81	S^{2-}	0.06238	16.03
Zn^{2+}	0.03059	32.69	SO_4^{2-}	0.02082	48.03

Note: mg/L = milligrams/liter; me/L = milliequivalent/liter; me/L = mg/L x factor; Factor = ionic charge/atomic or formula weight (Cl_f = 1/35.45 = 0.02821); mg/L = me/L x factor; Factor = atomic or formula weight/ionic charge (Cl_f = 35.45/1 = 35.45).

Relation Between Total Hardness and Calcium and Magnesium Values

The analytical value of the Total Hardness, as $CaCO_3$ should be equal to the sum of the determined calcium, as $CaCO_3$ and magnesium, as $CaCO_3$. The $CaCO_3$ forms of the two metals are converted from the analytical values of calcium, as Ca and magnesium, as Mg by calculation, as outlined in Chapter 10. If criteria is failed, reanalyzing the two metals and the Total Hardness is necessary to verify the wrong result(s).

Relation of the Different Forms of Nitrogens

The analytical group of nutrients includes the forms of nitrogen of greatest interest in waters and wastewaters, namely ammonia nitrogen (NH_3-N), nitrite nitrogen (NO_2^--N), nitrate nitrogen (NO_3-N), and organic nitrogen. To determine organic nitrogen, analyze Total Kjeldahl Nitrogen (TKN), which is the sum of the NH_3-N and organic nitrogen in the sample. When the analytical result of the NH_3-N is higher than the TKN value, the two tests should be repeated to rectify the problem. The Total Nitrogen content of the sample should be the sum of the concentrations of the nitrogen compounds. Otherwise, check the analytical and QC data or the analytical tests should be repeated until the correct result is achieved.

Relation of the Different Forms of Phosphorus

Phosphates in environmental samples are classified as ortho (water soluble), condensed, and organically bound phosphates, reported as PO_4-P. When the reported value for the total phosphates is different from the sum of the individual forms of phosphates, it is suggested that the analytical tests are repeated.

Chapter 12
Reporting Analytical Data

Report of analyses are the written records of analytical work. The content of the report should include all of the necessary information and should be clear and understandable for the peruser. The designer of the final report always has to remember that the reader frequently does not understand the issued results and needs more specification. Brief interpretation of the reported values is essential.

Chapters 10 and 11 have given an account of how analytical data is calculated and checked. Never issue any report until all of the checks have been completed and approved. All documentation used to approve and defend reported data must be collected and should be available and referenced so it can be found at any time it may be needed. The content of these documents should be detailed and clear enough to explain to the interested party how the final reported values have been generated.

12.1. Required Documentation

The documentation that is required to approve and defend reported data is as follows:

- **Documentation for sample collection and identification** — Sampling plan, Sample collection method, Sample identification, Chain-of-custody, Field notebook and documentation of field tests, Detailed field QC activities.
- **Documentation of the analytical performance** — Analytical method used, Method Detection Limit (MDL), Instrumentation (manufacturer, model, performance check, maintenance log), Calibration data (initial, continuing), Detailed analytical work (working papers, standards, reagents preparation, calculation), Date of the analysis and the name of the analyst (on working paper).
- **QA/QC documentation and data** — Analysis of blanks, Precision and accuracy data, QC charts, Acceptable ranges of accuracy and precision per parameter, Source of control samples, How spikes were prepared, How surrogates were prepared, Acceptance of clean ups.
- **Checks and validation of analytical data** — Documentation, how analytical data are checked and validated, Corrective actions (when applicable), Date and signature for approval of the reportable data of each parameter tested, Date and signature for the approval of the final analytical report.

12.2. Significant Figures in Analytical Reports

Numerical data are often obtained with more digits than are justified by its accuracy and precision. Report only such figures that are justified by the accuracy of the analytical method and do not use the common practice that all numbers in one column have the same number of figures to the right of the decimal point. The reported numbers belong to different parameters and they are determined by different analytical methods, so the significant figures will also be different. Significant figures and rounding off rules are discussed in Chapter 10.

Round Off

If an analytical result is, for example, 16.6 mg/L, the analyst should be certain about the "16" but uncertain about the "0.6". It may be 0.4, 0.5, 0.7, 0.8, because this number is not justified by the method. Therefore, the result is rounded off and reported as "17". When a result is generated for a certain method and it gives a value of "16.61" mg/l, it should be reported if these significant figures are justified by the method.

If a calculated value of a result is 2346 mg/L, but the analyst is not certain about the "46", the result should be reported as 2350 mg/L. In a number written as 5.000, it is understood that all the zeros are significant, or else the number would have to be rounded off to 5.00, 5.0, or 5, whichever is appropriate. If a result is reported as 360 mg/L, the recipient of the report should be certain that the zero is significant, because the zero cannot be deleted.

Calculations

In the practice, when calculation takes place, use significant figures as are present in the lowest significant number participate in the calculation (see Chapter 10). For example: 25 + 0.356 + 12.23 = 37.586 - report as 37, because the lowest significant number is two. 62 × 0.0022 × 52.16/1.245 = 5.71455743 - Report as 5.7, because the lowest significant figure is two.

12.3. Report with Confidence Interval

Reporting the calculated confidence limit of the measurement helps to estimate the reliability of the result. Standard deviation is calculated to three significant figures and rounded to two when reported as a data. Calculated confidence interval is rounded to two significant figures as reported. Usually the 95% confidence limit is used. For example:

Available data	= 24.3, 23.9, 24.1, 24.2, 23.8, 24.0, 24.2
Average ($\bar{x}$)	= 24.07
Standard deviation (s)	= 0.1799
Number of measurements (n)	= 7
Degree of freedom (n - 1)	= 7 - 1 = 6
"t" value from Student t table with 95% limit	= 2.517
Confidence limit (CI)	$= t\,s / \sqrt{n}$
CI	= 2.517 × 0.179 / 2.645 = 0.170
The result reported	= $\bar{x} \pm$ CI 24.07 ± 0.17

12.4. Units Used to Express Analytical Results

Units used to express analytical results depend on the analytical method used, the concentration of the analytes, and the matrices of the sample analyzed.

The most common unit used to express analytical results in environmental samples is *parts per million (ppm)* and *parts per billion (ppb)*. ppm is equal to milligram per liter (mg/L) or milligram per kilogram (mg/Kg). When the concentration is less than 0.1 ppm, it is more convenient to express the results as ppb, which is equal to microgram per liter (ug/L) or microgram per kilogram (ug/Kg).

If the concentration is greater than 10,000 ppm, the result is expressed in *percent (%)*. (1% is equal to 10,000 mg/L when the specific gravity is 1.00.)

If result is issued in ppm or percent by weight for *solid samples and liquids with high specific gravity,* a correction is necessary as follows:

$$\text{ppm by weight} = \text{mg/L}/\text{specific gravity}$$
$$\text{\% by weight} = \text{mg/L}/(10{,}000 \times \text{specific gravity})$$

Analytical results for *solid samples* report in milligram per kilogram or microgram per kilogram or as stated above in percent, according to the concentration value. The report must also include the statement that the result is on the "as is base", also called "wet base", of the solid, if the result is not corrected to dry, moisture free solid. Results reported as on the "dry base" mean that the value is corrected to dry, moisture-free solid. Calculations of these units are in Chapter 10.

Calculations related to water treatment require *analytical results in milliequivalent per liter (me/L)*. Also to check the accuracy of the analysis by cation-anion balance, the concentration of the ions must be in milliequivalent per liter. Conversion of milligram per liter values to milliequivalent per liter units is discussed in Chapter 11.

Color is reported in Color Unit (C.U.). Color Unit is the color produced by 1 mg platinum per liter in the form of chloroplatinate ion.

For water samples, *turbidity values* are reported as Nephelometric Turbidity Unit (NTU). This unit is related to the method, which uses nephelometric measurement. The method used to determine the odor in water samples is the Threshold odor test, and the unit used to express it is the Threshold Odor Number (T.O.N.)

The pH at which a water is just saturated with calcium carbonate ($CaCO_3$) is known as the pH of saturation, or pH_s. *The Langelier's Saturation Index (SI)* is defined as the actual pH minus pH_s. The SI is used as the unit for expression of the $CaCO_3$ saturation, or corrosivity test. A negative value indicates a tendency to dissolve $CaCO_3$ (the water is corrosive), and a positive value indicates a tendency to deposit $CaCO_3$ (the water is scale forming). When the value is 0 ± 0.2, the water is stable. The index is not directly related to corrosion, but deposition of a thin, coherent carbonate scale may be protective; therefore, the positive index is associated with the noncorrosive condition and the negative value may be a sign of corrosivity.

Conductivity is a numerical expression of the ability of an aqueous solution to carry the electrical current. The laboratory measurement of conductivity is usually a resistance, measured in ohms or megohms. Specific resistance is the resistance of a cube 1 cm on an edge. The reciprocal of the resistance is conductance. It measures the ability to conduct a current and is expressed in reciprocal of ohms or mhos. A more convenient unit in water analysis is micromhos (umhos) and the reported unit is micromhos per centimeter (umhos/cm). In the International System of Units (SI) the reciprocal of the ohm is the Siemens (S) and conductivity is reported as millisiemens per meter (mS/m).1 mS/m = 10 umhos/cm. To report results in SI units, umhos/cm divided by 10 to obtain mS/m. 1 umhos/cm = 1 uS/cm.

12.5. Units Used to Express Results for Microbiological Tests

The report on microbiological examination depends on the tests. Coliform tests by the *multiple tube fermentation procedure* are reported as the most probable number (MPN) index. The index of the number of coliform bacteria, more probably than any other number, would give the results shown by the examination. It is not an actual enumeration.

Using the *membrane filter procedure*, which permits the direct count of bacterial colonies, the report is the count/100 ml. For verified coliform counts, adjust the initial count based on the positive verification process and report as "verified Coliform count/100 ml". If confluent growth occurs, e.g., growth covering either the entire filtration area of the membrane or a portion of the membrane, and colonies are not discrete, report as "confluent growth coliforms" and request a new sample from the same place. In the next examination, samples have to be diluted for the proper size to avoid overgrowing bacterial density. If the total bacterial count (including coliform and non-coliform) exceeds 200 per membrane filter, or the colonies are not separated enough to count, report as "too numerous to count (TNTC)".

Total coliform bacteria maximum count is accepted as 80 colonies; therefore, it may be reported as estimated counts/100 ml >80 colonies, if 100 ml sample used for determination. On the other hand, all of the dilution should be counted. For example, if the plate is prepared from a 1:10 dilution sample, and is TNTC, it should be reported as estimated counts >800/100 ml. If the dilution was 1:1000, the bacterial count will be >80000/100 ml.

Fecal coliform bacteria is the same as mentioned above, except the accepted maximum count is 60 instead of 80 colonies. When no colonies are found on the membrane filter, report as <1 colony/100 ml.

The *heterotrophic plate count (HPC)*, formerly known as the Standard Plate Count (SPC), is a procedure for estimating the number of live heterotrophic bacteria. Colonies may arise from pairs, chains, clusters, or single cells and, therefore, report as "Colony-Forming Units (CFU)". Because the usual sample size for plating is 1 ml, HPC report as colony forming units (CFU)/ml. If an inoculated 1 ml sample has no colonies, report as CFU <1/ml. If plates from all dilutions of a sample have no colonies, use the lowest dilution plate for reporting; for example, if no colonies developed on the 1:100 dilution, the report is "estimated CFU <100/ml". Do not report as TNTC if the number of colonies per plates exceeds 300 colonies. The calculated values are based on the area of a plate, which is 57 cm^2 for disposable plastic plates, and 65 cm^2 for glass plates. Count colonies per square centimeter, and use the factor according to the area of the plate. Take consideration for any dilution and report the result as "estimated CFU/ml". When bacterial colonies on crowded plates are greater than 100/cm^2, the report will be estimated CFU >5700/ml, or CFU >6500/ml, corresponding to the plate used. If the plate belongs to any dilution, the dilution factor should also be considered. For example, if calculated colonies on one square was greater than 100, a plastic plate was used, and the plate was prepared from a 1:1000 dilution sample, the report will be: estimated CFU >5,700,000/ml.

Bacterial population in solid samples report as count/g sample.

Air density plates reports should contain the time at which the plate was exposed and CFU count per plate per minutes exposure, (CFU = Colony Forming Units). For example, CFU count is 10 colonies per plate per 15 minutes exposure.

12.6. Units Used to Express Results on Radioactivity Analyses

Results on radioactive analysis are reported in terms of picoCuries per liter (pCi/L) at 20°C. For samples with specific gravity that is significantly different from 1.000, the report is picoCuries per gram (pCi/g). (Curie is a unit quantity of any radioactive nuclide in which 3.7×10^{10} disintegration occurs per second; picoCurie is one trillionth, 10^{-12}, part of a Curie). If the quantity is 1000 to 1,000,000 pCi/L or pCi/g, use the nanoCurie per liter (nCi/L) or nanoCurie per gram (nCi/g) (nCi = 10^{-9} Ci = 1000 pCi).

For values higher than 1000 nCi, use microCurie per liter (uCi/l) or microCurie per gram (uCi/g). (1 uCi = 10^{-6} Ci = 10^{6} pCi).

12.7. Report Format

A simple table form is accepted for regular, routine reporting. It should be clear, easy to follow, and should contain all of the necessary information to evaluate the analytical report. For special, non-routine purposes, reports are more detailed, and a comprehensive QA/QC report should be attached to the analytical report.

Usual Form of Analytical Report

Title — It should be brief, but descriptive, to identify the goal of the analytical work.

Who requested the analysis — Identification of organization, or person for whom the work was done. Name, organization, address, work order, etc.

Report number — Laboratory Identification number (ID number) for the sample(s), as in the laboratory log book and, if applicable, the appropriate project number. The numbering system: calendar year/sequence number.

Date — Date report is completed.

Objective — Short statement about the reason the work was done.

Sample identification — Physical description of the sample, sampling area, possible photographs with the sampling area, and all the information related to the sample that may impact on the data.

Sampling details — Sampling procedures, sample type, sample preservation, sample collector, time and date of collection, chain of custody, sample field custody, and transportation.

Analytical methodology — Brief description of the methods and method references used for analyses. References should be specific, and provide all of the necessary information (revision date, reference number, etc.) Any modification from the original method should be stated.

Method Detection Limit (MDL) or Practical Quantitation Limit (PQL) — Each analysis should provide with the specified MDL or PQL.

Numerical values and units — Numerical values are always reported with the corresponding unit and with correct significant figures.

References — Any information related to previous analytical work on the same sample location. For example, part of a monitoring program, previously completed analytical work, and references to other reports on the same sample location.

Discussion — Interpretation of the results, recommendation of additional works, or for corrections. Any special observations related to the sample or for the analytical report that serve the objective.

Signatures — Signatures and titles of all persons responsible for the data and for the report.

Distribution list — Full distribution list for the report.

Attachment — QC report attached to the analytical data report. This report contains the basis of calibration, standardization, and statement where the documentation is found. Precision and accuracy data should be stated for each analytical performance in the analytical report with the acceptance limit.

The pages of the report should be properly numbered. The reader of the report must be certain that he/she has the full report for review. (For example, page 1 of 6 pages, or page 6 of 6 pages, etc.)

Part 4

Discussion of the Analyzed Parameters

Chapter 13
Introduction to Physical Properties

This chapter deals primarily with the introduction and general discussion of the physical properties of a sample. The typical, meaningful, and traditional parameters described here are temperature, pH, color, taste, odor, turbidity, conductivity, and solids. However, physical properties cannot be separated entirely from chemical composition. For example, calcium carbonate saturation is listed under physical properties, but it is dependant on chemical tests.

13.1. Electrical Conductance

Electrical conductance is a numerical expression of the ability of an aqueous solution to carry the electric current. This ability depends on the presence of ions, their total concentration, and on the temperature of the measurement. Because the test is a measure of the electric current carried by ionized substances, dissolved solids in the sample are basically related to this measure. Solutions of most inorganic acids, bases, and salts are relatively good conductors. Conversely, molecules of organic compounds that do not dissociate in aqueous solutions conduct the current very poorly, if at all. Conductivity is a useful test for quick determination of the mineral content of the sample. Conductivity is the reciprocal of the resistance. Since electrical resistance is measured in ohms, the electrical conductance is measured by its inverse, mhos. To express electrical conductance in environmental samples, the most convenient units are mmhos/cm (millimhos/cm) and umhos/cm (micromhos/cm) as discussed in Chapter 2 and Chapter 11. The centimeter usually indicated in the specific conductance is the standard distance of 1 cm apart from two electrodes of 1 cm^2. Temperature significantly influences the results. The standard reported temperature is 25°C.

13.2. Color

Color in water may result from the presence of metallic ions (iron manganese, copper, chromium), humus and peat materials, plankton, weeds and industrial wastes. When noticed in drinking water, color may cause psychological rejection and fear; there have been cases in which sudden change in color has stopped people from drinking contaminated water. Color is measured by a comparator, using color discs that give results in substantial agreement with the platinum-cobalt color standards. Reporting unit is "C.U." (Color units), the color produced by 1 mg/L platinum in the form of chloroplatinate ion.

13.3. Solids or Residues

The term residue refers to solid material suspended or dissolved in water or waste water. Residue may affect water or effluent quality adversely in a number of ways. Waters with high residue are generally of inferior palatability and may induce unfavorable physiological reaction in the consumers. A limit of 500 mg/L dissolved solids is desirable for drinking waters; highly mineralized waters are unsuitable for many industrial applications as well. Solid determinations are an important part in the control of water treatment processes. In drinking water, total dissolved solids are primarily made up of inorganic salts, mainly carbonate, bicarbonate, chloride, sulfate, nitrate, sodium, potassium, calcium, and magnesium. A major contribution to these constituents is the natural contact with rocks and soil, with a minor contribution from pollution.

Total Solids (TS) or Total Residue

Total solids are all solids contained in the water sample and are determined by evaporation and drying in an oven at a defined temperature.

Suspended Solids (SS) or Nonfilterable Residue

Suspended solids are not dissolved in the water sample. They are the portion of total residue retained by a standard glass fiber filter (Glass fiber filter, Whatman grade 934AH or Gelman type A/E or Millipore type AP40 or equivalents). The residue on the filter is dried to a constant weight at 103 to 105°C. If the suspended material clogs the filter and prolongs filtration, the suspended solids may be estimated from the difference between the total solids and the total dissolved solids.

Total Dissolved Solids (TDS) or Filterable Residue

Total dissolved solids are the portion of total residue that passes through a standard glass fiber filter (specification of the filter is as described for suspended solids). TDS are the solids dissolved in the water sample, and the value is approximately proportional to the measured conductivity of the water.

Volatile and Fixed Solids

"Fixed Solids" is the term applied to the residue of total, suspended, or dissolved solids after ignition in a muffle furnace for a specified time (one hour) and for a specified temperature (550 ± 50°C). The weight loss on ignition is the "Volatile Solids". The test does not precisely distinguish between organic and inorganic matter; it gives only estimated values.

Settleable Solids

Settleable solids are solids in suspension that can be expected to settle by gravity during a defined period of time, and report on either a volume (ml/L) or a weight (mg/L) basis. Values from this test are useful to evaluate sedimentation process during water and sewage treatments.

13.4. Salinity

Salinity is an important measurement in the analysis of certain industrial wastes, natural waters, and seawater. It is defined as the total solids in water after all carbonates have been converted to oxides, all bromides and iodides have been replaced by chloride, and all organic material has been oxidized. It is numerically smaller than the dissolved solids and usually reported as parts per thousands (o/oo) or grams per kilogram.

Associated terms are *chlorinity* (g/L Cl at 20°C), which includes chloride, bromide, and iodide, all reported as chloride, and *chlorosity*, which is the chlorinity multiplied by the water density at 20°C. A conversion table is available to convert chlorosity to salinity, based on an empirical relationship between salinity and chlorosity:

$$\text{Salinity o/oo} = 0.03 + [1.805 \times \text{Cl/L(20°C)} \times \text{density}]$$

In recent years, the conductivity and density (hydrometric measurement) methods have been used because of their high sensitivity and precision. Although conductivity has the greatest precision, it responds only for ionic solutes. Density is less precise, but responds to all dissolved solutes. Table 13-1 gives the corresponding densities (at 15°C) and salinities.

13.5. Temperature

Temperature readings are used in the calculation of various forms of alkalinity, in studies of saturation and stability with the respect of calcium carbonate saturation, in the calculation of salinity, and in general laboratory operations. In lymnological studies, water temperatures as a function of depth are often required. Elevated temperatures resulting from heated water discharges may have significant ecological impact. Identification of the source of the water supply is often possible by temperature measurement alone. Industrial plants often require data on water temperature for process use.

13.6. Turbidity

The clarity of a natural body of water is a major determinant of the condition and productivity of that system. Turbidity in water is caused by suspended matter, such as clay, silt, finely divided organic and inorganic matter, soluble colored organic compounds, and plankton and other microscopic organisms. Turbidity reporting units represent light-scattering and absorbing properties of suspended matter in water. The reportable unit is nephelometric Turbidity Unit (NTU). Turbidity measurement is not limited to the determination of small particles, but also related to the fact that these particles definitely tend to protect pathogens from disinfection treatment. Organic matter and microorganisms are presumed to protect pathogenic organisms from the bactericidal effect of disinfecting agents, while inorganic particles are not. The sample may also contain particles with toxic effect.

Table 13.1. Corresponding Densities and Salinities

Density	Salinity	Density	Salinity	Density	Salinity
0.9991	0.0	1.0048	7.3	1.0105	14.8
0.9992	0.0	1.0049	7.5	1.0106	14.9
0.9993	0.2	1.0050	7.6	1.0107	15.0
0.9994	0.3	1.0051	7.7	1.0108	15.2
0.9995	0.4	1.0052	7.9	1.0109	15.3
0.9996	0.6	1.0053	8.0	1.0110	15.4
0.9997	0.7	1.0054	8.1	1.0111	15.6
0.9998	0.8	1.0055	8.2	1.0112	15.7
0.9999	0.9	1.0056	8.4	1.0113	15.8
1.0000	1.1	1.0057	8.5	1.0114	16.0
1.0001	1.2	1.0058	8.6	1.0115	16.1
1.0002	1.3	1.0059	8.8	1.0116	16.2
1.0003	1.5	1.0060	8.9	1.0117	16.3
1.0004	1.6	1.0061	9.0	1.0118	16.5
1.0005	1.7	1.0062	9.2	1.0119	16.6
1.0006	1.9	1.0063	9.3	1.0120	16.7
1.0007	2.0	1.0064	9.4	1.0121	16.9
1.0008	2.1	1.0065	9.6	1.0122	17.0
1.0009	2.2	1.0066	9.7	1.0123	17.1
1.0010	2.4	1.0067	9.8	1.0124	17.3
1.0011	2.5	1.0068	9.9	1.0125	17.4
1.0012	2.6	1.0069	10.1	1.0126	17.5
1.0013	2.8	1.0070	10.2	1.0127	17.7
1.0014	2.9	1.0071	10.3	1.0128	17.8
1.0015	3.0	1.0072	10.5	1.0129	17.9
1.0016	3.2	1.0073	10.6	1.0130	18.0
1.0017	3.3	1.0074	10.7	1.0131	18.2
1.0018	3.4	1.0075	10.8	1.0132	18.3
1.0019	3.5	1.0076	11.0	1.0133	18.4
1.0020	3.7	1.0077	11.1	1.0134	18.6
1.0021	3.8	1.0078	11.2	1.0135	18.7
1.0022	3.9	1.0079	11.4	1.0136	18.8
1.0023	4.1	1.0080	11.5	1.0137	19.0
1.0024	4.2	1.0081	11.6	1.0138	19.1
1.0025	4.3	1.0082	11.8	1.0139	19.2
1.0026	4.5	1.0083	11.9	1.0140	19.3
1.0027	4.6	1.0084	12.0	1.0141	19.5
1.0028	4.7	1.0085	12.2	1.0142	19.6
1.0029	4.8	1.0086	12.3	1.0143	19.7
1.0030	5.0	1.0087	12.4	1.0144	19.9
1.0031	5.1	1.0088	12.6	1.0145	20.0
1.0032	5.2	1.0089	12.7	1.0146	20.1
1.0033	5.4	1.0090	12.8	1.0147	20.3
1.0034	5.5	1.0091	12.9	1.0148	20.4
1.0035	5.6	1.0092	13.1	1.0149	20.5
1.0036	5.8	1.0093	13.2	1.0150	20.6
1.0037	5.9	1.0094	13.3	1.0151	20.8
1.0038	6.0	1.0095	13.5	1.0152	20.9
1.0039	6.2	1.0096	13.6	1.0153	21.0
1.0040	6.3	1.0097	13.7	1.0154	21.2
1.0041	6.4	1.0098	13.9	1.0155	21.3
1.0042	6.6	1.0099	14.0	1.0156	21.4
1.0043	6.7	1.0100	14.1	1.0157	21.6
1.0044	6.8	1.0101	14.2	1.0158	21.7
1.0045	6.9	1.0102	14.4	1.0159	21.8
1.0046	7.1	1.0103	14.5	1.0160	22.0
1.0047	7.2	1.0104	14.6	1.0161	22.1

Table 13.1. Corresponding Densities and Salinities (continued)

Density	Salinity	Density	Salinity	Density	Salinity
1.0162	22.2	1.0215	29.1	1.0268	36.0
1.0163	22.4	1.0216	29.3	1.0269	36.2
1.0164	22.5	1.0217	29.4	1.0270	36.3
1.0165	22.6	1.0218	29.5	1.0271	36.4
1.0166	22.7	1.0219	29.7	1.0272	36.6
1.0167	22.9	1.0220	29.8	1.0273	36.7
1.0168	23.0	1.0221	29.9	1.0274	36.8
1.0169	23.1	1.0222	30.1	1.0275	37.0
1.0170	23.3	1.0223	30.2	1.0276	37.1
1.0171	23.4	1.0224	30.3	1.0277	37.2
1.0172	23.5	1.0225	30.4	1.0278	37.3
1.0173	23.7	1.0226	30.6	1.0279	37.5
1.0174	23.8	1.0227	30.7	1.0280	37.6
1.0175	23.9	1.0228	30.8	1.0281	37.7
1.0176	24.1	1.0229	31.0	1.0282	37.9
1.0177	24.2	1.0230	31.1	1.0283	38.0
1.0178	24.3	1.0231	31.2	1.0284	38.1
1.0179	24.4	1.0232	31.4	1.0285	38.2
1.0180	24.6	1.0233	31.5	1.0286	38.4
1.0181	24.7	1.0234	31.6	1.0287	38.5
1.0182	24.8	1.0235	31.8	1.0288	38.6
1.0183	25.0	1.0236	31.9	1.0289	38.8
1.0184	25.1	1.0237	32.0	1.0290	38.9
1.0185	25.2	1.0238	32.1	1.0291	39.0
1.0186	25.4	1.0239	32.3	1.0292	39.2
1.0187	25.5	1.0240	32.4	1.0293	39.3
1.0188	25.6	1.0241	32.5	1.0294	39.4
1.0189	25.8	1.0242	32.7	1.0295	39.6
1.0190	25.9	1.0243	32.8	1.0296	39.7
1.0191	26.0	1.0244	32.9	1.0297	39.8
1.0192	26.1	1.0245	33.1	1.0298	39.9
1.0193	26.3	1.0246	33.2	1.0299	40.1
1.0194	26.4	1.0247	33.3	1.0300	40.2
1.0195	26.5	1.0248	33.5	1.0301	40.3
1.0196	26.7	1.0249	33.6	1.0302	40.4
1.0197	26.8	1.0250	33.7	1.0303	40.6
1.0198	26.9	1.0251	33.8	1.0304	40.7
1.0199	27.1	1.0252	34.0	1.0305	40.8
1.0200	27.2	1.0253	34.1	1.0306	41.0
1.0201	27.3	1.0254	34.2	1.0307	41.1
1.0202	27.5	1.0255	34.4	1.0308	41.2
1.0203	27.6	1.0256	34.5	1.0309	41.4
1.0204	27.7	1.0257	34.6	1.0310	41.5
1.0205	27.8	1.0258	34.8	1.0311	41.6
1.0206	28.0	1.0259	34.9	1.0312	41.7
1.0207	28.1	1.0260	35.0	1.0313	41.9
1.0208	28.2	1.0261	35.1	1.0314	42.0
1.0209	28.4	1.0262	35.3	1.0315	42.1
1.1210	28.5	1.0263	35.4	1.0316	42.3
1.0211	28.6	1.0264	35.5	1.0317	42.4
1.0212	28.8	1.0265	35.7	1.0318	42.5
1.0213	28.9	1.0266	35.8	1.0319	42.7
1.0214	29.0	1.0267	35.9	1.0320	42.8

Density at 15°C. Salinity expressed as parts per thousand (0/00) or g/kg.

Chapter 14
Introduction to Metals

The effects of metals in water and waste water range from beneficial to troublesome to dangerously toxic. Some metals are essential, for example, calcium, magnesium, and others may adversely affect water consumers, waste water treatment systems, and receiving waters. Lead, mercury, and arsenic have proven to be significant pollutants and toxins. Some metals may be either beneficial or toxic, depending on their concentrations. As indicated by Paracelsus, the 16th-century Swiss physician who introduced lead and arsenic into pharmaceutical chemistry that the dose of a substance determines its toxicity. This notion lead to the controversy concerning the definition of hazardous environmental exposure to chemicals. Aluminum, zinc, and tin are common components of food storage and cooking containers and are usually considered nontoxic. More attention is being paid to aluminum, however, because high levels have been found in patients with certain mental disorders such as Alzheimer's disease. Copper is usually toxic only to those with special susceptibility known as Wilson's disease, a genetic disorder of copper metabolism. Iron is an essential nutrient, but is toxic in large doses. There are some toxicities that many metals have in common: one of these is termed metal fume fever. Exposure to freshly generated oxides of zinc or many other metals results in polymyalgia (generalized muscle pain), fever, chills, headache, nausea, vomiting, and weakness. Water distribution systems and house plumbing contribute to the presence of copper, lead, and cadmium at the consumer's tap. Cross-connections may introduce chromium, used for corrosion control in cooling towers. Arsenic, barium, cadmium, and chromium have been studied for potential carcinogenicity, and mercury and selenium have proven to be toxic. These metals have been selected and listed in the National Primary Drinking Water Regulations, mandated by the Safe Drinking Water Act (SDWA).

The Primary Drinking Water Standard contains the maximum contaminant levels for health affecting metals: arsenic (As), barium (Ba), cadmium (Cd), chromium (Cr), lead (Pb), mercury (Hg), and selenium (Se).

The Secondary Drinking Water Standard metals are copper (Cu), iron (Fe), manganese (Mn), silver (Ag), and zinc (Zn).

In mineral content analysis, the common metallic constituents are calcium (Ca), magnesium (Mg), sodium (Na), and potassium (K).

EPA Priority Pollutants listed 12 toxic metals as environmental concern: antimony (Sb), arsenic (As), beryllium (Be), chromium (Cr), copper (Cu), lead (Pb), mercury (Hg), nickel (Ni), selenium (Se), silver (Ag), thallium (Tl), and zinc (Zn).

EPTOX (Extraction Procedure Toxicity) and TCLP (Toxicity Characteristic Leachate Procedure) metals are: arsenic (As), barium (Ba), cadmium (Cd), chromium (Cr), lead (Pb), mercury (Hg), selenium (Se), and silver (Ag).

Metallic compounds may dissolve in water or may be connected in suspended materials in the sample. According to the determination, metallic compounds are defined in the following terms:

- Filterable (dissolved) metals — Those metals of an unacidified sample that pass through a 0.45 um membrane filter.
- Nonfilterable (suspended) metals — Those metals of an unacidified sample that are retained by a 0.45 um membrane filter.
- Total metals — The concentration of metals determined on an unfiltered sample after vigorous digestion, or the sum of the concentrations of metals in both filtrable and nonfilterable fractions.

14.1. Specifications and Health Effects of Selected Metallic Parameters

Aluminum (AL)

A silvery, light-weight, easily worked element that has excellent anticorrosion property, aluminum is nontoxic and a good thermal conductor. Most aluminum is produced by electrolyzing bauxite (an impure hydrated oxide ore). Aluminum is the most abundant metal found on the earth's crust, 8.1%. It is found in nature as aluminosilicates, such as clay, kaolin, mica, and feldspar. Uses of aluminum metal include building construction, aluminum paint, electrical uses, automobile highway signs, packaging, and containers. Bentonite and clays, natural aluminum minerals, are used in purification of water as coagulant, Al_3AlO_3 and $NaAl(SO_4)_2$ (alum). Aluminum compounds are used therapeutically to prevent hyperphosphatemia in renal disease, in antacid preparations, and as an antidote. Unprocessed food contains <10 mg/kg, although some vegetables and fruits may contain up to 150 mg/kg aluminum. High aluminum intake is also originated from packaging and processing food (aluminum cooking vessel, aluminum foil, etc.), and from aluminum containing antacids. Aluminum has been considered to be nontoxic. However, high aluminum levels have been found in some regions of the brains of patients who died of Alzheimer's disease. This started the investigation of the possible role of aluminum in the aging process and in Alzheimer's disease.

Antimony (Sb)

The symbol of antimony is Sb, from the old latin name of stibium. It belongs to the same periodic group as arsenic and resembles it chemically and biologically. Antimony is associated with the sulfide mineral stibnite. It is used for alloys and certain compounds are used for fireproofing textiles and for ceramics, glassware, and antiparasitic drugs. It is not known if antimony is required by the human body, but its toxicity appears with the same symptoms as arsenic. Practically no antimony has been detected in drinking waters; therefore, standards for this metal are not available.

Arsenic (As)

A silvery-white, very brittle, semi-metallic element, arsenic is notorious for its toxicity to humans. The most toxic are the trivalent compounds. The lethal dose is 130 mg. Accumulation in the body is expected to raise progressively at low intake level. Arsenic is used in bronzing, pyrotechnique, dye manufacturing, insecticides, poison, and medicine. As medication in low doses, arsenic is excellent to enhance growth. Groundwater may contain a higher concentration of arsenic that originated from geological materials. Sources of arsenic pollution are from industrial wastes, arsenic-containing pesticides, and smelting operation. These industrial processes (as well as coal burning) are responsible for the presence of arsenic in the atmosphere. The EPA has classified arsenic as "carcinogenic in humans by inhalation and ingestion, however, this chemical has a potentially essential nutrient value", and established the Maximum Contaminant Level (MCL) for drinking waters as 0.05 mg/L.

Barium (Ba)

Barium is a silvery white, slightly malleable metallic element and belongs to the alkaline earth metal group. It is found in nature as carbonate or sulfate salts in limestones, sandstones, and sometimes in soils. The source of barium pollution is from mining industries (oil, coal, drilling mud) and from combustion (aviation, diesel fuel). Barium is used in manufacturing paint, paper, ceramics, glass, oil-well drilling, rubber, linoleum, and X-ray diagnostic work. Acute exposure to barium results in gastrointestinal, neuromuscular, and cardiac effects. MCL in drinking water is 1.0 mg/L.

Beryllium (Be)

Beryllium is a member of the alkaline earth metal. It is very light with excellent thermal conductivity and has a high melting point. Most of its uses are based on these properties. A small quantity of beryllium is found in water and soil, and can be ingested via food. The results of a survey in the U.S. have shown beryllium in drinking water in trace quantities (average of 0.013 ug/L). There is some beryllium as air pollutant due to combustion, cigarette smoke, and beryllium processing plants. Only the water soluble salts of the metal (sulfates and fluorides) have acute effects that cause dermatitis, conjunctivitis, and (by inhalation) irritation of the respiratory tract. Chronic exposure to beryllium and its compounds may produce "berylliosis", a frequently fatal pulmonary granulomatosis. The toxic effect may be related to inhibition of enzyme activities. Animal studies indicate pulmonary cancer, osteosarcoma, rickets, anemia, and alveolar metaplasia, but carcinogenic effects on humans have not been confirmed. Since concentration of beryllium in water is minimal, there is no need to issue a standard.

Boron (B)

Boron is a metalloid with a unique chemistry. It belongs to the Group III elements, along with aluminum, gallium, indium, and thallium. Boron is produced from bedded deposits and brines, and concentration in seawater is high (5 mg/L). The major uses of boron include glass and ceramic preparation, cosmetics (soap and cleansers), food preservation (boric acid, sodium borate), herbicides, and is therapeutically used as an antiseptic. Boron is generally essential to plants, but concentrations of higher than 0.2 mg/L in irrigation waters is harmful for certain vegetation, and over 0.1 mg/L may damage some greenhouse plants. Drinking water generally contains less than 0.1 mg/L boron. High doses of boron cause circulatory problems, tachycardia, delirium, and coma. Extended ingestion of boron may develop into the sickness called "borism".

Cadmium (Cd)

A soft, blue-white metallic element, cadmium appears in nature in ores. It may contaminate water supply from mining, industrial operations, and leachates from landfill. It may also enter the distribution system as a corrosion product of galvanized pipes. High levels of cadmium in cigarette smoke contribute to air pollution. Exposure occurs in pigment workers and welders. Acute overexposure to cadmium fumes may cause pulmonary damage, while chronic exposure is associated with renal tubular damage and an increased risk of cancer of the prostate. MCL in drinking water is 0.005 mg/L.

Calcium (Ca)

Calcium is a silver-white, active alkaline earth metal, with the abundance of 3% in the earth crust, always found in combination with limestone, marble, and chalk ($CaCO_3$). Addition of water to calcium produces calcium hydroxide $Ca(OH)_2$. As a metal used in metallurgy, it becomes calcium sulfate, or gypsum $CaSO_4.2H_2O$ is widely used in construction, interior decoration, and as cement. Numerous calcium compounds have therapeutic uses, for example, as antispasmodic, diuretics, and antacid preparations and against low-calcium tetany. Calcium is an essential constituent of bone and teeth. Hypercalcemia (elevated blood calcium concentration) occurs in diseases such as hyperparathyroidism, sarcoidosis, malignancy, and Vitamin D poisoning (infants). Calcium toxicity signs and symptoms include anorexia, nausea, vomiting, dehydration, lethargy, coma, and death. Calcium salts are common in water. These salts may result from leaching of soil and other natural sources, or they may occur in sewage and industrial wastes. Excessive calcium levels in drinking water may relate to the formation of kidney or bladder stones in the human body, but no toxicity concern. An upper limit as a guide may be used, such as 75 mg/L in relation to hardness. Literature relating blood pressure to various drinking water parameters has recently been reviewed. According to this survey, high sodium and low calcium intake have been implicated as factors in the development of high blood pressure. Drinking water is a source of both of these minerals, although the contribution to total dietary intake is quite small.

Chromium (Cr)

Chromium is a grayish-white crystalline and is a very hard metallic element with a high resistance to corrosion. The earth's crust contains an average 125 ppm chromium. Trivalent and hexavalent chromium are both found in nature, but chromium trivalent (Cr^{3+}) predominates. Cr^{3+} may be nutritionally essential, but Cr^{6+} is highly toxic. It has a deleterious effect on the liver, kidney, and respiratory organs with hemorrhagic effects, and exposure to chromium hexavalent can cause dermatitis. Besides occupational dermatitis, chromate workers develop ulcerations and perforations of the nasal septum. Gastric cancers, presumably from excessive inhalation of dust containing chromium have also been reported. Chromium compounds are used in tanning, pigments, electroplating, as corrosion inhibitors, catalysts, and wood preservatives. The MCL in drinking water systems is 0.1 mg/L.

Cobalt (Co)

Chemically, cobalt belongs to the group of iron and nickel with similar properties. The main uses of cobalt are in alloys, glass, porcelain, and electroplating. Cobalt is a part of Vitamin B_{12} (Cyano-cobalamin), and is considered as an essential nutrient, but concentration as high as 1 mg/kg of body weight is noticed as health hazard. The quantity of cobalt in drinking water is negligible, so maximum contaminant level is not established. Cobalt exhibits toxic effects in humans on the thyroid, the heart, and the kidneys. Coffee and beer are considered potential sources of cobalt. Cobalt toxicity has been reported — in Belgium and Canada — with severe heart failure and about 40% mortality among heavy beer-drinkers when additional cobalt was added to the beer as a foam stabilizer.

Copper (Cu)

Copper is a reddish-brown, malleable, ductile, metallic element that is excellent for heat and electrical conductance. Elemental copper was one of the first metals used by humans. Bronze (copper hardened by alloying with tin) was the first "industrial metal". About half of the copper production is used as a conductor in electrical equipment due to its high conductivity. It is used in many alloys, as well as in plumbing and the manufacture of various parts and goods. Copper compounds are also used in pesticides, algicides, fungicides, insecticides, ceramics, and paints. Copper sulfate has been used as an emetic. Copper is essential in human nutrition, since it is important for enzymatic actions. MCL in drinking water is 1.0 mg/L.

Iron (Fe)

Iron is the fourth most abundant element in the earth's crust, and vital for both animal and plant life. Iron, and its carbon alloy steel, is the backbone of modern civilization. Iron is the central metal in the hemoglobin molecule, and is used as therapy for iron-deficiency anemia. Iron and its compounds are used as pigments, magnetic tapes, catalysts, disinfectants, tanning, and fuel additives. The pure metal is very reactive, and corrodes rapidly particularly in moist air (rust). Iron is an essential mineral, but is toxic in large doses. The iron content of environmental samples is a result of feeding aquifers, corrosion from pipes, leachates from acid mine drainage, and iron product industrial wastes. Iron exists as either ferrous (Fe^{++}) or ferric (Fe^{+++}) forms. It may be found in solution, colloidal state (subjected to organic matter), inorganic or organic iron compounds, or suspended forms. Ferrous and ferric iron are soluble in water, but ferrous iron is easily oxidized to ferric hydroxide, which is insoluble in water. Iron in water can cause staining of laundry and porcelain, and a bittersweet astringent taste is detectable in drinking water containing 1 mg/L iron. To prevent the formation of black iron deposit and iron bacterial growth, oxygen in the water should be higher than 2 mg/L and a free chlorine residual concentration of higher than 0.2 mg/L should be present. Maintaining a pH above 7.2 in a distribution system also helps to avoid high iron deposition. In the Secondary Drinking Water Regulation, the iron and manganese combined limit is 0.3 mg/L, based on aesthetic and taste consideration.

Lead (Pb)

Lead is a heavy, very soft, highly malleable, bluish-gray metallic element. Most lead is used as the metal and the sulfate salt in storage batteries, bullets, and radiation shielding. The use of many compounds, such as inorganic pigments and tetraethyllead for gasoline is decreasing. Rome was one of the first societies to have a piped water supply. The pipes were fashioned from lead. The latin word for lead is "plumbum". This root word gives us the chemical symbol of lead, Pb, as well as the name of the vocation concerned with water pipes-plumber. Lead is a toxic substance (it is particularly toxic to the nervous system) and children are especially susceptible to its effects. If lead is ingested, even in a tiny amount dissolved in drinking water, it is well-absorbed from the gastrointestinal tract and deposited in the central nervous system. When a sufficient amount accumulates, lead encephalopathy may be a severe result. Some historians have theorized that this may explain the fall of the Roman Empire. Obviously, only the upper-class citizens of Rome would have had water piped into their homes. According to the theory, their children drank the water and were exposed to the lead, putting them at high risk for lead neurotoxicity. Lead toxicity also may explain the bizarre behavior of some of the notorious roman emperors. The recent history of lead is equally fascinating. It is widespread in the environment. A person can be exposed to lead from a number of sources, including lead-based paint, air, soil, dust, food, and drinking water. When drinking water contains lead, a significant portion of the lead is absorbed when you drink it. Children absorb more lead than adults — about 50% of the ingested lead is absorbed into a child's body on average; however, between 10 and 15% of lead in the drinking water is absorbed in adult bodies. The presence of lead in the body is indicated by the level of lead found in the blood, expressed as ug of lead per deciliter of blood (ug/dL). Lead has no health benefits, and negatively affects virtually every system in the body. Although blood lead levels are on the 10 ug/L level, lead may contribute to decreased intelligence, decreased body growth, and poor posture and may also affect the nervous system. Lead is also a weak carcinogen, although data are not completed. Many children have suffered lead toxicity from ingestion of lead-based paint. Lead-based paint was used inside many homes until Congress passed the Lead-based-Poisoning Prevention Act in 1971. Since then, lead has not been added to paint used inside residences or on furniture or toys. Lead is also found in air, water, and soil. Concentration of lead in natural waters has been reported as much as 0.4 to 0.8 mg/L, mostly from mountain limestone and galena (PbS) deposits. Lead in surface and ground water ranges from trace amounts to 0.04 mg/L, averaging about 0.01 mg/L. Industrial and mining sources are responsible for some local high contamination levels, but most high level lead values detected in drinking waters are produced by corrosion in lead service pipes and house plumbing. The use of soft water of slightly acidic pH and the use of lead pipes in service and domestic water lines may cause high levels of lead in drinking water. In fact, the U.S. EPA estimates that infants dependent on formula may receive more than 85% of their lead from drinking water that contains lead. Lead in drinking water can be a significant source of lead intake if it is not controlled. Because scientific evidence indicates that lead has harmful effects even below the 10 ug/dL blood level, the U.S. EPA set the Maximum Contaminant Level Goal (MCLG) of drinking water lead level to zero, instead of the 0.02 mg/L limit. Lead, as a corrosion product in drinking water is associated with copper. Copper is needed for good health and has a beneficial effect in low levels. High levels may cause nausea and diarrhea. MCLG for copper is 1.3 mg/L. As with all drinking water standards, the new Lead and Copper Rule sets minimum standards. Final rules for lead and copper were published on June 7, 1991, Federal Register, 56:110:26460 by the U.S. EPA.

Lithium (Li)

Lithium is a soft, alkali metal, found mostly in aluminum silicates and in seawater at an average 11 mg/L concentration, or even higher in some spring water. Lithium is the lightest metal and it is used in metallurgy, the ceramic industry, air conditioning, and in the chemical industry. The biological function of lithium is still not clear. There is no regulation for lithium levels in drinking water.

Magnesium (Mg)

Magnesium is the lightest structural metal and its use is limited by its cost and flammability. It is the eighth abundant element in the earth's crust. Magnesium is a typical alkaline earth metal, similar to calcium. The major commercial sources are minerals (magnetite), rocks (dolomite), and seawater. It is used in flashlight photography, alloys, pyrotechnique, and as medicine (such as milk of magnesia and epsom salts [$MgSO_4.7H_2O$]). Magnesium, together with calcium, contribute to water hardness (see under inorganic nonmetallic parameters). New users of drinking water high in magnesium (500 to 1,000 mg/L) may experience a cathartic effect, but they usually become tolerant. Magnesium is essential for neuromuscular conduction, and is involved in many vital enzyme functions. WHO International Standards of Drinking

Water list the maximum acceptable level for magnesium as 50 mg/L and a maximum allowable level as 100 mg/L. Magnesium is non-toxic to humans and does not constitute a public health hazard because before toxic levels occur in water, the taste cannot be tolerated. Diets high in magnesium and low in calcium cause rickets.

Manganese (Mn)

Manganese is a very brittle metallic element that resembles iron but is harder and more complicated by the existence of six oxidation states (+1, +2, +3, +4, +6, +7). Manganese is found in many minerals as oxides, silicates, and carbonates. The major use of manganese is in iron alloys, dry cells, and other oxidizing chemicals, as potassium permanganate ($KMnO_4$). Human manganese toxicity has been shown on exposure to high levels in the air. Inhalation of large doses of manganese compounds, especially the higher oxides, can be lethal. Chronic manganese toxicity is well known in miners, millworkers, and others exposed to dust and fumes, and who drink well water with excessive manganese (often in mining villages). The usual signs and symptoms involve the central nervous system. Onset is insidious, with apathy, anorexia, and asthenia. Then comes the characteristic manganese psychosis, with unaccountable laughter, euphoria, impulsiveness, and insomnia, followed by over-powering somnolence. This may be accompanied by headache, leg cramps, and sexual excitement, followed by lethargy. Finally, speech disturbances, masklike faces, general clumsiness in movement, and micrographia occur. The symptoms are similar to Parkinson's disease. Although patients may be totally disabled, the syndrome is not lethal. Inhalation of manganese fumes can cause "manganese pneumonia", which can be fatal. Manganese is an essential trace element, but deficiency in humans has not been evaluated as a health hazard. The secondary drinking water standard for manganese (as recommended by the U.S. EPA) is 0.05 mg/L, and manganese and iron combined maximum level is 0.3 mg/L.

Mercury (Hg)

Mercury is a heavy, silver-white, liquid metal. It is the only metal that is liquid in normal temperature. It freezes at -38.9°C and boils at 357°C. This liquid range has led to its widespread use as the fluid in thermometers. The most common occurrence is in combination with sulfur, HgS, called cinnabar. Its symbol is Hg, which corresponds to Hydrargyrum and means " quick silver". Mercury and its compounds are very toxic. Mercury spills should be avoided in the laboratory because the vapor pressure of Hg (about 10^{-3} torr at room temperature) is sufficient to cause mercury poisoning if the vapors are inhaled for a long time of period. Soluble mercury compounds such as mercury(II)chloride ($HgCl_2$) are poisonous because they can quickly provide sufficient mercury to the body to cause death. By contrast, mercury(I)chloride (Hg_2Cl_2) is very insoluble in water and at one time — before the discovery of penicillin — it was used medicinally as a treatment for syphilis. Its insolubility prevents the body from absorbing lethal doses of mercury. Mercury is a cumulative poison, however, so even small amounts absorbed over extended periods can lead to serious medical problems. Metallic mercury — the type found in thermometers, sphygmomanometers, and other instruments — is not absorbed by the gastrointestinal tract and therefore not very hazardous if swallowed. However, it is absorbed in the lungs. Therefore, mercury vapors are hazardous, especially when heated. Thus, broken thermometers inside infant incubators can present severe problems for the babies inside when spilled mercury leaks down to the heating units. Metallic mercury is mixed vigorously with silver to form amalgam. This is what dentists use to fill teeth. If the amalgamator is open, significant mercury exposure may result. Mercury is also released when the dentist carves and shapes the amalgam. This presents more of an occupational hazard to the dentist and dental assistants, whose exposure is prolonged, than the patient, whose exposure is intermittent. In the 1950s, mercury was thought to be an innocuous water pollutant. Mercury was known to have been hazardous to miners and to the 19th century hat makers, who frequently developed tremors or “hatter’s shakes”, lost hair and teeth, and were called “mad hatters”. In the 1950s, an outbreak of mercury poisonings in Japan raised awareness of the hazard. Residents who ate seafood from Minamata Bay, which was contaminated with methyl mercury, developed numbness of the limbs, lips, and tongue. Muscle control was lost. Deafness, blurring of vision, clumsiness, apathy, and mental derangement also occurred. Of 52 reported cases, 17 people died and 23 were permanently disabled. Mercury is a by-product of manufacturing the plastic vinyl chloride, and it is also emitted in aqueous wastes of the chemical industry and incinerators, power plants, laboratories, and even hospitals. In streams and lakes, inorganic mercury is converted by bacteria into two organic forms. One of these, dimethyl mercury, evaporates quickly from the water. But the other, methyl mercury, remains in the bottom sediments and is slowly released into the water, where it enters organisms in the food chain and is biologically magnified (build up of chemical elements or substances in organisms in successively higher trophic levels). Fresh-water fish are particularly at risk for such contamination, especially those near chlor-alkali plants and paper plants where mercuric chloride is used as a bleach for the paper and is subsequently discharged into the water. Organic mercury compounds continue to be used as fungicides on seeds for crops. In one severe outbreak in New Mexico (1972), a family consumed a pig that had eaten contaminated seeds. Other outbreaks of acute mercury poisoning have resulted as well. Mercury is a nervous system toxin, causing tremor, ataxia, irritability, slurred speech, dementia, blindness, and death. The U.S. EPA established a maximum contaminant level (MCL) in drinking water of 0.002 mg/L.

Molybdenum (Mo)

Molybdenum is a lustrous, silver-white metallic element, mostly used in alloys, and particularly valuable to increase the quality of stainless steel. Molybdenum is an essential trace mineral in ruminants and plants. Deficiencies are unknown in humans so the diet must supply sufficient amounts to carry out its roles in enzyme functions. Molybdenum is also used in nuclear energy, electrical products, and the glass and ceramic industries.

Nickel (Ni)

A hard, silver-white, malleable, ductile metallic element mostly used in alloys and plating because of the resistance to oxidation. Nickel and its compounds have little toxicity, except for inhalation of nickel carbonyl (which is highly irritating to pulmonary tissue, sometimes causing death from pulmonary oedema). “Nickel itch” or contact dermatitis is most commonly seen in women, due to nickel sources such as costume jewelry, especially earrings. Chronic exposure to nickel is carcinogenesis and can cause cancer in the respiratory tract and the lungs. Nickel content of drinking water is negligible.

Potassium (K)

Potassium is a common element of the alkali metal group, similar to sodium, and it reacts vigorously with oxygen and water. The crustal abundance of potassium is

2.59%, and is never found free in nature. Potassium metal is used in organic synthesis, and several compounds have medicinal uses. Potassium salts such as hydroxide, nitrate, chloride, bromide, iodide, sulfate chromate, permanganate, and cyanide are very important. They are extensively used as fertilizers, in glass, and (in a limited way) in the chemical industry. Potassium is necessary for proper cardiovascular function. Natural potassium content contributes to the daily intake; meat, milk, and many fruits are good potassium sources. The normal daily intake is about 1.6 to 6.0 g. Sudden increase in intake (about 18 g/day for an adult) can cause acute poisoning (hyperkalemia), which may produce cardiac arrest. The condition is characterized by ECG changes, weakness, and flaccid paralysis. Variations of the sodium-potassium (Na:K) ratio in the diet may affect blood pressure, while increasing the ingestion of potassium increases the urinary sodium excretion and vice versa. Potassium levels in drinking water are not health threatening; therefore, neither primary nor secondary standard contaminant levels are available.

Selenium (Se)

Selenium is a nonmetal with properties and reactions similar to those of sulfur. It exists in many allotropic forms. It is used in electronics, glass, ceramics, pigments, some metal alloys, and rubber production. Other uses include as an insecticide and as a topical therapeutic agent for dandruff, acne, etc. Selenium is expected to be found in water from soil contamination in areas rich in selenium, or from industrial pollution. Selenium is nutritionally essential at low levels, and toxic in high levels. The best known biochemical function of selenium is as a component of the enzyme glutathione peroxidase, which prevents oxidative damage of important cell constituents. The U.S. EPA recommended MCL of selenium is 0.05 mg/L in drinking water.

Silver (Ag)

Silver is a brilliant white, hard, ductile, malleable, rare element, and is an excellent conductor of heat and electricity. Silver occurs in nature as native silver as well as in various minerals. It is mostly used in jewelry, alloys, photography, batteries, mirror production, electroplating, electrical contacts and conductors, and as a disinfectant. The major problem in humans from overexposure to silver is called argyria, which is characterized by a slate blue-gray coloration of the skin, mucous membranes, and internal organs. (A continuous daily dose of 0.4 mg of silver may produce argyria, according to WHO 1987.) Standards for silver in drinking water were introduced in 1962 as 0.05 mg/L because of the use as disinfectant. In 1989, the U.S. EPA removed silver from the primary drinking water standards, since health is not affected, and recommended a silver level of 0.09 mg/L as secondary drinking water standard.

Sodium (Na)

Sodium is the most familiar alkali metal (pH of the aqueous solution is alkaline), and it is soft at room temperature. Sodium is very reactive with water and oxygen; therefore, it is stored under oxygen-free liquids (for example kerosene). Major industrial uses of sodium compounds include manufacturing glasses, detergents, paper, and textiles, and in water treatment as a soda ash (Na_2CO_3) process for softening and increasing pH. It is also used in organic syntheses, sodium lamps, and photoelectric cells. Household bleach is a 5% solution of sodium hypochlorite (NaOCl), and sodiumbicarbonate is used for baking and is found in beverages and antacids. The most important compounds are sodium chloride (NaCL), sodium carbonate or soda ash (Na_2CO_3), sodium bicarbonate or baking soda ($NaHCO_3$), sodium hydroxide or caustic soda (NaOH), sodium phosphates (Na_3PO_4), and borax ($NaB_4O_7.10\ H_2O$). Sodium is a natural constituent of waters, but its concentration increases by pollution. Long-term excessive sodium intake is one of the many factors associated with hypertension. A high sodium-potassium ratio (Na:K) in the diet may be detrimental to persons with high blood pressure. Inadequate replacement of salt during sweating will lead to salt depletion, characterized by fatigue, nausea, giddiness, vomiting, and exhaustion. Sodium salts are extremely soluble in water and when the element is leached from soil or discharged to streams by industrial waste processes, it remains in solution. Sodium, if present in high amounts in drinking water, may be harmful to individuals with cardiac, renal, and circulatory disease. $Na_2SO_4.10\ H_2O$ known as Glauber salt is used as a laxative; therefore, when the high sodium level in water is due to sulfates, the water is not recommended for drinking purposes. Excess concentrations of sodium salts in drinking water can also be deleterious to animals. For example, well-water containing 7,000 mg/L of sodium has been known to be toxic to chickens (Tyler, 1949), and a threshold upper limit of 2,000 mg/L would cause problems for livestock. Sodium enters groundwater because of human pollution activities. The sodium content of groundwater in humid climates is on the order of 1 to 20 mg/L. The American Heart Association recommends a water intake concentration not higher than 20 mg/L, evaluating a total intake of 500 mg by diet, and considering 440 mg from food and 60 mg from water and other miscellaneous sources. Irrigation water with high sodium levels can bring a displacement of exchangeable cations (Ca^{++}, Mg^{++}) followed by replacement of the cations by Na. The ratio of Na^+ ions to total cation contents can be used as an assessment of the suitability of water for irrigation. The ability of water to expel calcium and magnesium by sodium can be estimated by determination of Sodium Absorption Ratio (SAR).

Thallium (Tl)

The crustal abundance of thallium is largely due to clays, soils, granites, and some sulfide minerals. Deep sea manganese nodules also contain a good quantity of thallium. Thallium belongs to the Group IIIA, boron-aluminum group. Except for boron, which is a metalloid, aluminum, gallium, indium and thallium are all metals. Gallium has a melting point of only 29.8°C, so the 37°C body temperature is high enough to cause the metal to melt in the palm of your hand. The density and atomic size of the group members increases going down the group; the density of thallium is 11.87 g/ml. Thallium is used mostly in electronic applications. It was previously used in rodenticides, cosmetics, and against fungal infections, but because of the high toxicity of the metal, these products are now banned in the U.S. Fresh water contains 0.01 to 14 ug/L, but seldom exceeds 30 ug/L. Freshwater animals may bioconcentrate thallium with a factor of <20, but a concentration factor of <700 has been reported for marine organisms. Thallium compounds are highly toxic; in humans, doses of 14 mg/kg and above are fatal.

Thorium (Th)

Thorium is a radioactive element that belongs to the Actinides series. Thorium is an important source of ^{233}U in the breeder reactor. Other uses of thorium are metallurgy, electronics, and catalysts. Thorium has radiological hazard and long-term exposure leads to malignant neoplasm. The standard set for thorium by the Commission of Radiological Protection is 0.5 mg/L for soluble thorium 232. The Primary Drinking Water Standard did not list the "zero" MCL for thorium.

Tin (Sn)

The symbol of tin refers to the name stannum. The stannous (+2 valency) compounds are more important than the stannic (4+ valency) ones. Tin is mostly used as plating metal for protection, alloys (tinfoil, solders, plates, tin cans), pigments, dyes, and heat stabilizers in PVC products. One oxygen containing compound of tin (bis-tributyltin oxide) is used in antifouling paints that are applied to boat hulls to prevent growth of marine organisms such as barnacles. However, its high toxicity to all forms of marine life has led to a ban on its use for this purpose. Tin is not found naturally in environmental samples; therefore, its presence is due to industrial pollution. Toxic effects of tin are reported from food with a concentration of higher than 250 mg/kg. Because the expected concentration of tin in water systems and natural raw waters is negligible, it was not necessary to regulate the tin level in drinking waters.

Vanadium (V)

Vanadium is a gray, relatively soft metal that is found in different minerals. Ferrovanadium is used to produce steels. It is also used in titanium-aluminum alloys. Vanadium salts have low oral toxicity and medium inhalation toxicity. Vanadium is possibly protective for the human body against atherosclerosis. If vanadium is a beneficial trace element, the daily contribution from drinking water is also useful.

Zinc (Zn)

Zinc is a bluish-white metallic element that is brittle and hard at room temperature, but malleable at 100 to 150°C. It is also a good electrical conductor. Major uses of zinc include zinc coatings to protect iron, steel, and brass. Zinc is also used in dry batteries, construction materials, and printing processes. High concentrations of zinc are found in the male reproductive system. It is highest in the prostate (about 850 mg/kg), semen and sperm (2,000 to 3,000 mg/kg), muscles, liver, kidney, pancreas, thyroid, and other endocrine glands. Zinc is also one important component of enzymes, and it is considered an essential trace element in human and animal nutrition. Acidic beverages made in galvanized containers may produce toxic levels of zinc concentration, and can cause nausea, vomiting, stomach cramps, and diarrhea. Excessive zinc intakes may inhibit copper absorption and lead to copper deficiency. Zinc has been positively correlated to lithium in drinking water and atherosclerotic heart disease studies. When zinc is detected in tap water, it is mostly a corrosion product from pipes. The U.S. EPA recommended MCL of zinc as secondary drinking water standard is 5 mg/L.

Chapter 15
Inorganic Nonmetallic Constituents

15.1. Acidity

Acidity of water is its quantitive capacity to react with a strong base to a designated pH. Acids contribute to corrosiveness and influence chemical reaction rates and chemical and biological processes. The measurement also reflects a change in the quality of the source of water. It is reported as $CaCO_3$. Strong mineral acids, weak acids such as carbonic and acetic acid, and iron and aluminum sulfates may contribute to acidity.

15.2. Alkalinity

Alkalinity of water is its quantitative capacity to react with a strong acid to a designated pH. Alkalinity is significant in many uses and treatments of natural and waste waters. Because the alkalinity of surface water is primarily the function of carbonate, bicarbonate, and hydroxide content, it is taken as an indication of the concentration of these constituents. The values of alkalinity are also contribute to the presence of borates, phosphates, silicates, and other bases in the analyzed sample.

Alkalinity in excess of alkaline earth metal concentrations is significant in determining the suitability of water for irrigation. Its measurement is also used in the interpretation and control of water and waste water treatment processes. It is reported as $CaCO_3$.

15.3. Bromide (Br^-)

Bromine belongs to the halogen family with fluorine, chlorine, and iodine. " Halogen" comes from greek roots meaning " salt producer" after the ability of halogens to combine directly with metals to make salts. Bromine was isolated first from salt marshes in France, and the French Academy name this brown, liquid element bromine from the greek "bromos" which is for "stench". Bromine is produced from its most common sources, naturally occurring brines, and it may occur in varying amounts in well supplies in coastal areas as a result of seawater intrusion. The bromide content of some ground water supplies has been ascribed to connate water. Industrial and oil-field brine discharges may contribute to the bromide found in some freshwater streams. Under normal circumstances, the bromide content of most drinking waters is negligible, seldom exceeding 1 mg/L. As all halogens, bromine is also an excellent disinfectant used as a sanitizer in water purification compounds (swimming pools) and fumigants. Presently, no standards have been issued by health authorities.

15.4. Carbon Dioxide (CO_2)

Surface waters generally contain less than 10 mg free carbon dioxide (CO_2) per liter while some groundwater may easily exceed that concentration. The carbon dioxide content of water may contribute significantly to corrosion. Recarbonation of the supply during the last stages of water softening is a recognized treatment.

15.5. Calcium Carbonate Saturation (Corrosivity)

The pH at which water is saturated with calcium carbonate ($CaCO_3$) is known as the pH of saturation, or pHs. **The Langelier's Saturation Index** is defined as the actual pH minus pHs. A negative index indicates a tendency to dissolve $CaCO_3$ (the water is corrosive) and a positive index indicates a tendency to deposit $CaCO_3$ (the water is scaleforming). In other words, water oversaturated with respect to $CaCO_3$ tend to precipitate $CaCO_3$, water undersaturated with $CaCO_3$ tend to dissolve $CaCO_3$, and water saturated or in equilibrium with $CaCO_3$ have neither precipitation nor dissolving tendencies; these waters are stable. Calculation of the Saturation Index (SI) is based on the values of pH, alkalinity, total dissolved solids (TDS), temperature, and hardness. A positive index is associated with noncorrosive conditions, whereas a negative index indicates the possibility of corrosion. An acceptable limit is ±0.2 SI and the MCL is "noncorrosive". Toxic metals as lead and cadmium are corrosion products from lead and galvanized pipes. Similarly, high iron, copper, and zinc content of water is also proportional to the corrosive property of the water.

15.6. Chloride (Cl^-)

Chloride, in the form of Cl^- ion, is one of the major inorganic anions in water and waste water, originated from natural minerals, seawater intrusion, and industrial pollution. In potable water, the salty taste produced by chloride concentrations is variable and dependent on the chemical composition of water. Some waters containing 250 mg/L chloride may have a detectable salty taste if the chloride is in the form of sodium chloride (NaCl), but chloride concentration of 1,000 mg/L may not cause a typical salty taste to waters if the salts are calcium and magnesium chlorides ($CaCl_2$, $MgCl_2$). Wastewaters usually have high chloride content caused by dietary sodium chloride (NaCl) in domestic wastewater, along the sea coast by the leakage of salty water into the wastewater system, and by pollution from industrial processes. Plants are more sensitive than humans to high chloride content; therefore, the recommended chloride content is 250 mg/L in drinking water, 100 mg/L for irrigation water, and 50 mg/L for industrial waters. The National Primary Drinking Water Regulations did not include chloride in the regulated health-related parameters. Drinking water chloride is only a small percentage of the total daily dietary chloride intake. High chloride concentration may harm metallic pipes.

15.7. Chlorine Residual (Cl_2)

Chlorine is the member of the Halogen group, as mentioned above in the discussion for bromine. It is a greenish-yellow gas and its name is originated from the greek "chloros" meaning yellow-green. Chlorine has an irritating, characteristic odor, which is detectable at 2 to 3 mg/L concentration, and is fatal at 1,000 mg/L in aqueous solution. Chlorine is used in the chemical industry and in the disinfection of drinking water. Chlorine is soluble in water according to the following equation:

$$Cl_2 + H_2O \rightarrow HOCl + H^+ + Cl^-$$

Hypochlorous acid (HOCl) is very unstable, and dissociate to

$$HOCl \rightleftarrows H^+ + OCl^-$$

Chlorine with its strong disinfecting power is applied to water to destroy disease-producing microorganisms. Chlorine applied to water in its elemental or hypochlorite form initially undergoes hydrolysis to form **free available chlorine** consisting of chlorine molecule (Cl_2), hypochlorous acid (HOCl), and hypochlorite ion (OCl^-). The relative proportion of these free chlorine forms depends on pH and temperature. Free chlorine reacts with ammonia and forms monochloramine, dichloramine, and nitrogen trichloride, called combined available chlorine. The presence and concentration of these compounds are also pH and temperature dependent. Beside the beneficial action of chlorination, it may produce adverse effects. When chlorine residuals combine with phenols, an objectionable taste and odor for drinking water may be produced. As disinfection by-products, cancer-causing Trihalomethanes (THMs) are formed by the reaction of natural organic compounds with chlorine. These substances are regulated by the U.S. EPA with the current standard for total THMs of 0.1 mg/L for systems serving more than 10,000 people (see THMs, Chapter 16).

15.8. Chlorine Demand

Chlorine demand is the quantity of chlorine that is reduced or converted to inert or less active forms of chlorine by substances in the water. Chlorine demand is defined as the difference between the amount of chlorine applied and the available free chlorine remaining in the end of the contact period. Chlorine consuming substances include ammonia, organics, cyanide, ferrous iron, manganous manganese, nitrite, sulfide, and sulfite ions. The chlorine demand varies with the amount of chlorine applied, contact time, pH and temperature. The residual chlorine in a chlorinated water supply is not rigidly proportional to the amount of chlorine added to the water, nor does it always increase with additional chlorine added. Usually water shows a continual rise in the chlorine residual until several ppm of chlorine are added. Then a definite decrease in residual chlorine covering a restricted range of increase dosages is seen. The point at which the residual chlorine starts to decrease is called the break point. The decreasing chlorine residual range is of short duration and is followed by a rise in residual chlorine nearly proportional to the rate of chlorine added. This latter rise in residual chlorine starts after the chlorine demand of the water has been completely satisfied. This method serves to determine the needed chlorine quantity and serves to avoid over-chlorination.

15.9. Chlorine Dioxide (ClO_2)

Chlorine dioxide is a deep yellow, volatile, unpleasant smelling gas, and is toxic under certain conditions. Concentrated chlorine dioxide gas and aqueous solutions of the gas are unstable and can decompose explosively. Because of this hazard, it is produced in relatively dilute solutions. Chlorine dioxide has been widely used as a bleaching agent in the paper and pulp industry, and has been applied to water to combat taste and odor due to phenolic-type wastes, algae, and to remove iron and manganese. The chief benefit of chlorine dioxide is that it does not form trihalomethanes when used as a disinfectant. However, other byproducts, such as chlorite and chlorate ions may be harmful.

15.10. Cyanide (CN^-)

Cyanide refers to all of the cyanide groups in cyanide compounds that can be determined as the cyanide ion, CN^-. The cyanide ion is a conjugate base of a weak acid, hydrogen cyanide, HCN, which is an extremely poisonous gas with an almond odor. Cyanide salts are equally lethal. The cyanide ion forms a complex ion with Fe^{3+} ion.

Fe^{3+} ions are essential in the operation of an enzyme system vital to every cell in the use of oxygen, but the enzyme system is deactivated when cyanide binds to Fe^{3+} ion.

Simple cyanide compounds are inorganic cyanide salts, such as potassium and sodium cyanide (KCN, NaCN) or other metal cyanides that are highly water soluble. In aqueous solutions, where the CN^- ion is present, HCN forms by hydrolytic reaction of CN^- ion. Organically bound cyanides, such as acrylonitrile (cyanoethene), are important industrial chemicals. Cyanides are used in plastics, electroplating, and metallurgy as well as in synthetic fibers and chemicals. In industrial finished wastewaters, cyanides are oxidized with chlorine to cyanate in alkaline media. The generally accepted industrial waste treatment of cyanide compounds is an alkaline titration.

15.11. Fluoride (F^-)

Fluorine is a pale, yellowish gas with a characteristic pungent odor. It belongs to the halogen family and is the most reactive nonmetallic element. Its existence in compounds has long been known, but its unusual chemical reactivity was launched by chemists to isolate fluorine itself. Its isolation won the Nobel prize for a French chemist in 1906. Even though hydrogen fluoride (HF) and all its salts are poisonous, the fluoride ion (F^-) in trace concentrations in drinking water helps growing bodies develop teeth that are particularly resistant to decay. Tooth enamel is mostly the mineral hydroxyapatite. When fluoride ion is available during the development of enamel, a much harder mineral tends to form instead — fluorapatite.

$$[Ca_3(PO_4)_2]_3 \cdot Ca(OH)_2 \qquad [Ca_3(PO_4)_2]_3 \cdot CaF_2$$

hydroxyapatite   fluorapatite

Hydroxyapatite contains hydroxide ions (OH^-), which are more avidly attacked by acids (produced by mouth bacteria feeding on sugars) than is the much weaker base, fluoride ion (F^-), in fluorapatite. Hence, enamel that includes fluorapatite resists decay better. A fluoride concentration of approximately 1.0 mg/L in drinking water effectively reduces dental cavities without harmful effects on health. Recent medical research has determined that regular ingestion of a small amount of sodium fluoride by pregnant women provides remarkable protection against tooth decay in babies long after birth. Discovery of the cavity fighting property of fluorides was made when the cause of spotted and darkened teeth of residents from several towns in the western U.S. was investigated. U.S. Public Health dentists soon found that along with the discolored teeth was an unusually low rate of tooth decay. This was attributed to the fluoride ion. At a concentration of 1.0 ppm in the drinking water, decay prevention occurs without discoloration, which begins at a concentration of more than 2.0 ppm. There is evidence to show that fluoride does have an influence on bone metabolism in all organisms and dental health in humans. A high fluoride level is harmful, causing fluorosis and bone problems. Fluoride inhibits many enzymes. In most cases, the enzymes affected contain a metal ion with which the fluoride combines to form a metal fluoride complex. This gives a general inhibiting effect on overall metabolism and the growth of cells. The adjustment of the fluoride content of drinking water to 1 ppm fluoride is in principle and in practice the most effective access to prevent dental cavities.

15.12. Total Hardness, as $CaCO_3$

Total hardness is commonly reported as an equivalent concentration of calcium carbonate ($CaCO_3$) hardness in water caused by calcium and magnesium salts. However, iron, manganese, and strontium also contribute to hardness if appreciable concentrations occur. Most of the calcium and magnesium present in natural water occurs as bicarbonates, sulfates, chlorides, and nitrates.

Temporary hardness is removed by boiling and is caused principally by the presence of bicarbonates of calcium and magnesium. *Permanent hardness* is mostly due to calcium sulfate, which is precipitated at temperature above 150°C (300°F). *Carbonate hardness* is due to the presence of calcium and magnesium normal carbonates and bicarbonates. *Noncarbonate hardness* includes the calcium and magnesium sulfates, chlorides, and nitrates. Sulfates are often the only noncarbonate hardness compounds present.

Hardness is expressed in terms of calcium carbonate. Alkalinity is also expressed in the same term, as was previously discussed. Thus, a report showing water to have a hardness of 100 ppm (mg/L) does not signify just what compounds cause the hardness but only that the hardness is equivalent to that produced by 100 ppm (mg/L) of $CaCO_3$.

The Relationship of Alkalinity as $CaCO_3$ and Total Hardness as $CaCO_3$

The relationship of alkalinity and total hardness may give the values of carbonate and noncarbonate hardness, expressed also as $CaCO_3$. If the normal carbonate and bicarbonate alkalinity expressed as $CaCO_3$ is greater than the total hardness, normal carbonates and bicarbonates of sodium or potassium are present. These compounds do not cause hardness and, in this case, the carbonate hardness would be equal to the total hardness. If the sum of the normal carbonate and bicarbonate alkalinities is equal to the total hardness, the carbonate hardness is also equal to the total hardness. If the sum of the normal carbonate and bicarbonate alkalinities is less than the total hardness, the carbonate hardness is equal to the noncarbonate hardness.

ALKALINITY > TOTAL HARDNESS		
Carbonate Hardness	=	Total Hardness
Noncarbonate Hardness	=	not present
ALKALINITY = TOTAL HARDNESS		
Carbonate Hardness	=	Total Hardness
Noncarbonate Hardness	=	not present
ALKALINITY < TOTAL HARDNESS		
Carbonate Hardness	=	Alkalinity
Noncarbonate Hardness	=	Alkalinity – Total Hardness

The most common unit for hardness is the expression of $CaCO_3$ equivalent in milligrams per liter. Other reporting systems are also used, such as French, German, and British Hardness Degree. 100 mg/L hardness, $CaCO_3$ is equal to 10 French hardness°, 5.6 German hardness°, and 7 British hardness°. The American Water Work Association indicates that "ideal" quality water should not contain more than 80 mg/L of hardness, $CaCO_3$. Hardness in many instances exceeds this level, especially where waters have contacted limestone or gypsum (200 to 300 mg/L is common). Large scale studies in Japan, U.S.A., England, Sweden, and Finland have shown that statistical interrelationships between cardiovascular diseases and certain ingredients, including hardness of the drinking water in the same region. Muss (1962) observed that lower death rates from heart and circulatory diseases occurred in states where public water supplies were higher in hardness. The theorized protective agents include calcium, magnesium, vanadium, lithium, chromium, and manganese. The suspect harmful agents include the metals cadmium, lead, copper, and zinc, all of which tend to be found in higher concentrations in soft water as a result of the relative corrosiveness of soft water.

15.13. Iodine (I_2)

Iodine is the last, and least active member of the halogen family.

Iodine was first prepared in 1811, by observing purple vapors rising from an extract of kelp ashes. The purple vapor condensed on a cold surface, forming nearly black crystals, and it was named after the greek "iodes", meaning "violet". Iodine has one isotope, ^{127}I. Many natural brines contain iodide ion, I^-. Iodine, I_2, is insoluble in water, but it dissolves readily in solution containing iodide ion, I^-, for example in the form of potassium iodide (KI). Iodine vapor irritates the eyes and mucous membranes. Iodine kills bacteria. The "tincture of iodine" is a solution of iodine in aqueous ethyl alcohol that is used as an antiseptic. Iodide ion is necessary for the production of thyroxine in the thyroid gland. Insufficient iodide ion in the diet leads to a condition known as goiter, which is an enlargement of the thyroid gland. To assure the presence of iodide ion in the diet, sodium and potassium iodide are added to table salt, which is sold as "iodized" salt. Higher concentrations may be found in brines, certain industrial wastes, and waters treated with iodine. Only ppb (microgram per liter) quantities of iodine are present in most natural waters.

15.14. Nitrogen Compounds

Nitrogen is a colorless, odorless, and tasteless gas that makes up 78% (v/v) of dry air. The largest use of nitrogen is in an oil field operation called enhanced oil recovery, where it helps to force oil from subterranean deposits. Another major use of nitrogen gas is in the metals and chemicals processing industry. Its low boiling point, 25°C (77°F) and chemical unreactivity make liquid nitrogen an ideal and inexpensive coolant in laboratory works, and manufacturers of frozen seafood, poultry, and other meat products use it for fast freezing. Nitrogen forms part of many essential organic molecules — notably, amino acids (the building blocks of proteins) and the genetic materials RNA (ribonucleic acid) and DNA (deoxyribonucleic acid). However, plants and animals cannot use nitrogen in the form of atmospheric nitrogen (N_2). To be usable, it must first be converted into ammonia (NH_3), or nitrate (NO_3^-). The conversion of atmospheric nitrogen into usable forms of nitrogen is called "nitrogen fixation," which is associated by bacterial activity. One nitrogen fixing bacteria, called Rhizobium, invades the roots of leguminous species (peas, beans, alfalfa, clover, and others). The roots form tiny nodules where the nitrogen fixation takes place. Bacteria species, as Azotobacter, are ready to fix atmospheric nitrogen directly in the soil. The fixed nitrogen is taken by plants and then synthesized into amino acids, proteins, DNA, and RNA. Animals, in turn, receive the needed nitrogen by eating plants and other animals. Nitrogen rich wastes from plants and animals will be decomposed by certain types of bacteria and converted to ammonia (NH_3). This is called "ammonification". Ammonia oxidizes to nitrite (NO_2) by oxidizing bacteria species (Nitrosomonas) and the nitrite is oxidized to nitrate (NO_3) by other oxidizing bacteria (Nitrobacter). The process is called "nitrification". The nitrates may be incorporated by the roots of plants and reused. Nitrates decompose to nitrites and convert into a gas, nitrous oxide (N_2O) by bacterial activity (Pseudomonas and others). This is then released into the atmosphere. The process is called "denitrification". In various forms, nitrogen travels from air to soil to plant to animal and then back to soil and atmosphere in a never ending cycle, as on Figure 15-1. Nitrogen compounds are the member of the inorganic nutrient group, and the wide variety of forms of nitrogen in the environmental samples have great interest. Major pollution sources of nutrients (nitrogen compounds and phosphorus) are surface and subsurface agricultural and urban drainage, animal waste runoff, as well as domestic and industrial waste effluents including sewage.

Organic Nitrogen, Total Kjeldahl Nitrogen (TKN)

It is defined functionally as organically bound nitrogen. Analytically, ammonia and organic nitrogen can be determined together and have been referred to as "Total Kjeldahl Nitrogen (TKN)". Organic nitrogen includes such natural materials as proteins and peptides, nucleic acids and urea, and numerous synthetic organic materials. Typical organic nitrogen concentrations vary from a few hundred ppb in some lakes to more than 20 ppm in raw sewage. Organic nitrogen value is calculated from the detected values of ammonia nitrogen and TKN. Organic nitrogen = TKN - NH_3-N

Total Oxidized Nitrogen

Total oxidized nitrogen is the sum of nitrate nitrogen (NO_3-N) and nitrite nitrogen (NO_2-N).

Nitrate-Nitrogen (NO_3-N)

In water supplies, nitrate-nitrogen owes its origin to several possible sources, including the atmosphere, legume plants, plant debris, animal excrement, and sewage, as well as nitrogenous fertilizers and some industrial wastes. Since the atmosphere consists of about 78% by volume of nitrogen, some of that present in

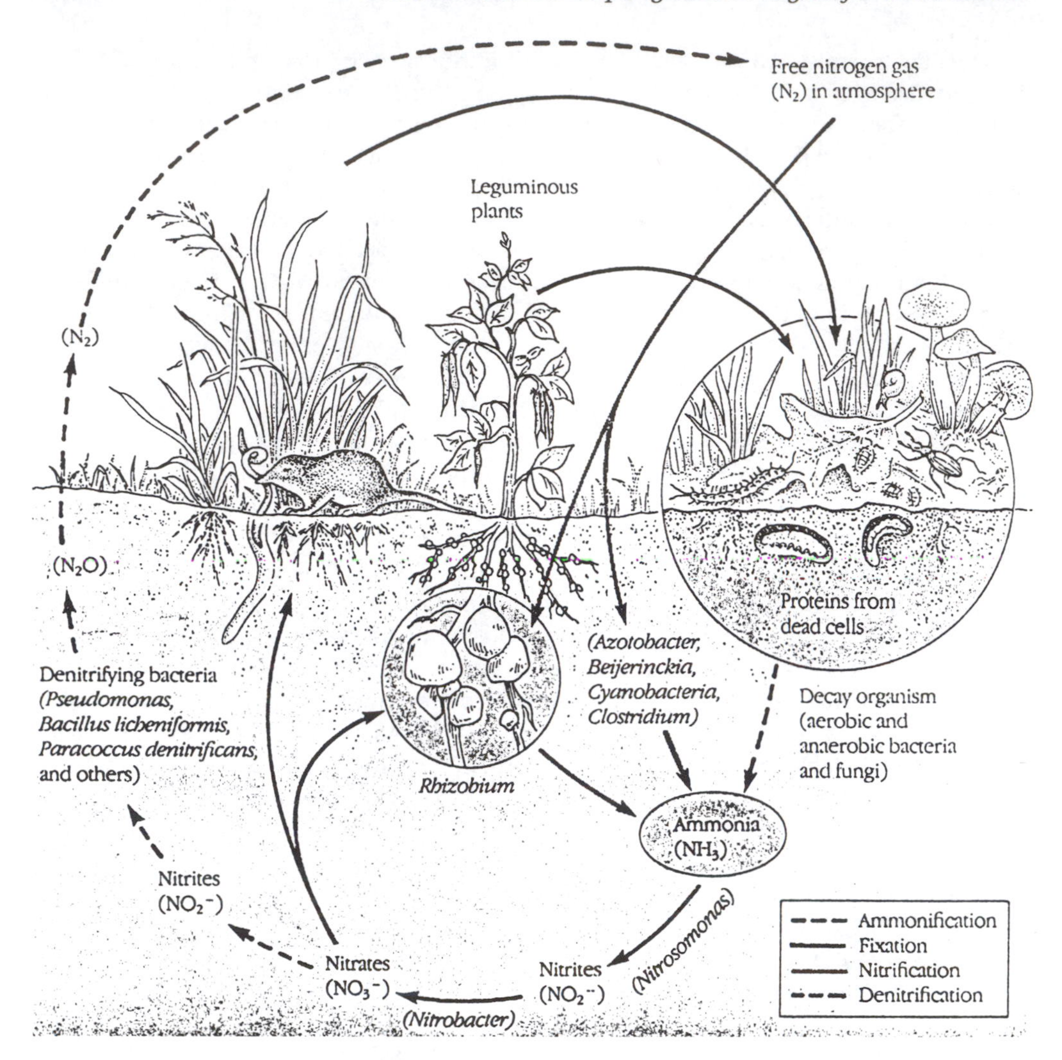

Figure 15.1. The nitrogen cycle.

water originates from this source, but most is generated by the decay of organic matter, and from industrial and agricultural chemicals. Bacteria decomposed organic matter, such as sewage and excrement, and these complex proteins, change to ammonia (NH_3), then nitrite (NO_2), and finally nitrate (NO_3). Since the various forms are highly soluble in water, they are easily leached downward from the soil by infiltrating water and may quickly reach the water table. Much of that present in the soil, however, is used by plants, since nitrate is a major nutrient for vegetation and is essential to all forms of life. In addition to decaying organic matter, fertilizers are a major source of nitrate in water supplies. The concentration of nitrate in both surface and groundwater can increase to alarming amounts. Nitrate concentration in water is reported either as nitrate (NO_3) or as nitrate nitrogen (NO_3-N). This system is confusing, so caution is urged when examining an analysis. For example, 45 mg/L of nitrate as NO_3 is equal to 10 mg/L nitrate-nitrogen as NO_3-N. To convert NO_3 to NO_3-N, multiply the NO_3 concentration by 0.23, and to convert NO_3-N to NO_3, multiply the NO_3-N concentration by 4.43.

For example:

$$\text{Nitrate as } NO_3 = 5.53 \text{ mg/L}$$
$$\text{Nitrate as } NO_3\text{-N} = 1.25 \text{ mg/L}$$

Nitrate-nitrogen (NO_3-N) concentration greater than 10 mg/L in drinking water has been known to cause infant methemoglobinemia, a disease characterized by cyanosis, a bluish coloration of the skin. The nitrate inactivate hemoglobin into methemoglobin, and in this form it becomes unfit to transport oxygen. The sickness is especially targeted on infants in the first three months. This disease may occur when a nursing child consumes either formula or milk directly from its mother that contains a large concentration of nitrate. Some evidence indicates that high concentrations of nitrate in drinking water for livestock has resulted in abnormally high mortality rates in baby pigs and calves and abortion in brood animals.

Shallow groundwater supplies are susceptible to pollution by nitrates. Shallow dug or bored wells, particularly, may contain excessive nitrates if they are improperly constructed so that surface water is allowed to flow into them or if they are near septic tanks, cesspools, or barnyard wastes. In cases such as these, the supply should be abandoned, because the pollutants might contain pathogenic organisms as well as other undesirable organic substances. If polluted by human excrement, the water supply might contain a relatively high concentration of chloride in addition to nitrates, and the presence of both provides a warning that the homeowner should have the supply checked for bacterial contamination. Furthermore, the concentration of nitrate in groundwater can fluctuate greatly from one season to the next. During the wet season, particularly while plants are still dormant, infiltrating rain or snowmelt may leach large amounts of nitrate to the water table. Most of it is either used by plants during the growing season or remains in the ground because of a soil-moisture deficiency, but after crops are harvested in the fall, nitrate may again be leached down to pollute shallow aquifers. In view of the cyclic nature of nitrate in shallow groundwater supplies, judgement must be used so that sampling is accomplished at the appropriate time. Nitrate is an essential nutrient for many photosynthetic autotrophs and in some cases has been identified as the growth-limiting nutrient.

Nitrite-Nitrogen (NO_2-N)

Nitrite is an intermediate oxidation state of nitrogen, both in the oxidation of ammonia to nitrate and in the reduction of nitrate. Such oxidation and reduction occur in wastewater treatment plants, water distribution systems, and natural waters. Nitrite can enter water supply systems through its use as a corrosion inhibitor in industrial process water.

A small amount of sodium nitrite ($NaNO_2$) is added to meats such as cold cuts and frankfurters because it is a good preservative. When the nitrite ion reaches the stomach, the high concentration of hydrochloric acid (HCl) there converts the NO_2^- to nitrous acid (HNO_2), which can react with secondary amines in the digestive tract to produce N-nitrosamine. N-nitrosamines are carcinogens. The problem is not easily solved and the question still exists: can we find another preservation technique for meat products, or accept the risk of using this chemical?

The reporting unit for nitrite is the same as for nitrate. Analytical reports should clearly state the expression of the unit, as nitrite (NO_2) or nitrite nitrogen (NO_2-N). The conversion factor is 3.3. For example:

$$\text{Nitrite as } NO_2 = 0.66 \text{ mg/L}$$
$$\text{Nitrite nitrogen, } NO_2\text{-N} = 0.20 \text{ mg/L}$$

Ammonia-Nitrogen (NH_3-N)

Ammonia concentration is generally low in groundwater because it adsorbs to soil particles and does not leach easily. Ammonia is mostly produced by decomposition of organic nitrogen containing compounds and by the hydrolysis of urea. During chlorination, ammonia combines with chlorine to form chloramine, as combined residual chlorine. Ammonia is naturally present in surface and groundwater.

Reporting form of ammonia is either ammonia (NH_3) or ammonia nitrogen (NH_3-N). The conversion factor to these units is 1.21. For example:

Ammonia as NH_3 = 0.50 mg/L

Ammonia nitrogen as NH_3-N = 0.41 mg/L

15.15. Dissolved Oxygen (DO)

Photosynthesis and respiration are fundamental to life on earth. The chemical processes involved may be simply expressed as set out below:

$$6\ CO_2 + 6\ H_2O \underset{\text{respiration}}{\overset{\text{photosynthesis}}{\rightleftarrows}} C_6H_{12}O_6 + 6\ O_2$$

These processes involve carbon dioxide and oxygen in the earth's atmosphere which (in aquatic systems) are dissolved in the water. Increased respiration, due to an increased amount of organic carbon in the water, results in changes in the pools of oxygen involved. The demand of dissolved oxygen in the water is comparatively small since oxygen has limited solubility in water, usually from about 6 mg/L to 14 mg/L. One of the most important sources of oxygen in waterways is oxygen in the atmosphere that dissolves in water at the water surface. The solubility of oxygen in water is affected by the partial pressure of oxygen gas, the temperature, and the salinity of the water. The solubility of oxygen in the water decreases with temperature and with increasing salinity. The solubility of oxygen in water exposed to water-saturated air at atmospheric pressure is shown in Table 15-1.

Many polluted waterways have little plant population, but in some cases the condition is suitable for aquatic plants. Plants can have a very important influence on dissolved oxygen content through photosynthesis and respiration. Photosynthesis occurs during the daylight and respiration by plants occurs at night. These processes cause a rise in dissolved oxygen during the day hours with the maximum in the afternoon. At sunset, production of the oxygen by plants is stopped and the dissolved oxygen content of the water starts to decrease because of the respiration of plants and aquatic organisms. Wastes, domestic and industrial, discharged into the waterways are usually rich in organic substances and have significant demands on oxygen in the water. Consequently, these wastes reduce the dissolved oxygen content of the water. The organic matter in these discharges contains organic carbon, but beside the organic carbon there are other elements, particularly nitrogen and sulfur. Complete oxidation of organic nitrogen leads to nitrates, and oxidation of sulfur ends in sulfates. In the case of lack of sufficient oxygen concentration, anaerobic microorganisms take over the degradation and decomposition of organic matter. Anaerobic respiration with sulfur and nitrogen containing organic matter leads to the formation of hydrogen sulfide (H_2S) and ammonia (NH_3), respectively. In natural systems, low dissolved oxygen concentration in water is usually associated with anaerobic respiration and

Table 15.1. Solubility of Oxygen in Water Exposed to Water-Saturated Air

	Chloride concentration in water mg/L				
	0	5,000	10,000	15,000	20,000
Temperature C°	Dissolved Oxygen mg/L				
0	14.6	13.72	12.90	12.13	11.41
1	14.19	13.35	12.56	11.81	11.11
2	13.81	12.99	12.23	11.51	10.83
3	13.44	12.65	11.91	11.22	10.56
4	13.09	12.33	11.61	10.94	10.30
5	12.75	12.02	11.32	10.67	10.05
6	12.43	11.72	11.05	10.41	9.82
7	12.12	11.43	10.78	10.17	9.59
8	11.83	11.16	10.53	9.93	9.37
9	11.55	10.90	10.29	9.71	9.16
10	11.27	10.65	10.05	9.49	8.96
11	11.01	10.40	9.83	9.28	8.77
12	10.76	10.17	9.61	9.08	8.58
13	10.52	9.95	9.41	8.89	8.41
14	10.29	9.73	9.21	8.71	8.24
15	10.07	9.53	9.01	8.53	8.07
16	9.85	9.33	8.83	8.36	7.91
17	9.65	9.14	8.65	8.19	7.78
18	8.45	8.95	8.48	8.03	7.61
19	9.26	8.77	8.32	7.88	7.47
20	9.07	8.60	8.16	7.73	7.33
21	8.90	8.44	8.00	7.59	7.20
22	8.72	8.28	7.85	7.45	7.07
23	8.56	8.12	7.71	7.32	6.95
24	8.40	7.97	7.57	7.19	6.83
25	8.24	7.83	7.44	7.06	6.71
26	8.09	7.69	7.31	6.94	6.60
27	7.95	7.55	7.18	6.83	6.49
28	7.81	7.42	7.06	6.71	6.38
29	7.67	7.30	6.94	6.60	6.28
30	7.54	7.17	6.83	6.49	6.18
31	7.41	7.05	6.71	6.39	6.08
32	7.28	6.94	6.61	6.29	5.99
33	7.16	6.82	6.50	6.19	5.90
34	7.05	6.71	6.40	6.10	5.81
35	6.93	6.61	6.30	6.01	5.71
36	6.82	6.51	6.20	5.92	5.64
37	6.71	6.40	6.11	5.83	5.56
38	6.61	6.31	6.02	5.74	5.48
39	6.51	6.21	5.93	5.66	5.40
40	6.41	6.12	5.84	5.58	5.33
41	6.31	6.03	5.76	5.50	5.25
42	6.22	5.94	5.68	5.42	5.18
43	6.13	5.85	5.60	5.35	5.11
44	6.04	5.77	5.52	5.27	5.04
45	5.95	5.69	5.44	5.20	4.98
46	5.86	5.61	5.37	5.13	4.91
47	5.78	5.53	5.29	5.06	4.85
48	5.70	5.45	5.22	5.00	4.78
49	5.62	5.38	5.15	4.93	4.72
50	5.54	5.31	5.08	4.87	4.66

Solubility of oxygen values are at a total pressure of 101.3 kPa (101.3 kPa = 1 atm). Units of pressure most frequently employed are atmospheres (atm), millimeters of mercury (mmHg), torr and pascals. 1 atm = 760 mmHg = 760 torr = 101,325 Pa = 101.325 kPa. To obtain results in milliliters oxygen gas per liter, multiply the Dissolved Oxygen mg/L by 0.70

the production of substances that are toxic to aquatic organisms. Hydrogen sulfide (H_2S) is probably the most important substance in this category. Based on these facts, organic discharges to a water body are due to the combined effects of reduced dissolved oxygen, organic enrichments, and the toxic effect of hydrogen sulfide. Depleted oxygen concentration in water systems leads to low oxygen intake by organisms, resulting in insufficient activities and death. Some "mobile" species can detect the decreased oxygen content and move into areas with more favorable oxygen content.

15.16. Phosphorus (P)

Phosphorus is present in nature only in its compounds. They occur in animal, plant, and mineral kingdoms, almost always as one or another variation of the phosphates. The name of the element is after the Greek roots phos "light" and phoros "bringing", which refers to its unique properties. Phosphorus has several allotropes, the two most known are white and red phosphorus. The white phosphorus is highly toxic, and at about 35°C spontaneously bursts into flame. In moist air and below 35°C, it slowly reacts with oxygen, which causes a glow. This phenomenon is called chemiluminescence, or phosphorescence. The red phosphorus is nontoxic and can be stored in air. As is well-known, both white and red phosphorus are involved in the chemistry of matches.

Phosphorus is found in living organisms as phosphates, which contain (PO_4^{3-}) ion. Phosphate is an important part of RNA and DNA, and phosphorus is found in cell membranes as phospholipid. The most important diphosphate and triphosphate in the body are called Adenosine-diphosphate (ADP) and Adenosine-triphosphate (ATP). They store and transfer energy. The body stores energy by forming P-O-P bonds. When it needs to use the energy, it hydrolyses, so the energy is given off. Tripolyphosphate ion as sodium salt ($Na_5P_3O_{10}$) is the chief phosphate in laundry detergents. The trouble with phosphates as laundry detergents is that they support the life of algae, so when phosphate containing wastewater gets into rivers, lakes, and bays, they become unfit for aquatic life (except algae) or for human recreation. Many areas, therefore, have banned the sale of phosphate-containing laundry products, since this is deemed much less costly than the removal of phosphates from wastewater.

Nutrients flow in cycles, as the carbon cycle (photosynthesis and respiration), nitrogen cycle, and the phosphorus cycle. Phosphate is slowly dissolved (leached) from rocks by rain and is carried to waterways. Dissolved phosphates are incorporated by plants and passed to animals in the food web. Phosphorus reenters the environment directly by animal excretum and by detritus decay. Each year large quantities of phosphates are washed into the oceans, where much of it settles to the bottom and is incorporated into the marine sediments. Sediments may release some of the phosphate needed by aquatic organisms, and the rest may become buried. Phosphate is a major component of fertilizer. By applying excess fertilizer, farmers may alter the phosphorus cycle.

In natural waters and wastewaters, phosphorus occurs as phosphates. These are classified as ortho-, condensed, organically bound, and total phosphates expressed as phosphorus.

Orthophosphates (PO_4-P) — Inorganic, water soluble phosphates, applied to agricultural or residential cultivated land as fertilizers and carried into surface waters with storm runoff.

Condensed phosphates (PO_4-P) — Polyphosphates are major constituents of many commercial cleaning preparations and laundry detergents. Condensed phosphates are used extensively in the treatment of boiler waters and added to some water supplies during treatment to prevent scale formation and inhibit corrosion.

Organic phosphates (PO_4-P) — Organic phosphates are formed primarily by biological processes. They are contributed to sewage by body wastes and food residues and can also be formed from orthophosphates in biological treatment processes or by receiving water biota.

Total phosphates (PO_4-P) — Include all the above mentioned forms of phosphorus and are reported as total phosphorus.

Phosphorus is essential to the growth of organisms and can be the nutrient that limits the primary productivity of a body of water. Phosphorus has been identified as the most critical growth factor in most waters. The discharge of wastewater, agricultural drainage, or certain industrial wastes to a water system may stimulate growth of photosynthetic aquatic micro- and macro-organisms. Sediments play a major role in the availability of phosphorus in many aquatic areas. A high proportion of the phosphorus is removed from the water by sorption onto sediment minerals.

The reporting form of phosphorus may be as phosphate (PO_4) or as phosphate phosphorus (PO_4-P). The conversion factor is 3.08.

For example:

$$\text{Total Phosphorus as phosphate } (PO_4) = 5.20 \text{ mg/L}$$

$$\text{Total Phosphorus, as phosphorus } (PO_4\text{-P}) = 1.69 \text{ mg/L}$$

15.17. Silica (SiO_2)

Silicon (Si) ranks next to oxygen in abundance in the earth's crust. The element silicon belongs to the carbon group (group IV) in the periodic table and, according to its properties, it is known as a metalloid. In most respects, metalloids behave as nonmetals both chemically and physically. However, in their most important physical property, electrical conductivity, they somewhat resemble metals. Metalloids tend to be semiconductors: they conduct electricity but not as well as metals. The property, particularly as found in silicon and germanium, is responsible for the remarkable progress in the field of solid-state electronics. The heart of small TV cameras, compact discplayers, calculators, and microcomputers is a microcircuit printed on a tiny silicon chip. Among the synthetic inorganic polymers, the most widespread uses are the silicons, as used in furniture and car polishes, waterproofing agent, lubricants, insulating electrical wire, and synthetic rubber.

Silicon is commonly found in combination with oxygen, which forms silica, SiO_2. Silica contributes about 75% of the crustal material, and occurs in the minerals, called silicates. The common mineral quartz, in its many varieties, is essentially pure silicon dioxide, SiO_2. Silicon and oxygen combine readily with many other major elements such as potassium, magnesium, sodium, iron, calcium, and aluminum, called silicates. Silicates are found mostly in igneous rocks and soils. Clay is a great variety of aluminum silicates and it is important in fertile soils, and usually formed by the weathering or alteration of aluminum-bearing rocks. The aluminum ions in clay help to hold such anions as nitrate, sulfate, and phosphate ions, which are also needed by plants. Clay is used in ceramic raw materials and pottery.

Silica is not readily dissolved by water. The origin of silica in groundwater is attributed to the chemical weathering of silicate minerals. Silica does not contribute to the hardness of the water. It is, however, an important constituent of the incrusting material or scale formed by many waters. As deposited, the scale is commonly calcium or magnesium silicate. Silicate scale cannot be dissolved by acids or other chemicals that are used for chemical treatments. Silica in water is undesirable for a number of industrial uses, because it forms difficult-to-remove silica and silicate scales in equipment, particularly in high-pressure steam boilers. Silica can be removed by distillation or by the use of strongly basic anion-exchange resins in the deionization process. Some water treatment plants use reverse osmosis, or precipitation with magnesium oxide in the lime softening process. Silica concentration in natural waters is generally small, ranging between 1 and 30 mg/L, but concentration in excess of 100 mg/L is not uncommon. Silica may be dissolved from nearly all rocks, but usually only in small concentrations, because of its slow solubility in the normal pH range of natural waters. In very acid hot springs or highly alkaline springs, silica concentrations may exceed several hundred milligrams per liter. Silica is widely used in making glass, silicates, ceramics, abrasives, enamels, petroleum products, and metal work, so it may appear in certain waste waters. Silicates are also used in water treatment systems as coagulants and corrosion inhibitors.

15.18. Sulfate (SO_4^{2-})

Many sulfate compounds are readily soluble in water. Most of them originate from the oxidation of sulfate ores, gypsum ($CaSO_4.2H_2O$) and anhydrite ($CaSO_4$), oxidation of pyrite (fool's gold), which is an iron sulfide (FeS_2), and the existence of industrial wastes. The sulfate content of groundwater generally does not exceed 100 mg/L. Higher sulfate content is expected in groundwater close to the deposits in sedimentary rocks. These deposits may include sodium chloride and other chloride salts. Groundwater may contain other minerals, such as magnesium sulfate (Epsom salt, $MgSO_4.7H_2O$) and sodium sulfate (Glauber's salt $Na_2SO_4.10H_2O$). These salts in sufficient quantities are responsible for a bitter taste to water, which is detectable over 500 mg/L. For people not accustomed to drinking high-sulfate water (over 600 mg/L), these salts may act as a laxative. It is mainly the reason that the MCL for sulfate is 250 mg/L. Since sulfate compounds are quite mobile in aquatic systems, there are few circumstances where concentrations will naturally decrease. One exception, however, is sulfate reduction by bacteria. Sulfate reducing bacteria derive energy from the oxidation of organic compounds, obtaining oxygen from sulfate ions. The bacterial reduction of sulfate produces H_2S gas as a by-product. If sufficient iron is present under moderate reducing conditions, iron sulfides may also be precipitated.

Much of the H_2S gas escapes directly into the atmosphere if the reduction is by bacteria in the soil. Sulfate is one of the major dissolved constituents in rain, where it usually appears in concentration of less than 5 mg/L. The amount is related to natural and human pollution of the atmosphere. This soluble substance may be derived from dust particles, oxidation of sulfur dioxide (SO_2) and hydrogen sulfide (H_2S) gases released from the decomposition of organic material, volcanic discharges, forest fires, and bacterial decay. Anthropogenic (human pollution) sources of the sulfur oxides are of major concern, because they are concentrated in urban and industrial regions, causing local levels to be quite high. About 70% of anthropogenic sulfur dioxide comes from electric power plants, most of which burn coal. In the atmosphere, SO_2 reacts with oxygen and water to produce sulfuric acid (H_2SO_4) as a toxic pollutant with far-reaching effect. Acid rain and snow are formed when two pollutant gases, sulfur and nitrogen oxides, combine with water. Sulfur oxides form sulfuric acid, nitrogen oxides form nitric acid. Both are powerful acids. They may accumulate in clouds and fall from the sky in rain and snow. The process is called "wet deposition".

Sulfur and nitrogen oxide gases also form sulfate and nitrate particulates. These may settle out from the atmosphere onto surfaces, as plants and other solid surfaces. These particles are called "dry deposits", and they combine with water to form acids.

15.19. Sulfide (S^{2-})

Sulfide is often present in groundwater, especially in hot springs. Groundwater that contains dissolved hydrogen sulfide (H_2S) gas is easily recognized by its rotten-egg odor. As little as 0.05 mg/L of hydrogen sulfide in cold water is noticeable and the odor from 1 mg/L is definitely offensive. Its common presence in waste waters comes partly from the decomposition of organic matter, sometimes from industrial wastes, but mostly from bacterial reduction of sulfate. Hydrogen sulfide escaping into the air from sulfide-containing waste water causes odor nuisance. Hydrogen sulfide is very toxic and has claimed the lives of numerous people working in sewers. It reacts with metals to form insoluble sulfides, and makes corrosivity a problem for all metallic pipes. It has caused serious corrosion of concrete sewers by biologically oxidizing to H_2SO_4 on the pipe wall.

15.20. Sulfite (SO_3^{2-})

Sulfite ions (SO_{32}^{-}) may occur in boilers and boiler feed waters treated with sulfite for dissolved oxygen control, in natural waters or waste waters as a result of industrial pollution, and in treatment plant effluent dechlorinated with sulfur dioxide. Excess sulfite ion in boiler waters is deleterious because it lowers the pH and promotes corrosion. Control of sulfite ion in waste water treatment and discharge may be environmentally important, principally because of its toxicity to fish and other aquatic life and its rapid oxygen demand.

Chapter 16
Organic Pollutants

Analyses of organics are made to evaluate possible effects of health of water consumers, measure the efficiency of waste treatment processes, and assess the quality of the receiving waters. Determination of some organic compounds, including pesticides and trihalomethanes in drinking water, is mandatory. The list of controlled organics is lengthy, both for drinking water and waste discharges. The analysis of organic matter can be classified into the next general types:

Total Amount of Organic Matter Present

- Demands:
  - Biochemical oxygen demand (BOD)
  - Chemical oxygen demand (COD)
  - Total organic carbon (TOC)
- Oil and grease, total
- Petroleum hydrocarbons
- Surfactants
- Tannin and lignin
- Total organic halogens (TOX)

Individual Organic Compounds or Groups of Compounds

- Volatile compounds
  - Volatile or purgeable halocarbons
  - Volatile or purgeable aromatics
  - Acrolein and acrylonitrile
- Extractables
  - Phenols, benzidines, phthalate esters
  - Nitrosamines, pesticides and herbicides
  - Polychlorinated biphenyls (PCBs),
  - Nitroaromatics and isophorone, nitrosamines
  - Polynuclear Aromatic Hydrocarbons (PAHs)
  - Dioxin, ethylene dibromide (EDB)

16.1. Determination of Total Amount of Organic Matter Present

Demands

The deleterious effects of pollutants can be generally related to three environmental factors: (1) excess plant production or nutrient enrichment, (2) deoxygenation, and (3) toxic physiological effects on organisms. Nutrients, such as nitrogen and phosphorus-containing substances, and the importance of dissolved oxygen in aquatic systems are discussed in Chapter 15. Deoxygenation is caused by the addition of organic matter, rich in chemical energy (carbohydrate, protein, and fat) to a water body. Therefore, the measurement of oxygen demand of an aquatic system is one easy and rapid way to detect the degree of pollution by organic matter. These oxygen-demanding processes can be measured by the biochemical oxygen demand (BOD), the chemical oxygen demand (COD), and the total organic carbon (TOC) content of the water.

Biochemical Oxygen Demand (BOD)

BOD determines the oxygen required to oxidize the organic matter in the sample, through the agency of microorganisms (bacteria). The test consists of the determination of DO prior to and following a period of incubation at 20°C. The incubation period is usually 5 days. If the oxygen demand of the sample is greater than the available DO, a dilution is made. Dilutions that result in a residual DO of at least of 1.0 mg/L and a DO uptake of at least 2.0 mg/L after the 5 days incubation produce the most reliable results. For wastes and sewage having an unknown oxygen demand, it is necessary to make up a number of dilutions in order to be sure that one will meet the requirements.

BOD may be *carbonaceous demand*, which measures the oxygen utilized for degradation of organic materials and also some inorganic material, such as sulfides and ferrous iron. BOD also may measure the oxygen used to oxidize reduced forms of nitrogen. This is called *nitrogenous demand*. The result of the BOD test is the sum of the carbonaceous and the nitrogenous demand. Addition of nitrification inhibitor to the sample provides a reliable measure of the carbonaceous demand. When inhibiting the nitrogeous oxygen demand, the report should be Carbonaceous Biochemical Oxygen Demand (CBOD) and when nitrification is not inhibited, report results as BOD_5.

Chemical Oxygen Demand (COD)

The COD is a measure of the oxygen equivalent of the organic matter content in the sample, which is sensitive to strong oxidizing agent. COD can be related empirically to BOD, organic carbon, or organic matter. Since it is a more rapid test than BOD, it has been largely replaced by the BOD determination because it does not give results that are comparable to those obtained from biological oxidation processes that occur in nature.

Total Organic Carbon (TOC)

Organic carbon represents many different compounds and oxidation states. It is a direct and convenient expression of the total organic content, from which the oxygen demand can be estimated. It is a more convenient and direct expression of total organic content than either BOD or COD, but the relevance to natural systems is not as clear as that of the BOD test. It does not replace BOD or COD testing.

Total carbon (TC) includes the inorganic carbons (IC) and the TOC. IC is a very large part of the TC. The determination of TC and total inorganic carbon (TIC), with the estimation of TOC by difference is common.

Oil and Grease

"Oil and Grease" is any material recovered as a substance soluble in trichlorotrifluoroethane, including biological lipids (animal and vegetation origin) and mineral hydrocarbons (petroleum origin). Certain constituents measured by the oil and grease analysis may influence wastewater treatment systems, and if present in excessive amounts, they may interfere with water treatment efficiency. A knowledge of the quantity of the total oil and grease in wastewaters is a helpful basis for the operation and evaluation of the treatment system. The presence of oil and grease in environmental samples may be the sign of possible oil pollution; further investigation would be warranted.

Total Recoverable Petroleum Hydrocarbons (TRPHs)

Oil and grease is composed primarily of fatty matter from animal and vegetable sources and hydrocarbons of petroleum origin. The separation of these materials is useful to specify the major source of the determined oil and grease content. Petroleum is a complex, naturally occurring mixture of organic compounds that is produced by the incomplete decomposition of biomass over a geologically long period of time. Petroleum compounds can occur in a gaseous form that is often called natural gas, as a liquid called crude oil, and as a solid and semisolid asphalt or tar associated with oil, sands, and shales. Hydrocarbons are the most important constituent of petroleum, and may be classified into three main groups: aliphatic, (open chain compounds), cyclic (ring structure compounds), and aromatic hydrocarbons (compounds contain benzene ring). During the refining of crude oil, the various hydrocarbon products are separated by fractional distillation at specific temperature. Typical yield of products are natural gas, gasoline, kerosene, heating oil, jet and rocket fuels, lubricating oils, waxes, bunker fuel for ships, and fuel for electrical utilities. Oil pollution can be caused by the spillage of crude oil or any of its refined products, and from the continuous discharge of contaminated process waters. Petroleum and hydrocarbon toxicity is a well-known and well-studied phenomenon.

Surfactants

Detergents contain chemicals called surfactants that drastically lower the surface tension of water. Surfactants may be classified into three major groups: anionic, cationic, and nonionic. The most commonly used surfactants belong to the anionic alkyl aryl sulfonate type group, and are called Alkyl Benzene Sulfonates (ABS). The most widely used anionic surfactant is Linear Alkyl Benzene Sulfonate (LAS). LAS is not a single compound, but may be comprised of any or all of 26 related compounds with various chain lengths and isomers.

Surfactants enter waterways through treated and untreated sewage containing domestic or industrial waste water and urban runoff. Surfactants are concentrated in bottom sediments. This leads to high concentrations of surfactants in sediments in areas receiving surfactant-containing wastewater. Concentration of 2.3 g/kg of sediment (dry weight) have been observed in some areas. The most important process in natural systems is microbial degradation. Many soil and aquatic organisms have the capacity to attack surfactants. In general, surfactants interact with membranes and enzymes, and inhibition of growth on plants and fish has been noted. Damage to respiratory epithelium of fish is well-known, as is the damage of other sensitive external organs.

Anionic surfactants have the ability to make the reaction between the anion of the surfactant and the methylene blue cation. Due to this ability, anionic surfactants are called Methylene Blue Active Substances (MBAS) and they are used for the measurement of the surfactant concentration in environmental samples. Soaps do not respond in the MBAS method. The form of reporting the analytical results for surfactants is "MBAS, calculated as LAS mg/L or ug/L".

Tannin and Lignin

Lignin is a plant constituent, and most commonly discharged to the waste effluent from manufacturing paper-pulp. Tannin is also a plant component, and may enter into waters as a pollutant through the waste of tanning industries. Tannin is an important additive to boiler waters to prevent scale formation.

Total Organic Halogens (TOX)

This measurement is an estimation of the total quantity of halogenated organic compounds in the sample. The presence of these substances is indicative of synthetic chemical contamination. The positive TOX result indicates the need for identification and quantification of specified substances by more complicated analyses. Halogenated compounds that contribute to the test are: trihalomethanes (THMs), organic solvents, halogenated pesticides, polychlorinated biphenyls (PCBs), and chlorinated aromatics.

Phenols

Phenols are compounds containing an OH^- group connected directly to an aromatic ring. Phenol itself is a solid that is fairly soluble in water. An aqueous solution of phenol is called carbolic acid. This solution is one of the earliest antiseptics because it is toxic to bacteria. Today it is no longer used directly on patients because it burns the skin, but it is still used to clean surgical and medical instruments. Phenol is a common and important industrial chemical and enters into the environment in wastewater and spills connected with its use in resins, plastics, adhesives, and other uses. If it is released into the environment, its primary removal is biodegradation which is generally rapid (days). Despite its high solubility and poor adsorption on soil, biodegradation is sufficiently rapid so that most groundwater is generally free of this pollutant. The exception would be in the cases of spills where high concentrations of phenol destroy degrading microbial populations.

Chlorination of phenol-containing waters may produce odorous and objectionable-tasting chlorophenols. The standard value of 0.001 mg/L as total phenol concentration based on taste resulting from chlorination and not because of toxicological concern.

Phenols are determined as total phenols, without differentiation of phenolic compounds by a more simple analytical method. A gas chromatographic (GC) method is applied to determine a wide variety of phenols in relatively low concentration (EPA method 604). The list of these phenolic compounds is as follows:

- 4-Chloro-3-methyl phenol
- 2-Chlorophenol
- 2,4-Dichlorophenol
- 2,4-Dimethylphenol
- 2,4-Dinitrophenol
- 2-Methyl-4,6-dinitrophenol
- 2-Nitrophenol
- Pentachlorophenol
- Phenol
- 2,4,6-Trichlorophenol

16.2. Analyses of Individual Organic Compounds or Groups of Compounds

Recent advances in the ability to analyze for extremely low concentrations of contaminants in water have shown that man-made organic chemicals occur in many groundwater supplies. Because many synthetic organic chemicals are widely used for industrial and domestic purposes, normal use and irresponsible disposal practices have permitted their widespread introduction to groundwater. Nine volatile synthetic organic chemicals have been recognized to be dangerous to human health: trichloroethylene, tetrachloroethylene, p-dichlorobenzene, 1,1-dichloroethane, vinyl chloride, 1,1,1-trichloroethane, 1,2-dichloroethane, benzene, and carbon tetrachloride. These chemicals are not naturally occurring, are extremely stable in aqueous systems, and are carcinogens. Some organics, even in very small concentrations, have been shown to be genotoxic, affecting the DNA molecule in humans. The U.S. EPA established the list of 123 priority pollutants including 113 organic compounds. Conventional treatment techniques do not remove organic pollutants. Only two methods are shown to be successful: aeration (air stripping) and adsorption using granular activated carbon (GAC).

Volatile Organic Compounds (VOCs)

Lightweight organic compounds that vaporize easily, VOCs are also part of the man-made synthetic chemicals, and are generally used in solvents and degreasing compounds as raw materials. VOCs in groundwater-derived water supplies represent a significant hazard to public health. Many VOCs have shown evidence of animal and human carcinogenicity, mutagenicity, or teratogenicity. VOCs are also quite persistent in groundwater because of their low reactivity and may be transported for long distance. They are also used in both domestic and industrial applications that can result in subsurface disposal (degreasing operation, septic system cleaning, pest and weed control, dry cleaning, fumigation). Oftentimes, to clean out a filled septic system, typically a gallon or less of cleaning fluid (trichloroethylene, methylenechloride, benzene) is flushed down the toilet once every one to two years. Therefore, septic discharges represents one of the most common sources of toxic organic contamination. VOCs as water and wastewater pollutants are represented by three groups:

1. Halogenated VOCs or Purgeable Halocarbons
2. Aromatic VOCs or Purgeable Aromatics
3. Acrolein and Acrylonitrile

Halogenated VOCs or Purgeable Halocarbons

Trihalomethanes (THMs)

Trihalomethanes (THMs) are compounds suspected to be cancer-forming when ingested in sufficient quantities. They are formed by the reaction of natural organic matter, methane (CH_4), with chlorine. For THMs or Total THMs (TTHMs) the following compounds were identified:

- Trichloromethane or Chloroform
- Bromodichloromethane
- Dibromochloromethane
- Tribromomethane or Bromoform

Chloroform is the trihalomethane that occurs in greatest quantity in finished waters. Because of the danger of chlorination, alternative disinfection methods (chlorine dioxide, ozone, UV radiation) are proposed. These alternatives depend on financial problems because of new equipment, installation, and training for operation. The U.S. EPA proposed maximum contaminant level goals (MCLG) for trihalomethanes are zero.

Other VOCs (Halogenated, Aromatic, Acrolein)

The list of monitored VOC compounds includes halogenated aliphatic (chainlike molecules), aromatic (contains benzene ring), and acrylonitrile (cyanoethene) the base of polyacrylonitrile polymers. The list of these health-threatened VOCs (including THMs) are listed and analyzed according to EPA methods 601, 602, and 603 for water samples and 8010, 8015, 8020, and 8030 in solid samples.

Bromodichloromethane	
Bromoform	cis-1,3-Dichloropropene
Bromomethane	trans-1,3-Dichloropropene
Carbon tetrachloride	Methylene chloride
Chlorobenzene	1,1,2,2-Tetrachloroethane
2-Chloroethylvinylether	Tetrachloroethene
Chloroform	1,1,1-Trichloroethane
Chloromethane	1,1,2-Trichloroethane
Dibromochloromethane	Trichloroethene
1,2-Dichlorobenzene	Vinyl chloride
1,3-Dichlorobenzene	Benzene
1,4-Dichlorobenzene	Ethylbenzene
1,1-Dichloroethane	Toluene
1,2-Dichloroethane	Xylene
1,1-Dichloroethene	BTX (benzene, toluene, xylene)
trans-1,2-Dichloroethene	Acrolein
1,2-Dichloropropane	Acrylonitrile

Benzidines

Benzidine is 1,1-Biphenyl-4,4-diamine released into the environment through emissions and wastewater production by manufacturing azo dyes. It absorbs to sediments and soils, and bioconcentrate in fish. Human exposure is primarily occupational.

3,3-Dichlorobenzidine (DCB) may be released as emissions or in wastewater during its production or use as an intermediate in the manufacture of pigments. Regulations in the United States require that DCB and its salts be used in isolated or closed systems to reduce its release. It adsorbs tightly to soils, sediments, and particulate matter, and bioconcentrates in fish. Exposure is mostly occupational. Benzidines are analyzed by GC method, according to EPA 605 method.

Phthalate Esters (Phthalic Acid Esters)

When a carboxylic acid and an alcohol are heated in the presence of an acid catalyst (usually HCl or H_2SO_4), the product is called an ester. Phthalate esters are the esters of the phthalic acid. Phthalate esters are liquids used as plasticizers (addition to synthetic polymers give them flexibility and workability). Phthalate esters enter into the environment in air emissions, wastewater effluent, and solid waste products. Analysis of phthalate esters in waters and solid samples is stated in EPA 606 and 8060 methods, respectively.

Compounds measured by these methods are:

Bis(2-ethylhexyl) phthalate	Diethyl phthalate
Butyl benzyl phthalate	Dimethyl phthalate
Di-n-butyl phthalate	Di-n-octyl-phthalate

Nitrosamines

Amines are organic relatives of ammonia, derived by replacing one, two, or all three hydrogens of ammonia with organic compounds. In primary amine one hydrogen, in secondary amine two hydrogens and in tertiary amine three hydrogens are substituted by organic groups. When secondary amine reacts with nitrous acid (HNO_2), the product is nitrosamine. Nitrosamines are known to be dangerous carcinogens. Nitrosamines can be formed in the environment by the reaction of amines with oxides of nitrogen or with sodium nitrite that has been added to meats as a preservative. When the nitrite (NO_2^-) ion reaches the stomach, the high concentration of hydrochloric acid (HCl) there converts the NO_2^- to HNO_2, which can react with secondary amines in the digestive tract to produce N-Nitrosamine. The analytical procedure of nitrosamines is in EPA 607 method.

Pesticides

Pesticides are substances that are used against insects, weeds, fungi, mites, nematodes, and rodents. Now the use of pesticides is an integral part of increased food production and spoilage prevention of harvested, stored foods. Pesticides also save human lives by disease control.

Classification of Pesticides Based on Their Use

- *Fungicides* are used to protect crop plants and animals from fungal pathogens.
- *Herbicides* are used to kill weeds.
- *Insecticides* are used to kill insect pests and vectors of deadly human diseases such as malaria, yellow fever, trypanosomiasis plague, and typhus.
- *Acaricides* are used to kill mites, which are pests in agriculture, and ticks, which can carry encephalitis of humans and domestic animals.
- *Molluscicides* are used against snails and slugs, which can be harmful for citrus groves, vegetables, and flowers.
- *Nematicides* are used against nematodes, parasites of the roots of crop plants.
- *Rodenticides* are used to control rodents (rats and mice).

Classification of Pesticides According to Their Chemical Structure

Pesticides include a vast array of natural and synthetic substances of widely different chemical natures.

Inorganic pesticides — This group includes compounds of various toxic elements, predominantly arsenic, copper, lead, and mercury. The most known inorganic pesticides are “Bordeaux mixture”, used as a fungicide with copper-based active ingredients. Various arsen compounds are used as herbicides, but lead arsenate and calcium arsenate are known as pesticides.

Organic pesticides — Chemically diverse group, some of them are produced from nature, but most are synthetic organic compounds.

Natural organic compounds — Natural organic compounds are extracted from certain plants as nicotine from tobacco, or pyrethrum produced from the daisy-like Chrysanthemum. One other example of a natural pesticide is an alkaloid extracted from a tropical plant, called strychnine, and used as a rodenticide.

Synthetic organometallic compounds — Most important in this category are the organomercurials.

Chlorinated hydrocarbons or organochloride compounds — They have low water solubility, high lipophilicity, and are persistent in the natural environment. They bioaccumulate in individual organisms and may biomagnify in food chains. Prominent subgroups of these chemicals are as follows.

DDT (dichloro-diphenyl-trichloroethane) — It was introduced in 1939 by the Swiss chemist, Paul Muller, who won a Nobel Prize in 1948 for that discovery. DDT was viewed as the savior of humankind. The use of DDT to control mosquitoes during and just after the Second World War reduced the incidence of malaria rapidly. The global production of DDT peaked in 1970. The relatives of DDT are DDD (dichloro-diphenyl-dichloro-ethane) and methoxychlor, and its breakdown product is DDE (dichloro-diphenyl-ethane). Residues of DDT and its relatives are persistent, and they typically have a half life of about 10 years in the environment. Because of the recognition of the widespread and persistent environmental contamination by DDT and its breakdown products, their ability to biomagnify, and the important nontarget damages that they caused, most industrialized countries banned the use of DDT after the 1970s (except some less developed countries where DDT was used for control of the mosquito vectors of disease). DDT was the first pesticide to which a large number of insect pests developed resistance. Because of the arised ineffectiveness of DDT, its use will eventually be shortened and it will be totally replaced by other insecticides.

Lindane (Gamma-Benzene hexachloride, BHC) — Lindane is the gamma isomer of benzene hexachlorocyclohexane, and the active insecticidal constituent of benzene hexachloride. It is used as an insecticide on hardwood logs and lumber, seeds, vegetables and fruits, woody ornamentals, livestock, and pets (external parasite control).

Cyclodienes — Are a group of highly chlorinated cyclic hydrocarbons that are used as insecticides. Important members are chlordane, heptachlor, aldrin, and dieldrin.

Chlorophenoxy acid herbicides — The most effective and widely used herbicides are the phenoxyacids, such as 2,4-Dichlorophenoxyacetic acid (2,4-D) and 2,4,5-Trichloro-phenoxyacetic acid (2,4,5-T, Silvex). Phenoxyacids have limited persistence in the natural environment, are moderately water soluble, are nonbioaccumulative, and do not biomagnify. "Agent Orange" is a 50-50 mixture of 2,4,5-T and 2,4-D. Agent Orange may have been the cause of serious medical problems that developed in soldiers and villagers throughout Vietnam, such as increase in miscarriages and birth defects. Studies have shown that 2,4-D is not toxic and it is virtually not restricted. The health defect is related to the contamination of 2,4,5-T by dioxin. However, other studies indicate that 2,4,5-T causes an increase in hydrogen peroxide in cells, which, in turn, alters the DNA and may lead to cancer.

Organophosphorus compounds — Their general properties differ markedly from the chlorinated hydrocarbons. Many members have limited persistence in the natural environment, are water soluble, are nonbioaccumulative, and do not biomagnify in food chains. They all have the same chemical structure and a nearly identical mode of action. Except for malathion and a few others, most of these agents are highly toxic. Humans exposed to even low levels may suffer from drowsiness, confusion, cramps, diarrhea, vomiting, headaches, and breathing difficulty. Higher levels cause severe convulsions, paralysis, tremors, coma, and death. Organophosphorus insecticides are fundamentally attributed to their inactivation of the enzyme, cholinesterase. This enzyme breaks down the neurotransmitter, acetylcholine, at the synapse. Insecticides that belong to this group are malathion, parathion, methyl parathion, phosphamidon and diazinon. Humans are mostly exposed to these pesticides in the air in the vicinity of agricultural crops that are sprayed or at production facilities. Occupational exposure to the chemicals will result mainly from dermal contact and inhalation of contaminated air. Dermal contact with treated surfaces such as leaves of sprayed crops is also important. Federal law limits the use of these chemicals to certified applicators or persons under their direct supervision. Protective clothing, goggles, and respirators must be worn and workers must keep out of treated areas for 48 hours after a field has been treated. General exposure may result from the consumption of contaminated food. Minor exposure may occur from the consumption of contaminated water. Reasons for the persistent popularity of organophosphates are their variety, which can be used by farmers to reduce the emergence of resistant strains of insects, and their biodegradability.

Carbamate pesticides — In general, carbamates are less toxic than organophosphates, although symptoms of acute poisoning are essentially similar. Clinical carbamate poisoning is nearly identical to that seen with organophosphates except that carbamates do not penetrate the central nervous system very well and thus produce only limited effects there. Members of this class include aldicarb (Temik), propoxur (Baygon), and carbaryl (Sevin). They are also very popular among farmers, gardeners, and foresters.

Triazine herbicides — Triazine herbicides are used in corn and other crops, and as soil sterilants. Prominent examples are atrazine, hexazinone, and simazine. They have low chronic toxicity with no significant increase in cancer evidence. However, the studies on this have been limited.

Other pesticides — There are literally thousands of other pesticides in addition to the ones discussed already. They include, for example, paraquat, used for marijuana fields, so smokers of marijuana may be exposed unknowingly to paraquat. The symptoms of inhalation include sore throat, nosebleed, and cough. Pentachlorophenol has been used as a wood preservative on utility poles and in log homes.

Rachel Carson's classic work, *Silent Spring* (1962) has called the pesticide situation to national attention. The U.S. EPA licenses pesticides, and the Food and Drug Administration (FDA) sets tolerance levels for concentration of residues of pesticides. By the early 1980s, several incidents of groundwater contamination resulting from field application of pesticides had been confirmed. Over 200 pesticides are in common use, and a wide range of hydrogeologic conditions can affect the susceptibility of groundwater to contaminate, as well as surface waters. Educating farmers and, indeed, individual households that pesticides are biologically active agents with potentially wide-ranging effects is mandatory. Besides the health effects, use of pesticides creates other problems by destroying beneficial species. Also, species develop resistance, so the surviving insects reproduce and eventually form a new population that can be killed only by large doses or new pesticides. Although many environmentalists support banning pesticides, this view is impractical. Pesticides are indispensable for agriculture and will remain in use; however, pesticides will play the much smaller role in agriculture as other, less harmful agents, such as pheromones, hormones, and natural chemical repellants. Because many synthetic organic chemicals are widely used for industrial and domestic purposes, normal use and irresponsible disposal practices have permitted their widespread introduction to groundwater. Detection of pesticide residues are in EPA methods 608, 614, 619, 630, and 632.

Ethylene Dibromide (EDB) or 1,2-Dibromoethane

Ethylene dibromide (EDB) is a highly volatile, clear, colorless, heavy liquid with a distinctive odor. It is a common industrial chemical that is used as a gasoline additive, but sometimes it is used to kill insects in fruit and grain and to fumigate buildings for termites. It is sometimes used by farmers to treat the soil to kill small destructive worms called Nematodes. Most uses of EDB have been banned in the U.S. from 1984, because of its carcinogenic and mutagenic effect. The half life of this chemical is 8 years. In consideration that EDB has been found in drinking water. The U.S. EPA proposed a MCLG of "zero". At the present time, there is no approved method to remove EDB from drinking water.

Dioxins

The term "dioxin" is commonly used to describe a family of substituted dibenzo-p-dioxins. They are a class of substances that are formed not as commercial products but as a result of contamination of merchandized chemical products in the environment. They occur in the environment in very low concentration and with very high toxicity effect. Particularly one member of this family, 2,3,7,8-tetrachlorodibenzo-p-dioxin (TCDD) produced substantial considerations. It is an unavoidable contaminates of the chlorophenoxy herbicide 2,4,5-trichlorophenoxy-acetic acid (2,4,5-T). This compound usually was used in combination with 2,4-dichlorophenoxyacetic acid (2,4-D), which contains no dioxin. They were used to control growth of weeds along roads, utility rights-of-way, and commercial forests. Their controversial uses were as weapons in the Vietnam War. (as discussed under "Chloro-phenoxy acid herbicides"). A combination of 2,4,5-T and 2,4-D, called Agent Orange, was sprayed on jungles and trails for years. As a result of the spraying, thousands of people were exposed to Agent Orange and its dioxin. At this time, the effect of dioxin was unknown. Later studies reached the conclusion that dioxins were extremely toxic and very stable compounds. The dioxin that contaminated 2,4,5-T is TCDD. This compound is capable of inducing three enzyme systems and suppress immunity, it is a potent carcinogen, and its teratogenic effect is also approved. The lethal doses for TCDD is 0.6 ug/kg, compare that with 2,4,5-T which is 381 mg/kg (635,000 times). The use of 2,4,5-T has been banned because of the dioxin contamination. 2,4-D contains no TCDD, therefore, it is still in use. In addition to environmental distribution in pesticides, dioxins can be formed as combustion products from burning vegetation treated with 2,4,5-T and 2,4-D. It may also be formed by the thermal decomposition of chlorophenol used to treat paper and wood or pesticide treated wastes. Wastes from chemical manufacturing can contain dioxins that may be dispersed during disposal to landfills. From the landfills, the groundwater and enter surface waters may be contaminated.

Polychlorinated Biphenyls (PCBs)

PCBs were made from biphenyl, a naturally occurring compound extracted from petroleum. The biphenyl is chlorinated at any or all of its carbon atoms in the presence of a catalyst, usually iron. PCBs are heavy, nonflammable, stable oils, with high boiling points and low electrical conductivity. These properties make them ideal coolants for electrical transformers, as well as superior hydraulic fluids and heat transformer oils. The production of PCBs is the chlorination of biphenyls to give the variety of products with different chlorine content. From the possible number of the 210 isomers are 102 commercial products that are known. These products are graded according to their chlorine content. By chemical structure, PCBs are similar to the organochlorine insecticides and, therefore, similar isolation techniques and analytical methodologies are used in the detection of these substances (EPA 608). In the U.S., Monsanto Co. was the only manufacturer of PCBs, with a trade name of "Aroclor". Each product was accompanied by a four digit code number, where the first two digits indicated the product, (12 = PCB, 54 = PCT, 25 and 44 = PCB and PCT mixture), and the last two digits indicated the chlorine content in percent. For example, Aroclor 1221 indicated that the product was PCB and its chlorine content was 21%. PCBs are used all over the world, produced by different companies and under different trade names. For example, in Germany a Bayer product was produced with a trade name of "Clophen". In France, Prodolec was produced under the name "Phenoclor".

The first reported incident related to the toxicity of PCBs was in Japan in 1968, where PCBs were used as a cooling fluid by manufacturers of rice oil. There was a leak or cross connection in the system, and the rice oil become heavily contaminated by PCBs, causing sickness (called oil sickness), characterized by chloracne, hyper-pigmentation of the skin and mucous membranes, liver diseases, fatigue, headache, oedema, menstrual disorders, and birth defects. Laboratory study indicates that PCBs cause hepatocellular carcinoma in research animals. Exposure to PCBs is virtually universal in industrialized countries. The chemical stability and fat solubility of PCBs causes them to bioaccumulate through the food chain. Very high levels were found in fish because of the dumping of used transformer oils into the rivers (ocean fish do not seem to be similarly contaminated). Consumers of fresh-water fish in North America are at high risk of absorbing PCBs into their bodies, where they are stored in the fat. The biological half-life of PCBs is years, and it is not clear what the long term risks of PCBs bioaccumulation. Human breast milk and cow milk can also be contaminated.

The ubiquitousness and insidiousness of PCB contamination is illustrated in the following incident in 1979. An unused transformer at a slaughter house was accidently punctured. The PCBs leaked out and entered the plant's grease and waste meat recycling system. The waste was sold to a large chicken farm as a chicken feed. Over the next several months, the eggs were sold and used by bakeries and mayonnaise preparation industries. After the contamination was discovered, millions of dollars of food was recalled from across the country and destroyed.

Shenandoah Stables was a large and successful quarter horse ranch in Moscow Hills, Missouri. In 1971 their problems started. To control dust, they sprayed 1000 gallons of waste automobile oil onto the acreage. Soon dozens of dead sparrows were found. Then dogs and cats on the farm became sick, lost their fur and dehydrated. After about a month many cats and dogs died. From the 85 horses, 43 died during one year. Horses had been bred on the ranch. Most pregnancies ended in spontaneous abortions. Of those born alive, all died within a few months. The owner and his two daughters also became sick. Tests on the oil applied to the farm showed high levels of dioxin and PCBs. A lot of other similar problems were also discovered.

The incidents pointed out the need for continuous inspection and measurement of the food supply. The major entry of PCBs into the environment results from vaporization during burning, leaks, disposal of industrial fluids, and disposal in dumps and landfills. Therefore, the Toxic Substances Control Act (TSCA) regulates the disposal of PCBs, which are being phased out of use in this country.

Nitroaromatics and Isophorone

When an aromatic compound is heated with a mixture of concentrated nitric acid, one of the hydrogens of the ring is replaced by a nitro group, NO_2^-. Aromatic nitrocompounds are very common. During environmental pollution control, the following nitroaromatic compounds are analyzed according to EPA method 609:

2,4-dinitrotoluene or 1-methyl-2,4-dinitrobenzene
2,6-dinitrotoluene or 1-methyl-2,6-dinitrobenzene
Isophorone
Nitrobenzene

2,4-dinitrotoluene and 2,6-dinitrotoluene — It is used in the manufacture of toluene diisocyanate for the production of polyurethane plastics, in the production of military and some commercial explosives and as a propellant additive. It is also used in the manufacture of azo dye intermediates, and in organic synthesis of toluidines and dyes. They may enter the environment in wastewater from the processes in which it is made and used.

Nitrobenzene — Nitrobenzene is produced in large quantities and may be released to the environment in emission and wastewater during its production and use. Since 98% of nitrobenzene is used to produce aniline in five regions of the country, industrial releases will be fairly localized. Nitrobenzene is also produced by the photochemical reaction of benzene with nitrogen oxides. Human exposure is primarily occupational by inhalation and dermal contact.

Isophoron — Isophoron is used as a solvent for a large number of natural and synthetic polymers, resins, waxes, fats, oils, and pesticides, in addition to being used as a chemical intermediate. It is released in the environment from a wide variety of industries, from the disposal of many different products, and during the application of some pesticides.

Polynuclear Aromatic Hydrocarbons (PAHs)

Although many compounds have now been found to be carcinogenic, the first to be discovered were a group of fused aromatic hydrocarbons, or polynuclear aromatic hydrocarbons, PAHs. PAHs are by-products of petroleum processing or combustion. They contain two or more benzene rings and have many diverse origins. Pyrolysis and incomplete combustion of organic matter yields PAHs in the products. Soot consists of a large number of linked aromatic rings and is also often produced.

About 200 years ago, an English surgeon found that cancer of the scrotum and other parts of the body among young boys who worked as chimney sweeps in London was very common. Today it is known that these cancers were caused by fused aromatic hydrocarbons present in the chimney soot. The polycyclic aromatic hydrocarbon, benzo(a)pyrene is found naturally in coal, petroleum, tar, and some wood. Benzo(a)pyrene is also a constituent of air pollution from combustion of gasoline, and it is also a component of tobacco smoke. Many of these compounds are highly carcinogenic at relatively low levels. Some of these dangerous substances are:

Acenaphtene	Chrysene
Acenaphtylene	Dibenzo(a,h)anthracene
Anthracene	Fluoranthene
Benzo(a)anthracene	Fluorene
Benzo(a)pyrene	Indeno (1,2,3)pyrene
Benzofluoranthrene	Naphthalene
Benzo(ghi)perylene	Pyrene
Benzo(k)fluoranthene	

Haloethers

Ethers are compounds with a formula of R-O-R´, where R and R´ can be the same or different organic groups. The chief chemical property of ethers is that, like the alkanes, they are virtually inert and do not react with most reagents. It is this property that makes some of them such valuable solvents. Haloethers are analyzed according to the EPA 611 method for the following compounds:

2-Chloroethyl-ether	4-Bromophenyl-phenyl-ether
2-Chloroisopropyl-ether	4-Chlorophenyl-phenyl-ether

Organic Analytical Groups

Analytical group	EPA method no.
Water and wastewaters	
Purgeable Halocarbons	601
Purgeable Aromatics	602
Acrolein and Acrylonitrile	603
Phenols	604
Benzidines	605
Phthalate Esters	606
Nitrosamines	607
Organochlorine Pesticides/PCBs	608 and 617
Nitroaromatics and Isophorone	609
Polynuclear Aromatic Hydrocarbons (PAHs)	610
Haloethers	611
Chlorinated Hydrocarbons	612
2,3,7,8-Tetrachlorodibenzo-p-dioxin	613
Organophosphate Pesticides	614 and 622
Triazine Pesticides	619
Dithiocarbamate Pesticides	630
Carbamate and Urea Pesticides	632
Solid wastes	
Halogenated VOCs	8010
Non-Halogenated VOCs	8015
Aromatic VOCs	8020
Acrolein, Acrylonitrile, Acetonitrile	8030
Phenols	8040
Phthalate Esters	8060
Organochlorine Pesticides/PCBs	8080
Nitroaromatics and Cyclic Ketones	8090
Chlorinated Hydrocarbons	8120
Organophosphorus Pesticides	8140
Chlorinated Herbicides	8150
Polynuclear Aromatic Hydrocarbons	8310

Method References: 40 CFR Part 136, Appendix A; EPA SW 846 Test methods for Evaluating Solid Waste, 1986, Third edition

Parameters per Analytical Methods

EPA 601

Bromodichloromethane	1,1-Dichloroethane
Bromoform	1,2-Dichloroethane
Bromomethane	1,1-Dichloroethene
Carbon tetrachloride	1,2-dichloroethene
Chlorobenzene	1,2-Dichloropropane
Chloroethane	cis-1,3-Dichloropropene
2-Chloroethylvinyl ether	trans-1,3-Dichloropropene
Chloroform	Methylene chloride
Chloromethane	1,1,2,2-Tetrachloroethane
Dibromochloromethane	Tetrachloroethene
1,2-Dichlorobenzene	1,1,1-Trichloroethane
1,3-Dichlorobenzene	1,1,2-Trichloroethane
1,4-Dichlorobenzene	Trichloroethene
	Vinyl chloride

EPA 602

Benzene	1,4-Dichlorobenzene

Chlorobenzene	Ethylbenzene
1,2-Dichlorobenzene	Toluene
EPA 603	
Acrolein	Acrylon
EPA 604	
4-Chloro-3-methylphenol	2-Nitrophenol
2-Chlorophenol	4-Nitrophenol
2,4-Dichlorophenol	Pentachlorophenol
2,4-Dimethylphenol	Phenol
2,4-Dinitrophenol	2,4,6-Trichlorophenol
2-Methyl-4,6-Dinitrophenol	
EPA 605	
Benzidine	3,3-Dichlorobenzidine
EPA 606	
Bis(2-ethylhexyl)phthalate	Diethyl phthalate
Butyl benzyl phthalate	Dimethyl phthalate
Di-n-butyl phthalate	Di-n-octyl-phthalate
EPA 607	
N-Nitrosodimethylamine	N-Nitrosodi-n-propylamine
N-Nitrosodiphenylamine	
EPA 608	
Aldrin	Endrin
Alpha-BHC (Benzene-hexachloride)	Endrin aldehyde
Beta-BHC	Heptachlor
Delta-BHC	Heptachlor epoxide
Gamma-BHC (Lindane)	Toxaphene
Chlordane	PCB-1016
4,4-DDD	PCB-1221
4,4-DDE	PCB-1232
4,4-DDT	PCB-1242
Dieldrin	PCB-1248
Endosulfan I	PCB-1254
Endosulfane II	PCB-1260
Endosulfan sulfate	
EPA 609	
2,4-Dinitrotoluene	Isophorone
2,6-Dinitrotoluene	Nitrobenzene
EPA 610	
Acenaphtene	Dibenzoanthracene
Acenaphtylene	Fluoranthene
Anthracene	Fluorene
Benzoanthracene	Indenopyrene
Benzopyrene	Naphthalene
Benzofluoranthene	Phenantrene
Benzoperylene	Pyrene
Chryse	
EPA 611	
Bis(2-chloroethyl)ether	4-Bromophenyl-phenyl-ether
Bis(2-chloroethoxy)methane	
Bis(2-chloroisopropyl)ether	4-Chlorophenyl-phenyl-ether

EPA 612	
2-Chloronaphtalene	Hexachlorobutadiene
1,2-Dichlorobenzene	Hexachlorocyclopentadiene
1,3-Dichlorobenzene	
1,4-Dichlorobenzene	Hexachloroethane
Hexachlorobenzene	1,2,4-Trichlorobenzene
EPA 613	
2,3,7,8-Tetrachlorodibenzo-p-Dioxin	
EPA 614	
Azinphos methyl	Ethion
Demeton	Malathion
Diazinon	Parathion ethyl
Disulfoton	Parathion methyl
EPA 619	
Ametryn	Secbumeton
Atraton	Simetryn
Atrazine	Simazine
Prometon	Terbutylazine
Prometryn	Terbutryn
EPA 630	
Amobam	Niacide
Busan 40	Polyram
Busan 85	Sodium diethyl-dithiocarbamate
Ferbam	
KN Methyl	Thiram
Mancozeb	ZAC
Maneb	Zineb
Metham	Ziram
Nabam	
EPA 632	
Aminocarb	Methiocarb
Barban	Methomyl
Carbaryl	Mexacarbate
Carbofuran	Monuron
Chlorpropham	Monuron-TCA
Diuron	Neburon
Fenuron	Oxamyl
Fenuron-TCA	Propham
Fluometuron	Propoxur
Linuron	Siduron
	Swep

Chapter 17
Microbiological Parameters and Radioactivity

Environmentally transmitted infectious disease constitutes the oldest environmental health problem. Environmental spread of infectious diseases, food- and water-borne illness, and soil-borne parasitic infections still occur each year. In the less industrialized countries, infectious disease (much of them environmentally transmitted) remains by far the greatest cause of morbidity and mortality. The annual death and disability of hundreds of thousands of people in the developing countries originated from water-borne infections.

Drinking water may be obtained from either surface sources (lakes, streams, surface waters) or underground sources (groundwater). A variety of processes is used to treat municipal water supplies to be safe for drinking purposes, and the success of the treatment should be checked on the continuous basis. Drinking water may be contaminated with infectious organisms because the water is untreated, deficiencies are present in the treatment system, deficiencies exist in the distribution system, such as cross-connection between sewage and water lines, or the properly functioning treatment system is unable to remove such contaminating agent, for example Giardia cysts. Microbiological testing determines the sanitary quality of environmental samples and gives the degree of contamination. Tests for detection and enumeration of indicator organisms, rather than of pathogens, are used. The coliform group of bacteria is the principal indicator of suitability of water for domestic and other uses. The significance of the coliform group density is the criteria of the degree of pollution and thus of sanitary quality. The environmental areas covered include (1) all waters, fresh, estuarine, marine, shellfish-growing, agricultural, groundwater, surface water, finished water, recreational water and industrial processing water; (2) all wastewaters of microbiological concern, domestic waste effluent, industrial wastes, shellfish-processing, and agricultural wastes; and (3) other areas of the environment, air, sediment, soil, sludge, oil, leachates, and vegetation.

Examination of routine bacteriological samples cannot be regarded as providing complete or final information concerning sanitary conditions surrounding the source of any particular sample. The results of the examination of a single sample from a given source must be considered inadequate. The final evaluation must be based on examination of a series of samples collected over a known and protracted period of time.

17.1. Heterotrophic Plate Count (HPC)

The test was formerly known as the "Standard Plate Count" determination. It provides an approximate enumeration of live heterotrophic bacteria. It may yield useful information about bacterial quality and may provide supporting data for further investigations. Colonies may arise from pairs, chains, clusters, or single cells, all of which are included in the term "colony forming units" (CFU), which is used to report the result.

17.2. Coliform Bacteria Group

Total Coliform Group

The density of the Coliform group is the principal indicator of suitability of a water for domestic, dietetic, or other uses. The significance of the tests and the interpretation of the results are well-authenticated and have been used as a basis for standards of bacteriological quality of water supplies. There are several criteria for an indicator organism. The most important criterion is that the organism be consistently present in human feces in substantial numbers so that its detection will be a good indication that human wastes are entering the water. The indicator organisms should also survive in the water at least as well as the pathogenic organisms would. Therefore, the presence of any significant number of coliform is evidence that the water is contaminated by fecal material, and any pathogens that leave the body through the feces can be present. Total Coliform group includes all of the aerobic and anaerobic, gram-negative, nonspore-forming, rod-shaped bacteria that ferment lactose in 24 to 48 hours at 37°C. The definition includes the genera: Escherichia, Citrobacter, Enterobacter, and Klebsiella.

Fecal Coliform Bacteria, *Escherichia Coli* (*E. coli*)

The fecal coliforms are part of the total coliform group. The major species is *Escherichia coli*, a species indicative of fecal pollution and the possible presence of enteric pathogens. The total coliform group includes organism of fecal and nonfecal origins. This test makes it possible to separate the members of the coliform group found in the feces of various warm blooded animals from those from other environmental sources. Organisms of fecal origin are detected at 44.5 ± 0.2°C incubation by using an enriched lactose medium. The fecal coliform test is applicable to investigations of stream pollution, raw water sources, wastewater treatment systems, bathing water, seawater, and monitoring general water quality. For potable waters, the recommended test is the detection of the total coliform group.

Fecal Streptococcus

The terms "fecal streptococcus" and "Lancefield's Group D Streptococcus" have been used synonymously. The members of the groups are Streptococcus faecalis, S. faecium, S. avium, S. bovis, S. equinus, and S. gallinarum. They all give a positive test with Lancefield's Group D antisera. The normal habitat of fecal streptococci is the gastrointestinal tract of warmblooded animals. Fecal coliform/fecal streptococcus ratios may provide information on possible sources of pollution. (FC/FS ratio). The ratio greater than 4.4 was considered indicative of human fecal pollution, and the ratio of less than 0.7 was the sign of nonhuman pollution. Presently, the FC/FS ratio is not recommended for use for such differentiation. The fecal streptococci are valuable pollution indicators in the study of rivers, streams, lakes, and marine systems, especially when used with fecal coliform bacteria.

Klebsiella

Klebsiella genus is included in the coliform group and may be associated with coliform regrowth in water distribution systems. It is also the major component of coliform population of industrial wastes containing high bacterial nutrients, for example in paper mill, textile, sugar cane, farm production, and other food processor industrial systems. These wastewaters contain great quantities of carbohydrates and can support these organisms in the effluent and receiving waters.

17.3. Pathogenic Microorganism

A wide variety of enteric pathogenic organisms may occur in environmental samples. Human pathogens that are transmitted in water and the effects they cause are as follows:

- Salmonella typhi — typhoid fever
- Other Salmonella species — salmonellosis (gastroenteritis)
- Shigella species — shigellosis (bacillary dysentery)
- Vibrio cholera — cholera
- Escherichia coli — gastroenteritis
- Yersinia enterocolitica — gastroenteritis
- Campylobacter fetus — gastroenteritis
- Hepatitis A virus — hepatitis
- Poliovirus — poliomyelitis
- Giardia intestinalis — giardiasis
- Balantidium coli — balantidiasis
- Entamoeba histolytica — amoebic dysentery

A more serious problem is that several pathogens are even more resistant than coliforms to chemical disinfection, in their order of increasing resistant to chlorination, coliforms, viruses, and protozoan cysts. It has been found that chemically disinfected water samples that are free of coliforms are often contaminated with enteric viruses. The cysts of Giardia lamblia are so resistant to chlorination that eliminating them by this method is probably impractical. Mechanical methods are probably necessary, such as filtration and flocculation to remove colloidal particles, because the microorganisms are trapped mostly by surface adsorption in the sand beds.

Pathogenic Bacteria

Among pathogenic microorganisms that can be demonstrated in waste water and under certain conditions, surface and groundwater, the most common and important are: Salmonella, Shigella, enteropathogenic *Escherichia Coli*, Leptospira, and Vibrio Cholera. Routine examination of water and wastewater for pathogenic microorganisms is not recommended because very well-equipped laboratories with well-trained personnel are needed.

Salmonella

The Salmonella bacteria (named for Daniel Salmon) are gram-negative, facultatively anaerobic, motile rods that ferment glucose to produce acid and gas. Their normal habitat is the intestinal tracts of humans and many animals. Salmonella causes salmonellosis, with a usually moderate fever accompanied by nausea, abdominal pain, cramps, and diarrhea. The severity of the infection depends on the number of Salmonella digested. Meat products are particularly susceptible to contamination by

Salmonella. If these products are mishandled, the bacteria can grow to infective numbers very quickly. The sources of the bacteria are the intestinal tract of many animals, and meat can be contaminated during processing. Poultry, eggs, and egg products are often contaminated by Salmonella. The organisms are generally killed by normal cooking. Consumption of raw, unpasteurized milk can also be a good source of salmonellosis.

Salmonella typhi

The organism is a species of the Salmonella family, and it produces typhoid fever. Before the days of proper disposal, drinking water treatment, and food sanitation, typhoid was a very common disease, and was responsible for many of the deaths. After recovery, many people remain carriers of the bacteria, and continue indefinitely to shed the organism.

Shigella

Shigellosis, or bacillary dysentery, is a severe form of diarrhea caused by the bacteria Shigella, named after the japanese microbiologist, Kiyoshi Shiga. Shigella produce an exotoxin that inhibits protein synthesis. In the intestine, Shigella produce tissue destruction and cause severe diarrhea with blood and mucus in the stool. While most shigellosis epidemics are food-borne or spread by person to person contact, it may be caused by contaminated drinking water. Water-borne shigellosis may result from accidental interruption of water treatment, untreated water supply, contamination of well water, cross connection between contaminated water pipe and potable water supply lines.

Vibrio cholera

During the 1800s, the bacterial infection called Asiatic cholera crossed Europe and North America in repeated epidemics. Today, cholera is endemic in Asia, particularly in India, and very rare in Western countries. The spread of cholera in 1970 may reflect the lack of international quarantine enforcement by some countries having primitive public water supplies and inadequate sanitary regulations. The organism that causes the sickness is Vibrio cholera, a slightly curved, gram-negative rod with a single polar flagellum. It grows in the small intestine and produces an enterotoxin that causes the intensive secretion of water and minerals. Because of the loss of water and electrolytes (12 to 20 liters daily), the results are shock, collapse, and often death. In the United States, the waters of the Gulf and Pacific coasts support the population of the Vibrio cholera that has been the cause of a number of outbreaks of gastroenteritis, usually associated with ingestion of contaminated seafood.

Enteropathogenic Escherichia coli

Escherichia coli is the most common and well-known microorganism in the human intestinal tract. It is usually harmless, but some species produce toxins that cause diarrhea. One group of the pathogenic *E.coli* group, the enterotoxigenic *E. coli*, causes "traveler's diarrhea", and the common infant diarrhea in developing countries.

Yersinia

Yersinia enterocolitica and Yersinia pseudotuberculosis are gram-negative organisms and intestinal inhabitants of many domestic animals and are often transmitted in meat and milk. Both organisms have the ability to grow at refrigerator temperature (4°C). The name of the caused sickness is yersiniosis and its symptoms are diarrhea, fever, headache, and abdominal pain.

Campylobacter

Campylobacter are gram-negative, spirally curved rods that cause gastroenteritis with mild symptoms. Like Salmonella, they are common in the intestinal tract of certain animals, especially sheep and cattle. The second most common causal agent of diarrhea is Campylobacter.

Legionella pneumophila

A type of pneumonia was identified as legionellosis (Legionnaire's disease), which received attention in 1976, when a series of deaths occurred among the members of the American Legion who had attended a meeting in Philadelphia. A total of 182 persons became sick with pneumonia at this meeting, and 29 died. Close investigation identified an aerobic, gram-negative, rod shaped, unknown bacterium, with polar, subpolar, and lateral flagella as the causative agent of the tragedy. Now the bacterium is called Legionella pneumophila, and the disease is legionellosis, which is commonly characterized by very high fever, cough, and general symptoms of pneumonia. Recent studies show that the organism can grow well in the water of air conditioning cooling towers, which might explain some epidemics in hotels, business districts, and hospitals. The organism is transmitted via the airborne route and it is ubiquitous in a moist environment. It was also isolated from contaminated potable water distribution systems and from lakes, streams, reservoirs, and wastewaters. Legionella is resistant to chlorine and can survive in waters with low chlorine levels.

Viruses

Virus is the latin word for poison. Viruses are found as parasites in all types of organisms. Viruses cause diseases for humans, animals, and plants. They also infect fungi, bacteria, and protozoa. Viruses differ in size from bacteria. Although most are quite a bit smaller than bacteria, some of the larger viruses are about the same size as some very small bacteria; the range of the diameter for viruses is 20 to 300 nm. Viruses are not normal flora in the intestinal tract, and they are excreted only by infected individuals. These viruses are transmitted from person to person, and they are present in domestic sewage, which after treatment enters waterways to become part of the rivers and streams that are very often the sources of drinking water. Ground water may also be polluted. More than 100 different enteric viruses are recognized. The most common viruses excreted with feces are the hepatitis A viruses and the Norwalk-type agents that can cause acute infectious nonbacterial gastroenteritis.

Hepatitis A

The hepatitis A virus (HAV) is the causating agent of the infectious hepatitis. After a typical entrance via the oral route, the hepatitis virus multiplies in the intestinal tract and then spreads to the liver, kidney, and spleen. The virus is shed in the feces and can also be detected in the blood and urine. Recent studies indicate that the virus might also be carried in oral secretions. The virus probably survives for several days on surfaces, such as cutting boards. The hepatitis A virus is very resistant to disinfectants such as chlorine at the concentration used in drinking waters. Oysters that live in contaminated waters are also a good source of infection. The incubation time of two to six weeks makes it difficult to find the source of the infection. The symptoms are loss of appetite, nausea, diarrhea, abdominal discomfort, fever, and in some cases jaundice, with yellowing of the skin and the whites of the eyes and dark urine, which is typical of liver infections. Persons at risk for exposure to the virus and travelers in high risk areas can be given immune globulin, which gives protection for several months.

Norwalk Agent

A number of viruses, such as the polioviruses, echoviruses, and coxsackieviruses, are transmitted by the feces-oral route. About 90% of the viral gastroenteritis are caused by rotavirus (named after its wheel shaped, rota means wheel) or by the Norwalk agent (named after the Outbreak in Norwalk, Ohio in 1968). The symptoms are nausea, diarrhea, and vomiting with a low fever for a few days.

Pathogenic Protozoa

Protozoans are one-celled, eucaryotic organisms that inhabit water and soil and feed upon bacteria and small particulate nutrients. From the large number of protozoans, only a few are pathogenic.

Giardia lamblia

One group of the protozoans is called Flagellate. Flagellates are typically spindle-shaped, with flagella projections from the front end. An example of a flagellate that is a human parasite is Giardia lamblia. The parasite is found in the small intestine of humans and other mammals. It is passed out of the intestine and survives as a cyst before being ingested by the next host. It is the cause of a prolonged diarrheal disease in humans called giardiasis. The disease often persists for weeks and is characterized by weakness, nausea, weight loss, and abdominal cramp. The parasite sometimes occupies so a large place of the intestinal wall, that they disturb absorption. Most outbreaks in the United States are transmitted by contaminated water supplies. The problem is that the cyst is not sensitive to the regular disinfection by chlorine in the concentration recommended for drinking water. The best treatment is the filtration of the water supply to remove the cysts.

Entamoeba histolytica

The amoebas move by extending usually blunt, lobelike projections of the cytoplasm called pseudopods. Any number of pseudopods can flow from one side of the amoeba cell, and the rest of the cell will flow toward the pseudopods.

Amoebic dysentery or amoebiasis is found worldwide and is spread by food and water contaminated by cysts of the protozoan amoeba Entamoeba hystolitica. Although stomach acid (HCl) can destroy vegetative cells, it does not affect the cysts. It causes severe dysentery and the feces contain blood and mucus (the vegetative form feed on red blood cells). In severe cases, the intestinal wall is perforated and invasion of other organs (liver) is possible. The major source of amoebiasis is drinking water contamination by sewage, oral-fecal route, and consumption of uncooked polluted vegetables.

17.4. Iron and Sulfur Bacteria

Iron Bacteria

The specific form iron takes in water depends on the amount of oxygen in the water and the pH. In natural groundwater systems where oxygen concentrations are low or absent and the pH is from 6.5 to 7.5, the iron occurs primarily as dissolved ferrous ions (Fe^{2+}). Ferrous iron is unstable when it comes in contact with oxygen. It changes to ferric iron (Fe^{3+}) and precipitates as ferric hydroxide ($Fe(OH)_3$). Ferric oxides or oxyhydroxides come out of solution and coat surrounding surfaces. Iron bacteria

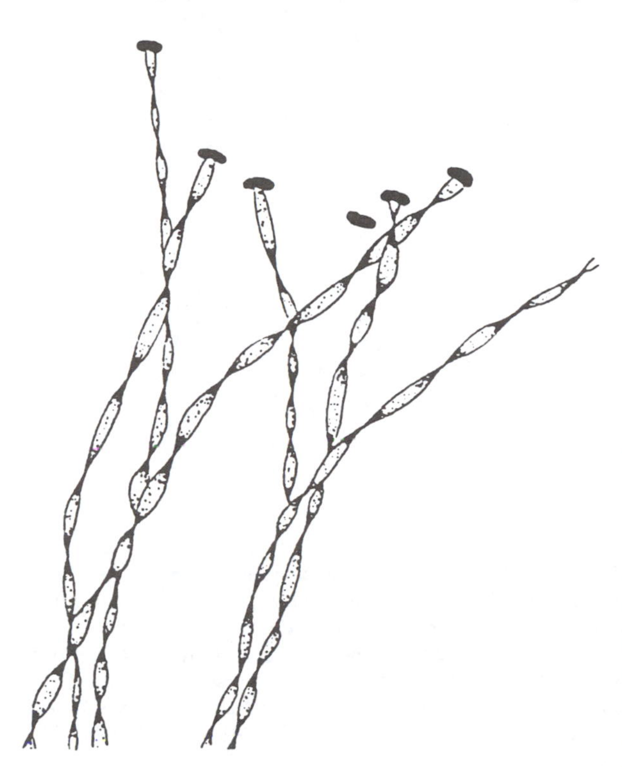

FIGURE 17-1. Common iron bacteria: Gallionella ferruginea. Gallionella is a common enzymatic iron bacteria, composed of twisted bands resembling a ribbon or chain. It can be recognized by the twisted stalks and the bean-shaped bacterial cell at the end of the twisted stalks. A precipitate of inorganic iron on and around the stalks often blurs the outlines. The average cells are 0.4 to 0.6 um in width by 0.7 to 1.1 um in length.

widely occur in wells open to the atmosphere when sufficient iron and/or manganese are present. The principal forms of iron bacteria plug wells by enzymatically catalyzing the oxidation of iron, using the energy to promote the growth of treadlike slime, and large amounts of ferric hydroxide in the slime. Precipitation of the iron and rapid growth of the bacteria create a voluminous material that quickly plugs pipes or screen pores of the sediment surrounding the well bore. Sometimes the quick growth of iron bacteria can render a well useless within a few months. The large amount of brown slime so produced will impart a reddish tinge and an unpleasant odor to drinking water and may render the supply unsuitable for domestic or industrial purposes. Some bacteria that do not oxidize ferrous iron nevertheless may cause it to be dissolved or deposited indirectly. In their growth, they either liberate iron by utilizing organic radicals to which the iron is attached or they alter environmental conditions to permit the solution or deposition of iron. In consequence, less ferric hydroxide may be produced, but taste, odor, and fouling may be engendered.

Gallionella

Gallionella is a common enzymatic iron bacteria composed of twisted bands resembling a ribbon or chain. It can be recognized by the twisted stalks and the bean shaped bacterial cell at the end of the twisted stalks (Figure 17.1).

Crenothrix, Clonothrix, and Leptothrix

Crenothrix, clonothrix, and leptothrix are filamentous iron bacteria. (Figure 17.2. and 17.3.)

Sulfur Bacteria

Sulfur bacteria is bacteria that oxidizes or reduces sulfur compounds.

Sulfate Reducing Bacteria

Sulfate reducing bacteria grows anaerobically and reduces sulfate to hydrogen sulfide, H_2S. The sulfate reducing bacteria contributes greatly to tuberculations and galvanic corrosion of water mains. It can cause taste and odor problems in water.

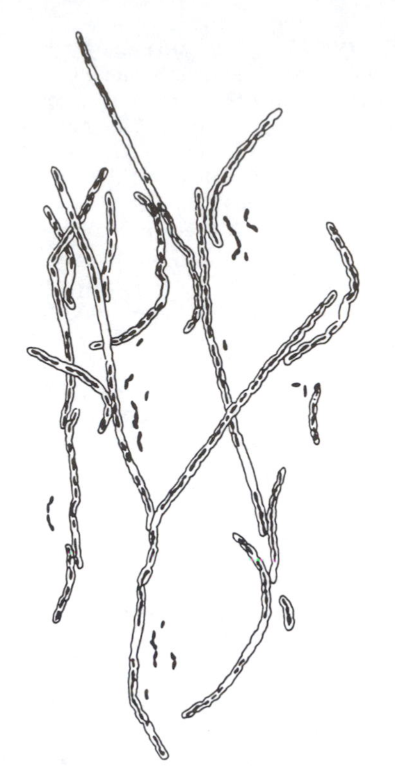

FIGURE 17-2. Filamentous iron bacteria: Spherotilus natans. Cells are within the filaments. Some of them free "swarmer" cells. Filaments also show areas devoid of cells. Individual cells within the sheat may vary in size, varying 0.6 to 2.4 um in width by 1.0 to 12.0 um in length; most strains are 1.1 to 1.6 um wide by 2.0 to 4.0 um long. (*Standard Methods for the Examination of Water and Wastewater*, APHA-AWWA-WPCF, 17th Ed., 1989, p. 9-117)

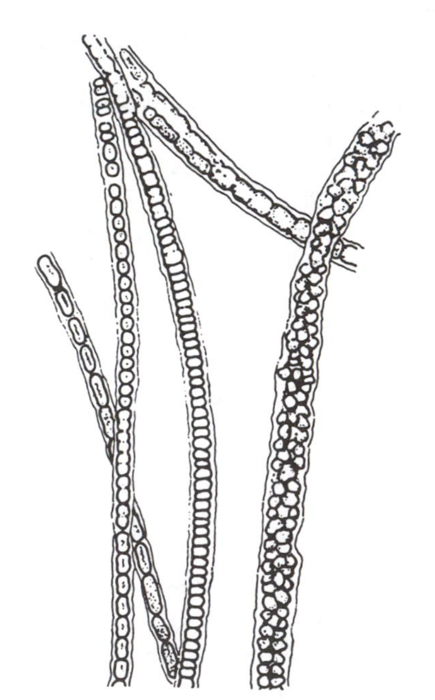

FIGURE 17-3. Filamentous iron bacteria: Crenothrix polyspora. Cells within the filaments show variations of size and shape, giving the name "polyspora". Young growing colonies are usually not encrusted with iron or manganese. Older colonies often exhibit empty sheats that are heavily encrusted. Cells may vary considerably in size: rod-shaped cells average 1.2 to 2.0 um in width by 2.4 to 5.6 um in length; cocoid cells of "conidia" average 0.6 um in diameter. (Cocoid = spherical or ovoid shaped; "conidia" = means dust, spores freely detach from a chain and float in the air-like dust). (*Standard Methods for the Examination of Water and Wastewater*, APHA-AWWA-WPCF, 17th Ed., 1989, p. 9-116)

FIGURE 17-4. Filamentous sulfur bacteria: Thiodendron mucosum and Beggiatoa alba. Thiodendron mucosum. Portion of a colony, with branching mucoid filaments. Individual cells are within the jelly-like material of the filaments. Beggiatoa alba. Filaments are composed of a linear series of individual rod-shaped cells that may be visible when not obscured by light reflecting from sulfur granules. (*Standard Methods for Examination of Water and Wastewater*, APHA-AWWA-WPCF, 17th Ed., 1989, p. 9-120)

Photosynthetic Green and Purple Sulfur Bacteria

Photosynthetic sulfur bacteria grows anaerobically in the light and uses H_2S as a hydrogen donor for photosynthesis. The sulfide is oxidized to sulfur or sulfate.

Aerobic Sulfur Oxidizer Bacteria

Aerobic sulfur oxidizer bacteria oxidizes reduced sulfur compounds aerobically to obtain energy for chemi-autrophic growth.

Sulfur Bacteria Produces Sulfuric Acid

Sulfuric acid produced by sulfur bacteria has contributed to the destruction of concrete sewers and acidic corrosion of metals. (Pictures of some common sulfur bacteria are shown in Figures 17.4. and 17.5).

17.5. Actinomycetes

Actinomycetes are filamentous bacteria, but their morphology resembles that of the filamentous fungi, and some actinomycetes sometimes further resemble molds. Actinomycetes are very common inhabitants of soil, lake, and river mud. Actinomycetes may cause problems in wastewater treatment by developing a thick foam during activated sludge process.

Streptomyces

The best known genus of actinomycetes is Streptomyces, which produces a gaseous compound called geosmin that gives fresh soil its typical musty odor. Traces of this compound can create a disagreeable odor to water and give a muddy flavor to fish.

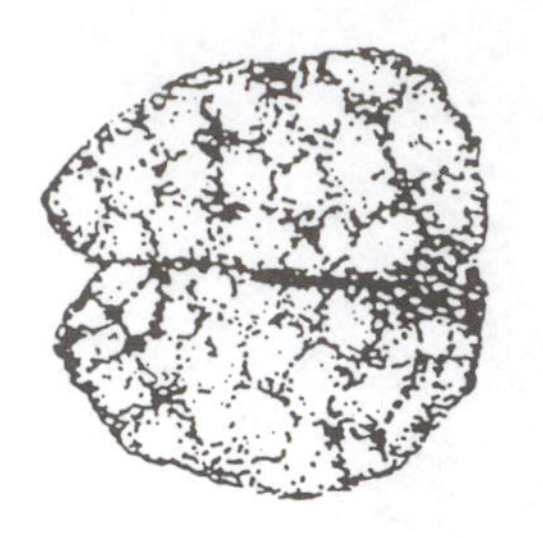

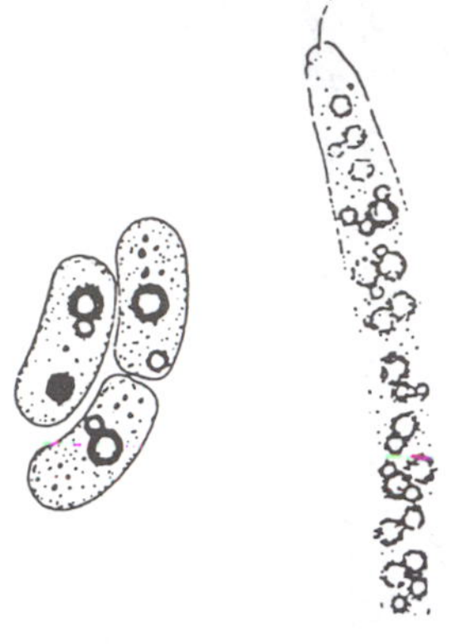

FIGURE 17-5. Nonfilamentous and Photosynthetic sulfur bacteria. Dividing cell of colorless, nonfilamentous sulfur bacteria, *Thiovolum majus*, containing sulfur granules. Cells may measure 9 to 17 um in width by 11 to 18 um in length and are generally found in nature in a marine littoral zone rich in organic matter and hydrogen sulfide. Photosynthetic purple sulfur bacteria. Left: *Chromatium okenii*. Right: *Thiospirillum jenense*. Photosynthetic sulfur bacteria used hydrogen sulfide (H_2S) and produce granules of sulfur (S). These purple and green sulfur bacteria most frequently occur in waters containing hydrogen sulfide (H_2S). Green sulfur bacteria are small, ovid or rod shaped. Purple sulfur bacteria are large, generally stuffed with sulfur globes. (*Standard Methods for Examination of Water and Wastewater*, APHA-AWWA-WPCF, 1989, pp. 9-119, 9-120)

17.6. Fungi

Fungi include yeasts and molds. Yeasts are unicellular and molds are multicellular filamentous organisms. Fungi are heterotrophic organisms, and may be found wherever nonliving organic matter occurs. The association between fungal density and organic matter suggests that fungi may be useful indicators of pollution. For example, certain species use pentose sugars; therefore, this species is a good indicator of pulp and paper mill wastes. Other species live only in warmer waters; therefore, they are indicators of thermal pollution. Some pathogenic fungi live in recreational waters, such as pools and beaches, and are also found in beach sands and shower stalls.

Aspergillus

Most common pathogenic fungi are Aspergillus. They cause pulmonary aspergillosis by breathing contaminated air.

Candida albicans

Candida albicans possibly live in wastewater effluent, and may pollute recreational waters and streams. A large number of females have vaginal candidiasis from contaminated baths and beaches. Vaginal candidiasis causes irritation, severe itching, and a thick, yellow discharge with a yeasty odor.

17.7. Radioactivity

Radioactivity in water and wastewater originates from natural and artificial sources. Fortunately, naturally occurring radioactive contamination of groundwater is not a widespread problem, but in certain areas radioactive elements have adversely affected groundwater. Groundwater usually contains nucleids of the uranium and thorium series.

There are four naturally occurring radium (Ra) isotopes — Ra^{222}, Ra^{224}, Ra^{226}, and Ra^{228}. Radium is one of the decay products of uranium (U^{238}).

Radon-222, daughter product of Ra^{226}, is a radioactive gas and naturally occurs in rocks and soil.

The development of nuclear science and its application to power development, industrial operations, and medical uses require that attention be given to formulating techniques to assess the resultant environmental radioactive contamination.

Measurement techniques are not difficult to devise because radiation counting equipment of high sensitivity, selectivity, and stability is fairly commonplace.

The Maximum Contaminant Levels (MCLs) were established by the EPA in the National Interim Primary Drinking Water Regulations and the International Commission of Radiation Protection (ICRP), with MCLGs of zero for Ra^{226}, Ra^{228}, natural uranium, radon, and gross alpha particle.

Part 5

Environmental Regulations

Chapter 18
Regulations and Standards

18.1. Introduction to Environmental Laws

Pollution is any physical, chemical, or biological alteration of air, water, or land that is harmful to living organisms. Environmental science relates to the physical and chemical changes in the environment through contamination and modification, and to the technology of how to control and correct these changes. Environmental deterioration started with the increasing population and industrialization. In almost every country, the dominant problems are air, water, and soil pollution; decrease in farmable land; the danger in radiation; the growth of the quantity of solid wastes; depletion of energy resources; and the continuously increasing number of toxic pollutants in the environment. It was not until 1968, however, that much of the rest of the world become aware of the tragedy. Over the years, laws have been enacted to protect the environment and human health. Numerous laws were developed, and today the U.S. has the world's most comprehensive and toughest set of environmental laws and regulations.

One of the most significant advances in environmental protection was the National Environmental Policy Act (NEPA) in 1969. NEPA declares the national policy calling on the federal government and describes the impacts in the Environmental Impact Statement (EIS). Drafts of the EIS are available for the public and review by federal agencies before the commencement of the project. NEPA established the Council on Environmental Quality. The Council publishes an Environmental Quality annual report on the environmental protection efforts of the federal government. It also develops new environmental policies and recommends them to the president.

In 1970, the Environmental Protection Agency (EPA) was established. The EPA was directed to accomplish the Federal Water Pollution Control Act and the Clean Air Act. Now EPA manages many of the environmental protection laws that are issued from Congress. Current responsibilities of the EPA include research on the health and environmental impacts of a wide range of pollutants as well as the development and enforcement of environmental standards for pollutants outside of the workplaces. Later, individual states began drafting legislation to regulate environmental problems in their own states.

Other federal agencies include the following:

- Occupational Safety and Health Administration (OSHA) with the responsibilities of research and enforcement of worker safety and health laws.
- National Institute for Occupational Safety and Health (NIOSH) to identify substances that pose potential health problems and recommend exposure levels to OSHA.
- Food and Drug Administration (FDA) for research and enforcement of laws to protect consumers from harmful foods, drugs, and cosmetics.
- Health Services Administration (HSA) with the responsibilities of familial planning and community health programs.
- Health Resources Administration (HRA) for research, planning, and training, and collection of statistics on health in the U.S.
- National Institute of Health (NIH) with the charge of cancer studies through the National Cancer Institute (NCI), studies on radiation and other environmentally related diseases, through the National Institute of Environmental Health Services (NIEHS).
- Centers for Disease Control (CDC) is the leader in epidemiological studies on diseases.
- National Oceanic and Atmospheric Administration conduct research and monitoring of the oceans and atmosphere, and modeling the better impacts of air and water pollution.

The final decision of a long term discussion of the federal government and state position in environmental issues is to develop federal standards while allowing the states to manage their own programs as long as they are at least as stringent as the ones set up by federal law. For example, states have their own programs and management for the Clean Air Act (CAA), Surface Mining Control and Reclamation Act (SMCRA), and the Resource Conservation and Recovery Act (RCRA). These acts also provide money to assist the states in these responsibilities.

18.2. Groundwater Contamination and Protection

Nearly half of the population of the U.S. and 90% of the population in rural areas depend on groundwater as their primary source of water. (Formation and properties of groundwater is in Chapter 3.) The frightening reality is that groundwater is not safe from contamination. Groundwater contamination is a national problem and the sources of potential contaminants are everywhere. Once contaminated, groundwater is difficult and sometimes impossible to clean. Contaminants remain concentrated in slow moving plumes (concentrated contaminant masses), and are typically present for many years. We are all concerned about what contaminated groundwater can do to our health. Among the early Greeks, Hippocrates recognized the danger of contaminated drinking water and recommended that the water be filtered and boiled. Many famous characters of history have fallen victim to waterborne diseases. Among them are King Louis VIII of France, Charles X of Sweden, Prince Albert of England, his son Edward VII, and his grandson George V. George Washington was also known to have suffered from water-caused dysentery, as did Abigail Adams, wife of Zachary Taylor, second president of the U.S. Louis Pasteur's daughters are said to have died from typhoid fever (USDA 1955). Contamination of groundwater by organic and inorganic chemicals, radionuclides, and microorganisms has occurred in every state and is being detected with increasing frequency. Public health concerns also arise because some groundwater contaminants are linked to cancers, liver, kidney and nervous system damages, and birth defects. Concern among medical experts is great because some fear that there is no threshold level for most contaminants, that is, there is no level free from risk of cancer or other problems. Detecting groundwater pollution is expensive and time consuming. Numerous test wells must be drilled to sample water and determine the rate and direction of flow. It is very easy to miss a tiny stream of pollutants that flow through one portion of a large aquifer. For example, liquids that do not readily mix with water may travel along the top or bottom of the aquifer in thin layers that are difficult to detect. Florida's groundwater resource is a clean, relatively inexpensive, and readily available source of drinking water. Florida's unique geology (a thin, permeable soil that covers a high water table), porous limestone formation, high rainfall, and high potential for salt water intrusion make the groundwater resource extremely vulnerable to contamination. Additional detection of contaminated groundwater areas are associated with increased efforts to monitor known problems, locate unknown problems, and monitor potential problems. Unfortunately, corrective and preventive actions will involve millions of dollars or more.

Sources of Groundwater Contamination

According to the Congressional Office of Technology Assessment (OTA), in 1984, the sources of groundwater contamination belonged to six categories:

- Subsurface percolation: e.g., septic tanks and cesspools
  Injection wells: hazardous wastes, non-hazardous wastes
  Land application: irrigation by wastewater, wastewater byproducts (sludge), hazardous wastes, non-hazardous wastes
- Landfills: industrial non-hazardous wastes, industrial hazardous wastes, municipal sanitary landfills
  Open dumps
  Residential waste disposal
  Surface impoundments: hazardous and non-hazardous wastes
  Waste tailings
  Waste piles: hazardous and non-hazardous wastes
  Materials stockpiles
  Graveyards and animal burial
  Aboveground storage tanks: hazardous and non-hazardous waste, non-waste
  Underground storage tanks: hazardous and non-hazardous waste, non-waste
  Containers: hazardous and non-hazardous waste, non-waste
  Open burning and detonation sites
  Radioactive disposal sites
- Pipelines: hazardous and non-hazardous waste, non-waste
  Materials transport and transfer operations: hazardous and non-hazardous waste, non-waste
- Irrigation practices
  Pesticide applications
  Fertilizer applications
  Animal feeding operations
  De-icing salt applications
  Urban runoff
  Percolation of atmospheric pollutants
  Mining and mine drainage: surface mine, underground mine
- Production wells

Oil and gas wells
Geothermal and heat recovery wells
Water supply and other wells (monitoring and exploration wells)
Construction excavation

- Groundwater-surface water interactions
Natural leaching
Saltwater intrusion

Landfills

Landfill is defined as "any land area dedicated or abandoned to the deposit of urban solid waste regardless of how it is operated or whether or not a subsurface excavation is actually involved" (USEPA, 1973). The function of a landfill or residuals repository is to permanently keep the waste from migrating back into the environment (surface water, groundwater, or air). The old system just dropped the wastes into the ground. Landfills contaminate groundwater because they cannot offer an impermeable barrier to the toxic substances. Chemicals in landfills escape their containers, mix with other substances, and can cause dangerous chemical reactions. Rain or runoff water wash out these dangerous toxic substances from the landfill toward the water table. Liners do not give proper protection, animals can dig through it, and the plastic can be dissolved by the wastes, etc. The modern, new technology requires standards and effective work from a landfill. The effectiveness of a system depends on local geology and meteorology to keep hazardous materials from migrating away from the landfill. Geologic features may enhance effectiveness by providing a subsurface water barrier (usually clay) beneath the deposited material. This inhibits the movement of waste materials in the subsurface area for long periods of time. Meteorology determines how much water may enter the landfill prior the final cap being placed on the site. In addition, the RCRA requires certain minimum construction standards, such as double-synthetic liners with double leachate collection systems over an impermeable clay soil layer. Periodic maintenance of fill cover is also recommended. If the cover is seeded by vegetation, it should be done by high transpiration crops, and intensive irrigation is not advised. Landfills can be municipal, sanitary, industrial, or hazardous waste. Regional, newly constructed landfills should replace numerous small dumps that endanger groundwater quality.

What is leachate?

Leachate is contaminated water that is produced when water percolates through wastes from inland deposit sites. The leachate picks up different contaminants that migrate various distances based primarily on the magnitude of the landfill operations, physical, chemical, and biological properties of the contaminants, and hydrogeological and soil conditions around the site. If leachates migrate from the site, they contaminate groundwater and surface waters.

Septic Tank Systems

Septic tank systems that are properly designed, constructed, maintained, and located represent an efficient and economical sewage disposal alternative without threatening ground water resources. However, poor design, improper construction and maintenance, and bad locations have led to groundwater pollution. Septic tank systems have a design life of about 10 to 15 years. The deterioration of these systems may threaten groundwater quality. For septic tanks, the most commonly used degreaser and tank cleaner chemicals are organic solvents (such as trichloroethylene), which results in organic pollution to groundwater. If a system is operating correctly, waste water flows out from the tank through pipes and into the drain field where it will be treated by soil filtration and biological processes. Septage that is pumped from septic tanks, after the tank has reached its capacity, is full of bacteria, viruses, ammonia, and organic nitrogen. Illegal dumping of septage creates serious health hazards. Commercial and industrial septic systems in laundries, paint shops, hardware stores, restaurants, gasoline and service stations, laboratories, beauty shops, etc. create different pollution possibilities.

Cesspools are typically 5 to 6 foot diameter holes several feet below ground surface. The facilities were built to receive raw sewage directly from the house discharge systems. Thus, such systems caused raw sewage to move directly into groundwater areas thereby causing high potential pollution.

Pivies are common in rural areas where indoor water systems are not available. The water introduced to such systems is relatively small, therefore, problems associated with odors and disease-carrying insects are more dominant than groundwater contamination.

Surface Impoundments

Domestic and industrial surface impoundments are often referred to as pits, ponds, lagoons, and basins. Sizes vary from a few square feet to several thousand acres and they serve as disposal or storage of hazardous and nonhazardous wastes. The wastes are liquids or mixed solids-liquids. They are commonly used for treating municipal waste waters, animal feedlots, farms, and many industries. Most of them are unlined and, therefore, allow waste seepage. Impoundments caused hazardous emergencies and contaminated waters when heavy rains or broken dikes allowed overflow of wastes. Additionally, a large percentage of these impoundments are not monitored (no monitoring wells are available to check the effect of the impoundment on ground water quality).

Injection Wells

Industrial wastes are commonly injected directly into the ground. The shape and size vary from a shallow dug pit to a few feet deep wells to force liquid wastes to thousands of feet below the surface. They are very common because of the low cost. The hazard from these wells depends on the toxicity of the waste. Fortunately, many laws now require discharge permits and approved locations. The effects of industrial waste disposal on groundwater dangerously threatens this valuable resource across the world.

Abandoned Wells

Abandonment of oil and gas wells usually results when drilling operations are not successful, or the production of the well becomes low. Abandoned wells are powerful tools to pollute groundwater. The presence of unrecorded abandoned wells in areas subject to flooding introduces surface waters to fresh water aquifers. The presence of improperly abandoned wells and improper plugging of exploration holes are the main source of salt water intrusion into fresh water aquifers. Contamination by oil field brine deteriorates the quality of the groundwater.

Underground Storage Tanks

Underground storage tanks are often used in gasoline service stations, farms, store heating oil for homes, gasoline, and storage of hazardous chemicals and chemical wastes. One major category of these underground storage tanks is petroleum hydrocarbons, both aliphatic (open chain compounds) and aromatic (containing benzene ring). Gasoline that has seeped down to ground water will tend to float on top of the water table. Volatile components escape as vapors. Benzene, toluene, and xylene (BTX) can attach to the soil and some compounds are dissolved in the water. The main concern is caused by benzene because of its high carcinogenic effect. Many other compounds such as toluene, xylene, ethylene dibromide (EDB), and ethylene dichloride are priority pollutants or hazardous substances. We now know that the expected lifetime of a tank is a maximum of 15 to 20 years. Improper installation, loose fittings, corrosion of connecting distribution lines, as well as high velocity pumping can cause contamination problems. Prevention includes the use of fiberglass tanks, coated steel tanks, synthetic liners, and double-walled tanks.

Agricultural Wastes

Agriculture includes a variety of activities and many of them represent potential sources of groundwater contamination. The sources of agricultural pollutants are animal wastes, fertilizers, irrigation residues, and pesticides.

Animal feeding represents a group of contaminants that includes nitrogen compounds, phosphates, chlorides, heavy metals, and bacteria. The production of manure contaminates underlying aquifers with nitrogen compounds, salts, hormones, bacteria, viruses, and parasites. *Fertilizers* containing nitrogen, potassium, and phosphorus can also threaten ground water supply. Nitrogen compounds are the easiest to leach especially in the form of nitrates. Organic fertilizer (manure) is also a large part of nitrate pollution.

Irrigation also presents a source of ground water contamination. Irrigation water usually has high dissolved solids. Salts are not transpired by crops or evaporated from soil. Therefore, it builds up within the soil and leaches into the groundwater. When the soil becomes unproductive from the high salt content, the salt will flush by overirrigation and transfer down to the ground water system.

Chemigation is the process whereby pesticides and fertilizers are directly mixed by irrigation water. They pose a substantial threat to groundwater.

Land application is practically spreading waste sludge and waste waters over farm lands. If properly designed and operated, land application recycles nutrients and waters to the soil and aquifer. It is dangerous when heavy metals, toxic chemicals, nitrogen, and pathogens are in the sludge, or the waste water is treated improperly and the soil type and depth of the groundwater are not considered. Crops that will receive the waste water should be selected based on the nitrogen uptake ability.

Pesticides are mostly organic chemicals that could cause adverse health effects. A detailed discussion of pesticides is in Chapter 16.

Hazardous Waste Sites

Hazardous waste sites represent a variety of forms from waste piles to impoundments to landfills. The major concern with hazardous and toxic waste disposal is their health effects and the possibility of human exposure. The main concern should be emphasized on the uncontrolled hazardous waste sites. The Comprehensive Environmental Response Compensation and Liability Act (CERCLA), called "Superfund", issued in 1980, provides liability, compensation, cleanup, and emergency response for hazardous substances released into the environment. The mission of CERCLA (Superfund) is to clean up hazardous waste disposal mistakes of the past and to cope with emergencies of the present.

Mining

Coal mining can pollute water with coal particles, minerals, bacteria, high sulfate level, and others. Acid mine drainage is produced when iron- and sulfur-containing minerals, such as pyrite (sometimes called fool's gold [FeS]) and marcasite (called white iron pyrite) as part of the coal, contact with oxygen and water and produce sulfuric acid (H_2SO_4). The acidic flow releases dangerous minerals from soil, such as aluminum, arsenic, barium, cadmium, nickel, lead, and mercury. Copper, silver, gold, uranium, zinc, lead, and asbestos minings are even worse contaminants.

Cooling Waters

Cooling water is returned to the same aquifer from which it was originally pumped. The chemical quality of the water is unchanged from its original state, except an increase in temperature. The increased solubility of aquifer materials is due to this rise in temperature, but it is not significant, except in carbonate aquifers. Sequestering agents, such as complex polyphosphate-based chemicals added to the water to inhibit oxidation of iron, can become a source of contamination.

Surface Waters

Surface waters from open bodies (rivers, lakes) can enter aquifers where groundwater levels are lower than surface water level. Surface runoff in urban areas also contains many contaminants that are introduced to groundwater.

Storm Water Runoff

Storm water runoff is rich in inorganic, organic, bacteriological, and biological contents. Storm waters contain nitrates, lead, chromium, other heavy metals, pesticides, and wide variety of organic pollutants. During the winter, road salting is a potential contributor to chloride pollution of groundwater.

Atmospheric Precipitation

Raindrops pick up different air pollutants and become active contaminants when they and water that carries them reach the groundwater table.

Salt Water Intrusion

Fresh groundwater in coastal aquifers is discharged into the ocean. If the need of groundwater increased, the flow of groundwater toward the sea decreased or reversed. Seawater will move toward the aquifer and produce seawater intrusion. In Florida, the problem is related to permeable limestone aquifers, lengthy coastlines, and the desire of most inhabitants to live near coastal beaches. The main contributor to seawater intrusion in coastal aquifers is overpumping. If the pumping of freshwater aquifers reverses the gradient, the freshwater flow ceases and seawater then moves into the aquifer.

Ground Water Protection

Protection and appropriate management of the quality of groundwater resources are significant environmental issues. Development of laws and regulations serve to evaluate the quality of groundwaters, control pollution sources, and design remediation activities. Monitoring programs have been developed to collect and analyze data on groundwater quality, to find the actual source(s) and cause(s) of the pollution, and to compile geologic and hydrologic information. The objectives of the monitoring wells are to determine which pollutants are present and at what concentration and to give information on the distribution of contaminants. States design their own monitoring programs, and by summarizing the results on the water quality, build up their effective groundwater protection program and quality management.

Overview of Some of the Major Groundwater Management Regulations

Clean Water Act (CWA)

The CWA controls the discharge of toxic materials into surface streams. The act regulates the level of pollution by setting discharge limits and water quality standards. The concept of federal discharge permits was incorporated into the National Pollution Elimination System (NPDES). The EPA set up 34 industrial categories covering over 130 toxic pollutants that are discharged into surface waters. Discharges of these substances are required to use the best available technology (BAT) to achieve these limits. Toxic and hazardous waste discharges directly to a receiving body of water are regulated by NPDES permit, whereas discharges acceptable to an industrial or municipal sewer system are allowed without a federal permit. The EPA provides financial assistance to state water programs for the construction of waste water treatment plants, water quality monitoring, and enforcement. The CWA also includes guidelines to protect wetlands from dredge and fill activities.

Safe Drinking Water Act (SDWA)

It was established to protect ground waters and drinking water sources. The EPA established Maximum Contaminant Levels (MCLs) and Maximum Contaminant Level Goals (MCLGs) for each contaminant that may effect human health. SDWA includes over 83 contaminants, grouping as inorganic chemicals, synthetic organic chemicals, and microbiological and radiological contaminants. It also regulates the injection of liquid wastes into underground wells to assure that this disposal method does not damage the quality of groundwater and groundwater aquifers.

Resource Conservation and Recovery Act (RCRA)

The primary concern of RCRA is to protect groundwater supplies by creating a management system for hazardous wastes from the time it is generated until it is treated and disposed. The EPA also issued a list of toxic chemicals as Appendix VIII of its RCRA standards. Wastes containing chemicals on this list may be deemed hazardous wastes. RCRA Subtitle I is related to the Underground Storage Tank Program

Toxic Substances Control Act (TSCA)

The EPA has the authority to control the manufacture of chemicals. TSCA bans the manufacture of polichlorinated biphenyls (PCBs) and also controls the disposal of this chemical substance. 40 C.F.R. Parts 712-799 (C.F.R.= Code of Federal Regulations)

Federal Insecticide, Fungicide and Rodenticide Act (FIFRA)

The act controls the manufacture and use of pesticides, fungicides, and rodenticides (40 C.F.R. Parts 162-180). Example of cancelled registration chemicals include DDT, keptone, and ethylene dibromide (EDB).

Comprehensive Environmental Response, Compensation, and Liability Act (CERCLA) (Superfund)

It is designed to address the problems of cleaning up existing hazardous waste sites. CERCLA provides the EPA with "broad authority for achieving clean-up at hazardous waste sites" and finance the clean-up jointly by industry and the government (Superfund). According to CERCLA, substances which "when released into the environment may present substantial danger to the public health or welfare or the environment..." are hazardous substances. Through this definition, CERCLA establishes a list of substances which, when released in sufficient amounts, must be reported to the EPA.

In 1986, the Superfund Amendments and Reauthorization Act (SARA) was passed, pertaining to carcinogen testing and regulations. Section 121 requires that cleanup at Superfund sites "assures protection of human health and the environment". It provides authority and money to the EPA to act quickly in case of hazardous material spill.

Section 313 of Title III of SARA, which is the Emergency Planning and Right to Know Act, requires industrial and commercial facilities to report annually to the EPA on the type of hazardous substances they handle and any release of such compounds into various media such as air and water. With the approval of the EPA, the state provides the enforcement of a program.

Groundwater Classification

Groundwater classification refers to a comprehensive system to classify waters for different groundwater protection strategies. States can use a variety of protection mechanisms to implement classification systems. Examples are taken from the U.S. EPA "Selected State and Territory Ground Water Classification Systems" May 1985, Office of Ground Water Protection, Washington, DC.

Class GAA: Existing or proposed drinking water use without treatment.

Class GA: May be suitable for public or private drinking water use without treatment. The best usage of Class GA water is as a source of potable water supply. Class GA waters are fresh ground waters found in the saturated zone of unconsolidated deposits and consolidated rock or bedrock.

Class GB: May not be suitable for public or private use as drinking water without treatment since the groundwater is known or presumed to be degraded. High density housing, waste disposal, or industrial sites form the basis for designating the ground water as GB.

Class GC: May be suitable for certain waste disposal practices because land use or hydrogeologic conditions render this groundwater more suitable for receiving permitted discharges than for development as a public or private water supply. Often, these areas have suffered waste disposal practices that have permanently made the groundwater unsuitable for drinking without treatment.

Class GSA: The best usage of Class GSA waters is as a source of potable mineral waters, for conversion to fresh potable waters, or as a raw material for the manufacture of sodium chloride or its derivatives or similar products. Such waters are saline waters found in the saturated zone.

Class GSB: The best usage of class GSB waters is as a receiving water for waste disposal.

18.3. Drinking Water Standards

The correct definition of drinking or potable water is the water delivered to the customer that can be safely used for drinking, cooking, and washing. Potable water must meet the physical, chemical, bacteriological, and radiological quality standards criteria established by the regulatory agencies. Our sources of water supply are being endangered by hundreds of new chemicals and pollutants every year. Our ability to detect contaminants has been improving, and new, modern equipments make it possible to detect contaminants in a very small quantity (ppb or less). Very small amounts can be significant; for example, a microscopic virus can cause serious sickness. Drinking water quality is protected by laws and regulations that must be enforced.

The World Health Organization (WHO) is a specialized agency of the United Nations devoted to public health. WHO issued the "Guidelines for Drinking Water Quality" that "intended for use by countries as a basis for the development of standards which, if properly implemented, will ensure the safety of drinking water supplies. Although the main purpose of these guidelines is to provide a basis for the development of standards, the information given may also be of assistance to countries in developing alternative control procedures where the implementation of drinking water standards is not feasible." In the U.S., the EPA mandated to protect the environment.

To safeguard public drinking water supplies and to protect public health, the U.S. Congress established the Safe Drinking Water Act (SDWA) in 1974. It gave the federal government, through the EPA, power to establish regulatory standards, setting MCLs for drinking water parameters. Each state must adopt drinking water standards at least as strict as the national standard. A state may set stricter standards if it wishes. Each state must be able to carry out adequate monitoring and enforcement requirements. In no case will any exception or variance be granted if there is any risk to the public health. The Health Authority reviews, inspects, collect samples, monitors, and evaluates drinking waters on a continuing bases. The Centers for Disease Control have issued a statement about illnesses linked to polluted drinking water. This situation calls for a great deal of research and action on a continuous basis.

The Drinking Water Standards, established by the EPA, reflect the best scientific and technical judgement available.

The National Drinking Water Advisory Council, state and local agencies, and experts in the field developed the National Interim Primary Drinking Water Regulations (NPDWRs), which is still incomplete.

MCLs are enforceable standards. MCLGs (previously called Recommended Maximum Contaminant Levels [RMCLs]) are not enforceable levels. MCLs are set as close to the MCLGs as possible based on currently available treatment technologies.

In 1979, Trihalomethane compounds were added to the standards. In 1983, categories for primary regulations were developed. They are Volatile Synthetic Organic Chemicals (VOCs), Synthetic Organic Chemicals (SOCs), Inorganic Chemicals, Microbiological Pollutants, Radionuclides, and Disinfectant by-products including Trihalomethanes.

In 1985, MCLs for eight VOCs were published and 83 contaminants were on the list. In 1986, SDWA was renewed and amended. Amendments set mandatory deadlines for the regulation of key contaminants, established treatment technologies, increased enforcement powers, and provided new authorities to promote protection of groundwater resources. In 1988, the first Drinking Water Priority List (DWPL) was finalized and published (Table 18-5 and 18-6). Until 1989, the EPA issued MCLs for 83 contaminants; 25 more were added through 1991. (A list of the 83 contaminants is on Table 18-7.)

Under the revised SDWA, it will be easier for the EPA to ensure that the states take enforcement action swiftly and effectively. Effective enforcement is vital to the success of the amended SDWA.

The most recently finalized regulation is the Phase V Rule, which was published in the Federal Register on July 17, 1992. Although the effective date of the rule is January 17, 1994, the planning and monitoring programs for drinking water quality should begin in 1993. In this rule, 23 compounds are regulated, 5 inorganic and 18 synthetic organic substances (9 pesticides and 3 volatiles). Inorganic compounds (IOCs) should be monitored annually in surface water systems, and every three years in groundwater systems. Cyanide monitoring is required only for vulnerable systems, because the sources of cyanide are industrial processes. VOCs need to be checked quarterly for all systems by four samples. In cases when the result of VOC compounds are greater than 0.5 ppb, the requirement of quarterly checking remains. If no VOCs are detected, the checks are reduced to once every three years. SOCs and pesticides must be monitored for all systems. Four quarterly samples are required every three years unless the system qualifies for decreased monitoring. Collection of confirmatory samples is a state option, and must be collected within 14 days of the original sampling. Systems that have failed the MCLs of these regulated parameters must be treated for acceptance by using the BAT.

Drinking water regulations fall into two basic categories: primary and secondary. Primary standards determine how clean drinking water must be to protect public health. The parameters are health related.

Secondary standards are not health related. They are intended to protect "public welfare" by offering unenforceable guidelines on the taste, odor, or color of drinking water, as well as certain other considerations.

The primary and secondary drinking water regulations as amended by the SDWA in 1974 is on Table 18.1. The next set of standards set for drinking water safety already contained 25 drinking water contaminants with interim MCLs,and the contaminants are divided by categories, as on Table 18.2. Table 18.3. has the list with MCLs and RMCLs as proposed in 1986. Phase V Rules with MCLs and MCLGs and health effects of the contaminants are on Table 18.8 and Table 18.9.

18.4. Surface Waters

Freshwater ecosystems fall into two categories: standing systems, such as lakes and ponds, and flowing systems, such as rivers and streams. Lakes and ponds are more susceptible to pollution because the water is replaced in a slow rate. A complete replacement of a lake's water may take 10 to 100 years or more, and during these years pollutants may build up to toxic levels. In rivers and streams, the flow of the water is fast and easily purges out pollutants. If the pollution is continuous, and distributed uniformly along their banks, the cleaning effect by purging does not work well. Rivers, streams, and lakes contain many organic and inorganic nutrients needed by the plants and animals that live in them. These nutrients in higher concentrations may become pollutants. Organic pollutants can come from feedlots, sewage treatment plants, and some industries (dairy products, meat products, etc.). The increased organic matter will stimulate the growth of the bacterial population. The increased number of bacteria will consume the organic matter, so help clean out pollution. Unfortunately, bacteria use up oxygen also and, therefore, they decrease the dissolved oxygen content of the water. The lack of dissolved oxygen will cause the death of fish and other aquatic organisms, and the aerobic (oxygen requiring) bacteria population will change to anaerobic (non-oxygen requiring) bacteria. Anaerobic bacteria will produce foul-smelling and toxic gases such as methane (CH_4) and hydrogen sulfide (H_2S). This process in rivers and streams occurs more readily in the hot summer months. When the organic pollutants are used up, and no more enter

Table 18-1. Primary Drinking Water Regulations (SDWA 1974)

Parameter	Maximum level
Arsenic	0.05 mg/L
Barium	1.0 mg/L
Cadmium	0.010 mg/L
Chromium	0.05 mg/L
Fluoride	4.0 mg/L
Lead	0.05 mg/L
Mercury	0.002 mg/L
Nitrogen, nitrate	10 mg/L
Selenium	0.01 mg/L
Silver	0.05 mg/L
Turbidity	1–5 NTU
Coliform, total	zero counts/100 ml
Endrin	0.0002 mg/L
Lindane	0.004 mg/L
Methoxychlor	0.1 mg/L
THMs	0.100 mg/L
Toxaphene	0.005 mg/L
2,4 D	0.1 mg/L
2,4,5 TP Silvex	0.01 mg/L
Radium 226 + 228	5pCi/L
Gross Alpha	15pCi/L
Gross Beta	50 pCi/L
Secondary Drinking Water Regulations	
Chloride	250 mg/L
Color	15 C.U.
Copper	1 mg/L
Corrosivity	Non-corrosive
Foaming agents	0.5 mg/L
Iron	0.3 mg/L
Manganese	0.05 mg/L
Odor	3 T.O.N.
pH	6.5-8.5 Units
Sulfate	250 mg/L
TDS	500 mg/L
Zinc	5 mg/L

Note: NTU = Nephelometric Turbidity Unit; pCi/L = picoCurie/Liter (definition is on Table 18-3.2.); 2,4-D = 2,4-Dichlorophenoxyacetic acid; 2,4,5-T = 2,4,5-Trichlorophenoxyacetic acid; C.U. = Color Unit; T.O.N. = Threshold Odor Number.

into the water body, oxygen levels return to normal. Returned oxygen level is replaced by oxygen from air and by oxygen released by plants during photosynthesis. Organic pollutants nourish bacteria and certain inorganic pollutants stimulate the growth of aquatic plants. These pollutants are called nutrients, and include nitrogen as ammonia and nitrate, and phosphorus as phosphates. The origin of these compounds is fertilizers, laundry detergents, and sewage treatment plants. If the levels of these nutritional compounds become high, the growth of plants can be enormous, covering lakes and rivers with thick algae and dense aquatic plants. Excessive plant growth disturbs fishing, swimming, boating, and navigation. When these plants die, they are degraded by aerobic bacteria. The lowered dissolved oxygen content of the water kills aquatic organisms, anaerobic bacteria growth, and produces odorous and toxic gases. So, inorganic pollutants cause the same problems in surface waters as organic contaminants do.

Classification of surface waters is based on their water quality, and on their use. The five main groups of surface waters are as follows:

Table 18-2. Drinking Water Standards (SDWA 1986)

Primary Standards	
	MCLs
Inorganics	
Arsenic	0.05 mg/L
Barium	1.00 mg/L
Cadmium	0.01 mg/L
Chromium	0.05 mg/L
Lead	0.05 mg/L
Mercury	0.002 mg/L
Nitrate-Nitrogen	10.0 mg/L
Selenium	0.01 mg/L
Fluoride	4.00 mg/L
Silver	0.05 mg/L
Turbidity	1-5 NTU
Organics	
Endrine	0.0002 mg/L
Lindane	0.004 mg/L
Methoxychlor	0.10 mg/L
Toxaphene	0.005 mg/L
2,4-D	0.10 mg/L
2,4,5-T (Silvex)	0.01 mg/L
Volatile Organics (VOCs)	
Trichloroethylene	0.005 mg/L
Carbontetrachloride	0.005 mg/L
Vinyl chloride	0.002 mg/L
1,2-Dichloroethane	0.005 mg/L
Benzene	0.005 mg/L
para-Dichlorobenzene	0.075 mg/L
1,1-Dichloroethylene	0.007 mg/L
1,1,1-Trichloroethane	0.20 mg/L
Total trihalomethanes	0.10 mg/L
Microbiology	
Total Coliform bacteria	zero counts/100 ml
Radionuclides	
Radium 226 and 228 (total)	5 pCi/L
Gross alpha particle activity	15 pCi/L
Gross beta particle activity	50 pCi/L
Secondary Standard	
Parameter	Recommended level
Chloride	250 mg/L
Color	15 C.U.
Copper	1.0 mg/L
Corrosivity	non-corrosive
Foaming agent	0.5 mg/L
Hardness, as $CaCO_3$	50 mg/L
Iron	0.3 mg/L
Manganese	0.05 mg/L
Odor	3 T.O.N.
pH	6.5-8.5 pH unit
Sulfate	250 mg/L
Total dissolved solids (TDS)	500 mg/L
Zinc	5.0 mg/L

Note: Recommended levels for these substituents are mainly to provide aesthetic and taste characteristics. Secondary drinking water regulations are not health related. They are intended to protect "public welfare" by offering unenforceable guidelines on the taste, odor, or color of drinking water.

Abbreviations: MCLs = Maximum Contaminant Levels; N.T.U. = Nephelometric Turbidity Unit; PCi/L = picoCurie per liter (definition is on Table 18-3.2.); C.U. = Color Unit; T.O.N. = Threshold Odor Number; 2,4-D = 2,4-Dichlorophenoxyacetic acid; 2,4,5-T = 2,4,5-Trichlorophenoxyacetic acid (Silvex).

Table 18-3. Primary Drinking Water Regulations with MCLs and MCLGs

EPA already has a head start on many of regulatory tasks mandated in the 1986 amendments to the Safe Drinking Water Act (SDWA). Maximum Contaminant Levels (MCLs) and Maximum Contaminant Level Goals (MCLGs) have been proposed for a whole range of drinking water contaminants.

	MCLs mg/L	MCLGs mg/L
Inorganics		
Arsenic	0.05	0.05
Barium	1.0	1.5
Cadmium	0.01	0.005
Chromium	0.05	0.12
Lead	0.05	0.02
Mercury	0.002	0.003
Nitrate-nitrogen	10.0	10.0
Selenium	0.01	0.045
Silver	0.05	—
Fluoride	4.0	4.0
Organics		
Endrin	0.0002	—
Lindane	0.004	0.0002
Methoxychlor	0.10	0.34
Toxaphene	0.005	0
2,4-D	0.10	0.07
2,4,5-T Silvex	0.01	0.052
Volatile organics		
Trichloroethylene	0.005	0
Carbon Tetrachloride	0.005	0
Vinyl Chloride	0.002	0
1,2-Dichloroethane	0.005	0
Benzene	0.005	0
para-Dichlorobenzene	0.075	0.075
1,1-Dichloroethylene	0.007	0.007
1,1,1-Trichloroethane	0.20	0.20
Bacterial		
Coliform bacteria	1-4/100 ml	0
Total trihalomethanes		
Bromoform, Chloroform, Dibromochloromethane Dichlorobromomethane	0.1 mg/L	0
Radionuclides		
Radium 225 & 228 total	5 pCi/L	0
Gross alpha particles activity	15 pCi/L	0
Gross beta particles activity	4 mrem/year	0
Turbidity	1–5 NTU*	0.1 NTU

**MCL for Turbidity depends on sampling technique*

Source: 40 Code of Federal Regulations (C.F.R.), part 141.

Abbreviations: MCLs = Maximum Contaminant Levels; MCLGs = Maximum Contaminant Level Goals; 2,4-D = 2,4-Dichlorophenoxyacetic acid; 2,4,5-TP = 2,4,5-Trichlorophenoxyacetic acid; pCi/L = picoCurie/liter. 1 Ci = 3.7×10^{10} disintegrations/sec. Because the multiplication factor for p (pico) is 10^{-12}, 1 pCi = 3.7×10^{-12} disintegrations/sec; mrem/year = millirem is one unit of radiation dose. rem = Roentgen Equivalent for Men. "rem" is the dosage of an ionizing radiation, that will cause the same biological effects as one roentgen of X-ray or Gamma-ray dosage. mrem is 1000 times less as rem. (A chest X-ray typically involves about 0.007 rem or 7mrem, and the U.S. federal standard for maximum safe occupational exposure is roughly 5000 mrem/year); NTU = Nephelometric Turbidity Unit.

Table 18-4. EPA Monitoring Requirements for Systems Using Groundwater

Monitoring for Regulated Contaminants

Inorganic chemicals

For inorganic chemicals, samples must be taken at three year intervals. If a level over the MCL is found, three additional samples must be taken within a month. The average of these four samples, if over the MCL, constitute a violation. For new systems, samples must be taken quarterly, to establish the appropriate monitoring rate for that particular system.

Organic chemicals

For organic chemicals, the frequency of monitoring is determined by the state, but is in no case less than three-year intervals. As with inorganic chemicals, once an excess is found, three additional samples must be taken within a month, and the average of these four readings determines compliance. For new systems, samples must be taken quarterly, to establish the appropriate monitoring rate for that particular system.

Volatile organic compounds

To establish the sampling rate, four samples must be taken, quarterly, for carbon analysis, to determine the vulnerability of the system to VOCs. (States can reduce subsequent monitoring if no VOCs are found in the first sample.) After this initial sampling, monitoring must be repeated anywhere from quarterly to once every five years, depending on whether VOCs were found in the initial samples and how likely they are to reoccur. Violations are determined by average of quarterly results.

Bacterial contaminants

The monitoring requirements for bacterial contaminants vary depending on the size of the water system; anywhere from 1 to 500 samples are required monthly. Violations of the MCL are calculated in a variety of ways, depending on the number of persons served and the sampling technique.

Total trihalomethanes (THMs)

Total Trihalomethanes must be monitored quarterly at four different sample points in the system. The annual average (or any 12 month average) determines compliance.

Radionuclides

Radionuclides must be monitored every four years, but much stricter requirements should be imposed by the states once any radiation level is found, or if the system is in the vicinity of mining activity. The average of four quarterly samples determines compliance.

Turbidity

Turbidity must be monitored daily. A violation exists if the average of two consecutive samples is in excess of 5 NTU or if the monthly average of daily readings exceeds 1 NTU.

Monitoring for Unregulated Contaminants

List 1. Monitoring Required for All Systems

Monitoring of these contaminants is required at least every five years, though states can monitor more frequently and/or can add contaminants to the list. All these compounds are VOCs.

Bromobenzene	1,1-Dichloroethane
Bromodichloromethane	1,1-Dichloropropene
Bromoform	1,2-Dichloropropane
Bromomethane	1,3-Dichloropropane
Chlorodibromoethane	1,3-Dichloropropene
Chlorobenzene	2,2-Dichloropropane
Chloroethane	Ethylbenzene
Chloroform	Styrene
Chloromethane	1,1,2-Trichloroethane
o-Chlorotoluene	1,1,1,2-Tetrachloroethane
p-Chlorotoluene	1,1,2,2,-Tetrachloroethane

Table 18-4. EPA Monitoring Requirements for Systems Using Groundwater (continued)

Dibromomethane	Tetrachloroethylene
m-Dichlorobenzene	1,2,3-Trichloropropane
o-Dichlorobenzene	Toluene
trans-1,2-Dichloroethylene	p-Xylene
cis-1,2-Dichloroethylene	o-Xylene
Dichloromethane	m-Xylene
List 2. Monitoring Required for Vulnerable Systems	
Ethylene Dibromide (EDB)	
1,2-Dibromo-3-Chloropropane (DBCP)	
List 3. Monitoring Required at the State's Discretion	
Bromochloromethane	
n-Buthylbenzene	
Dichlorodifluoromethane	
Fluorotrichloromethane	
Hexachlorobutadiene	
Isopropylbenzene	
p-Isopropyltoluene	
Naphthalene	
n-Propylbenzene	
sec-Butylbenzene	
tert-Butylbenzene	
1,2,3-Trichlorobenzene	
1,2,4-Trichlorobenzene	
1,2,4-Trimethylbenzene	
1,3,5-Trimethylbenzene	

Sources: 40 C.F.R. Part 141, 52 Federal Register 25710 (July 8, 1987)

Class I Waters: Potable water supplies
Class II Waters: Shellfish propagation or harvesting
Class III Waters: Recreation-propagation and maintenance of healthy, well-balanced population of fish and wildlife
Class IV Waters: Agricultural water supply
Class V Waters: Navigation, utility, and industrial use

Groundwater contamination by flow from surface water is well-known. Surface waters from open bodies (rivers, lakes) can enter into aquifers where groundwater levels are lower than surface water levels. The reverse situation, when groundwater contaminating surface water is also possible, occurs when the water table is higher or the surface water is lowered by pumping wells.

Monitoring, maintaining, and regulating the quality of surface waters is the responsibility of state programs. The Florida surface water program is the Surface Water Improvement and Management (SWIM) program, mandated for protection and restoration of rivers, lakes, bays, or other water bodies of regional or statewide significance in 1988.

18.5. Industrial Waters

Quality requirements for industrial uses vary widely according to potential use. Industrial process waters must be of much higher quality than cooling waters. For example, salt and brackish waters are commonly used as cooling waters (especially if they are used only once). Municipal supplies are generally good enough to satisfy the quality requirements of most process waters, exception of those waters used for boilers. Boiler waters are especially checked and treated for quality. The silica content

Table 18-5. SDWA Drinking Water Priority List (DWPL)

Inorganics	
Aluminum*	Strontium
Ammonia	Vanadium*
Boron	Zinc*
Cyanogen chloride	Chloramine
Molybdenum*	Chlorate
Ozone & byproducts	Chlorine
Silver*	Chlorine dioxide
Sodium*	Chlorite
	Hypochlorite ion
Organics	
Bromobenzene	1,1-Dichloropropene
Bromochloracetonitrile	1,3-Dichloropropene
Bromodichloromethane	2,4-Dinitrotoluene
Bromoform	Ethylenethiourea (ETU)
Bromomethane	Halogenated acids, alcohols, aldehydes, ketones, and other nitriles
Chloroethane	Isophorone
Chloroform	Methyl-tert-butyl ether
Chloromethane	Metolachlor
Chloropicrin	Metribuzin
Cyanazine	o-Chlorotoluene
Dibromoacetonitrile	p-Chlorotoluene
Dibromomethane*	2,4,5-T
Dibromochloromethane	1,1,1,2-Tetrachloroethane
Dicamba	1,1,2,2-Tetrachloroethane
Dichloroacetonitrile	Trichloroacetonitrile
1,1-Dichloroethane	1,2,3-Trichloropropane
1,3-Dichloropropane	Trifularin
2,2-Dichloropropane	
Microbiology	
Cryptosporidium	

**Seven contaminants removed from the list of "83" requirements and placed to the DWPL*
The first Drinking Water Priority List was published in 1988, 53 Federal Register 1901 (Febr.22.1988)

of water is very sensitive. Silica is an important constituent of the incrusting material or scale formed by many waters. As deposited, the scale is commonly calcium or magnesium silicate. Silicate scale cannot be dissolved by acids or other chemicals. Therefore, silica-rich water that is used in boilers must be treated. Sanitary requirements for waters used in processing milk, canned goods, meats, and beverages exceed even those for drinking water. Table 18.12 lists some quality criteria for industrial process waters.

18.6. Irrigation Water Quality

Water quality problems in irrigation include salinity and toxicity. *Salinity* affects crop production, because crop roots have great difficulty extracting enough water and nutrients from saline solution. Consequently, crop production is limited because sufficient water cannot reach the root zone. *Toxicity* is also a problem in maintaining good yield. Boron, chlorides, and sodium are common toxic substances.

Of particular consequence is the *ratio of sodium to calcium and magnesium*. When sodium-rich water is applied to soil, some of the sodium is taken up by clay and the clay gives up calcium and magnesium in exchange. Clay that takes up sodium becomes sticky and slick when wet and has low permeability. Then the dry clay

Table 18-6. Disinfectants and Disinfection By-Products

Disinfectants	Trihalomethanes
Chlorine	Chloroform
Hypochlorite ion	Bromoform
Chlorine dioxide	Bromodichloromethane
Chlorite	Dibromochloromethane
Chlorate	Dichloroiodomethane
Chloramine	
Ammonia	Halogenated acids, alcohols, aldeydes and ketones
Ozone	Monochloroacetic acid
	Dichloroacetic acid
Halonitriles	Trichloroacetic acid
Bromochloroacetonitrile	Chloralhydrate
Dichloroacetonitrile	2,4-Dicholorophenol
Dibromoacetonitrile	
Others	
Chloropicrin	
Cyanogen chloride	
3-chloro-4(dichloromethyl)	
5-hydroxy-2(5H)-furanone (MX)	

The list was the first target for the Drinking Water Priority List.

shrinks into hard clods that are difficult to cultivate. Even worse, *high concentration of sodium salts can produce alkali soils* in which little or no vegetation can grow. On the other hand, when the same clay carries excess calcium or magnesium ions, it tills easily and has good permeability. If an irrigation water contains calcium and magnesium ions sufficient to equal or exceed the sodium ion, enough calcium and magnesium is retained on clay particles to maintain good tilth and permeability. These waters serve well for irrigation. The sodium effect can be calculated by the *Sodium Absorption Ratio (SAR)* method.

$$SAR = Na/\sqrt{(Ca + Mg)/2}$$

Na, Ca and Mg values are expressed in milliequivalent per liter. Water with SAR values below 10 is sufficient for irrigation, and SAR values of 18 or higher are not recommended for irrigation. Table 18.12 contains analytical parameters needed to evaluate water used for irrigation and Table 18.11. contains the trace elements and their recommended maximum concentrations in irrigation waters.

18.7. Waste Characterization

The few characteristic properties that qualify a waste as a RCRA waste material are ignitability, corrosivity, reactivity, and toxicity.

Ignitability refers to the characteristics of being able to sustain combustion and includes the category of flammability (ability to start fires when heated to temperatures less than 60°C or 140°F).

Corrosive wastes may destroy containers, soil, and groundwater, or react with other materials to cause toxic gas emissions. Corrosive materials provide a very specific hazard to human tissue and aquatic life where the pH levels are extreme.

Reactive wastes may be unstable or have a tendency to react, explode or generate pressure during handling. Also, pressure-sensitive or water reactive materials are included in this category.

Table 18-7. List of the 83 Contaminants Regulated by SDWA

Inorganics	**Organics**	**Volatiles**
Aluminum*	Acrylamide	Benzene
Antimony	Adipates	Carbon tetrachloride
Arsenic	Alachlor	p-Dichlorobenzene
Asbestos	Aldicarb	1,2-Dichloroethane
Barium	Atrazine	1,1-Dichloroethane
Beryllium	Carbofuran	Trichlroethene
Cadmium	Chlordane	Vinyl chloride
Chromium	2,4-D	Chlorobenzene
Copper	Dalapan	cis-1,2-Dichloroethene
Cyanide	Dibromochloropropane	trans-1,2-Dichloroethene
Fluoride	Dibromomethane*	Methylene chloride
Lead	1,2-Dichloropropane	Tetrachloroethene
Mercury	Dinoseb	Trichlorobenzene
Molybdenum*	Diquat	
Nickel	Endothall	**Added parameters**
Nitrate	Endrin	Aldicarb Sulfoxide
Selenium	Epichlorohydrin	Aldicarb sulfone
Silver*	Ethylene dibromide (EDB)	Ethylbenzene
Sodium*	Glyphosate	Heptachlor
Sulfate	Hexachlorocyclopentadiene	Heptachlor Epoxide
Thallium	Lindane	Styrene
Vanadium*	Methoxychlor	Nitrite
Zinc*	PAH's	
	PCB's	**Radionuclides**
Turbidity	Pentachlorophenol	Beta particles &
	Phtalates	Photon activity
Microbiology	Pichloram	Gross Alpha Particle
Giardia lamblia	Simazine	Radium 236 & 228
Legionella	2,3,7,8-TCDD (Dioxin)	Radon
Heterotrophic	Toluene	Uranium
Plate Count	Toxaphene	Uranium
Total Coliform	2,4,5-TP (Silvex)	
Viruses	1,2,2-Trichloroethane	
	Vydate	
	Xylene	
	Xylene	

*The * marked items were removed from the list of 83 at 1/22/88 and placed on the Drinking Water Priority List (DWPL).*

Toxicity is a function of the effect of waste materials that may come into contact with water or air and be leached into the groundwater or dispersed in the environment. The toxic effects that may occur to humans, fish, or wildlife are the principal concerns.

Criteria for Hazardous Waste Evaluation

The criteria for hazardous waste evaluation are as follows:

Ignitability: Flashpoint <140°F or 60°C

Corrosivity: pH less than 2.00 or higher than 12.00

Reactivity: Reacts violently or generates pressure. The substance should be free from cyanide (CN) and sulfide (S)

Toxicity: Leaching tests Extraction Procedure (EP) Toxicity and Toxicity Characteristics Leachate Procedure (TCLP) parameters should meet MCL criteria

The list of the analyzed parameters with MCLs are on Tables 18.13 and 18.14.

Table 18-8. Primary Drinking Water Regulations and Monitoring Requirements

Parameter	MCLs mg/L	MCLGs mg/L	Monitoring requirements for system using ground water
Inorganics			
Arsenic	0.05	0.05	For inorganic chemicals, samples must be taken at three year intervals. If a level over the MCL is found, three additional samples must be taken within a month. The average of these four results, if over the MCL, constitutes violation. For new systems, samples must be taken quarterly, to establish the appropriate monitoring rate for that particular system.
Barium	1.0	1.5	
Cadmium	0.01	0.005	
Chromium	0.05	0.12	
Lead	0.02	0.015	
Mercury	0.002	0.003	
Selenium	0.01	0.05	
Silver	0.05	—	
Nitrate-N	10.0	10.0	
Fluoride	4.0	4.0	
Organics			
Endrin	0.0002	—	For organic chemicals, the frequency of monitoring is determined by the state but in no case is it less than three intervals. As with inorgainc chemicals, once an excess is found, three additional samples must be taken within a month, and the average of these four readings determines compliance. Establishing monitoring rate is the same.
Lindane	0.004	0.0002	
Methoxychlor	0.10	0.74	
Toxaphene	0.005	0	
2,4-D	0.10	0.07	
2,4,5-TP Silvex	0.010	0.052	
Volatile Organics (VOCs)			
Trichloroethylene	0.005	0	To establish the sampling rate, systems must take four quarterly carbon samples to determine vulnerability of systems to VOCs. (State can reduce subsequent monitoring if no VOCs are found in in first samples.) After this initial sampling, monitoring must be repeated anywhere from quarterly to once every five years, depending on whether VOCs are found in initial samples and how likely they are to occur. Violations are determined by average of quarterly sampling results.
Carbontetra chloride	0.005	0	
Vinyl chloride	0.002	0	
1,2-Dichloroethylene	0.005	0	
Benzene	0.005	0	
p-Dichlorobenzene	0.075	0	
1,1-Dichloroethylene	0.007	0.007	
1,1,1-Trichloroethane	0.2	0.2	
Total Trihalomethanes (THMs)			
Bromoform	0.10	—	Total Trihalomethanes monitored quarterly at four different sample points in the system. The annual average (or any 12 month average) determines compliance.
Chloroform			
Dibromochloromethane			
Dichlorobromomethane			
Turbidity	1.0 NTU	0.1 NTU	Turbidity monitored daily. A violation exists if the average of two consecutive samples is in excess of 5 NTU or if the monthly average of daily samples exceeds 1 NTU.

Table 18-8. Primary Drinking Water Regulations and Monitoring Requirements (continued)

Parameter	MCLs mg/L	MCLGs mg/L	Monitoring requirements for system using ground water
Bacteriology			
Total coliform count/100 ml	1	zero	The monitoring requirements for bacteriological contamination vary depending on the size of the water system; anywhere from 1 to 500 samples are required monthly. Violations of the MCL are calculated in a variety of ways, depending on the number of persons served and the sampling technique.
Radionucleides			
Radium 226 & 228 (total)	5 pCi/L	zero	Radionuclides must be monitored every four years but much stricter requirements should be imposed by the states once any radiation is found, or if the system is in the vicinity of mining activity. The average of four quarterly samples determines compliance.
Gross alpha particles activity	15 pCi/L	zero	
Gross beta particles activity	4 mrem/year	zero	

Based on 40 C.R.F. Part 141 11/13/85 and F.R. Vol.54, No.97.5/24/89

18.8. Air Pollution and Control

People have known for centuries that air carries "poisons". Coal miners used to take canaries with them into the mine because the death of the bird meant the presence of toxic gases. An important exposure route to hazardous materials is by air, and its effects frequently appear far away from the pollution sources. The atmosphere contains hundreds of air pollutants from natural and anthropogenic sources. These pollutants are called *primary pollutants.*

By using the energy from the sun, they react with one another or with water vapor in the air and produce dangerous new chemical substances called *secondary pollutants.* These reactions are called *photochemical reactions* because they involve sunlight and chemicals, resulting in a brownish-orange shroud of air pollution called *photochemical smog.* Secondary pollutants are ozone, formaldehyde, peroxyacylnitrate (PAN), sulfuric acid, and nitric acid (causes of acidic rains). The health risks are apparent, causing burning or itching eyes and irritated throats as acute health effects, or bronchitis, emphysema and lung cancer as chronic effects.

Air pollution control started in 1955; however, it was the Clean Air Act in 1970 that designed for a real control program, which was amended in 1975 and 1977. There are two broad regulatory classifications of air pollutants: criteria pollutants and non-criteria pollutants. *Criteria Pollutants* (normal air pollutants) are those for which there are established federal ambient air quality standards. They include gases in the form of nitrogen oxides, ozone, sulfur dioxide, and carbon monoxide, and solids in the form of particulate matter and lead (as a particulate).

Non-criteria Pollutants (toxic air contaminants) are those that do not have established federal ambient air standards. Non-criteria pollutants include practically every other compound or element that could have an impact on human health or the environment.

Table 18-9. Drinking Water Contaminants and Related Health Effects

Primary Standards	
Parameters	Related health effects
Inorganics	
Arsenic	Dermal and nervous system toxicity effect
Barium	Circulatory system effects
Cadmium	Kidney effects, hypertension, anemia
Chromium	Liver and kidney effects
Lead	Central and peripheral nervous system damage and kidney effects. Highly toxic for infants and pregnant women
Mercury	Central nervous system disorders, and kidney effects
Selenium	Gastrointestinal effects
Silver	Discoloration of the skin (Argyria)
Nitrate	Methemoglobinemia (Blue-baby syndrome)
Fluoride	Dental caries (if it is not available), Dental fluorosis (brownish discoloration of the teeth) and skeletal damage, if the concentration exceeds MCL
Organics	
Endrin	Nervous system and kidney effects
Lindane	Nervous system and liver effects
Methoxychlor	Nervous system and kidney effects
Toxaphene	Cancer risk
2,4-D	Nervous system and kidney effects
2,4,5-TP	Nervous system and kidney effects
Volatile Organics (VOCs)	
Trichloroethylene	Cancer
Carbon tetrachloride	Cancer
Vinyl chloride	Cancer
1,2-Dichloroethylene	Cancer
Benzene	Cancer
p-Dichlorobenzene	Cancer
1,1-Dichloroethylene	Cancer
1,1,1-Trichloroethane	Cancer
Trihalomethanes	
All the four types	Cancer
Bacteriology	
Total coliforms	Although not necessarily in themselves disease-producing organisms, coliforms can be indicators of organisms that cause assorted gastro-enteric infections, dysentery, hepatitis, typhoid fever, cholera, and other diseases
Radionuclides	
Radium 226 & 228	Bone cancer
Gross alpha particles activity	Cancer
Beta particle and photon radioactivity	Cancer
Unregulated synthetic organic chemicals	
Acrylamide	Cancer
Alachlor	Cancer
Aldicarb, aldicarb sulfoxide, and aldicarb sulfone	Nervous system effects
Dibromochloropropane (DBCP)	Cancer
1,2-Dichloropropane	Liver and kidney effects
Epichlorohydrine	Cancer

Table 18-9. Drinking Water Contaminants and Related Health Effects (continued)

Parameters	Related health effects
Ethyl benzene	Liver and kidney effects
Ethylene dibromide (EDB)	Cancer
Heptachlor	Cancer
Heptachloroepoxide	Cancer
Pentachlorophenol	Liver and kidney effects
Polychlorinated biphenyls (PCBs)	Cancer
Styrene	Liver effects
Toluene	Nervous system and liver effects
Xylene	Nervous system effects
Secondary Standard	
pH	Water should not be too acidic or too basic; must fall between 6.5 and 8.5 pH units
Chloride	Taste, corrosion
Copper	Taste, staining
Foaming agents	Aesthetic effects
Sulfate	Taste and laxative effect
Hardness	Taste. Possible relation between low hardness and cardiovascular diseases. Also an indicator of corrosivity (lead problems). Can damage plumbing and limit effectiveness of soaps and detergents
Zinc	Taste
Color	Aesthetic effects
Corrosivity	Aesthetic (taste, and color) and health related (lead)
Iron	Taste and color
Manganese	Taste and color
Odor	Aesthetic effects
Sodium monitoring	Hypertension

Ambient Air Quality Standard (AAQS)

The ambient air quality standard (AAQS) includes those concentrations of contaminants in the air above which adverse health effects occur. The ambient air standards are listed on Table 18.16.

The sources of air pollution are divided into two groups, namely *mobile sources*, including engines, usually associated with transportation, i.e., automobiles, airplanes, trucks, trains, ships, and *stationary sources* such as pipelines, factories, boilers, storage vessels, storage tanks, etc. These sources are classified as point sources (chimneys, stacks, etc.) and area sources (parking lots, industrial facilities). Pollutions are generally regulated by their sources.

The federal government has primary authority to regulate emissions from mobile sources. Regulations for automobile emission controls have become more stringent as increasingly effective technology has developed. The use of catalytic converters and unleaded gasoline has been a great step in the development of better air quality.

To regulate stationary sources, the EPA sets National Stationary Standards called New Source Performance Standard (NSPS). These are emission standards adopted by the federal government on an industry-specific basis for all new sources of air contaminant-emitting equipment or processes located anywhere in the U.S. These standards are controlled by local authorities under the jurisdiction of the respective state. Inspection and maintenance of vehicles for air emissions are also regulated by state laws.

Table 18-10. Quality Tolerances for Industrial Process Waters

Industry	Turbidity (NTU)	Color (C.U.)	Hardness $CaCo_3$ mg/L	Alkalinity $CaCO_3$ mg/L	Fe + Mn mg/L	Total solids mg/L	Others
Food products							
Baked goods	10	10	*	—	0.2	—	a
Beer	10	—	—	75–150	0.1	500–1000	a,b
Canned goods	10	—	25–75	—	0.2	—	a
Confectionery	—	—	—	—	0.2	100	a
Ice	5	5	—	30–50	0.2	300	a,c
Laundering	—	—	50	—	0.2	—	—
Manufactured products							
Leather	20	10–100	50–135	135	0.4	—	—
Paper	5	5	50	—	0.1	200	d
Paper pulp	15–50	10–20	100–180	—	0.1–1	200–300	e
Plastics	2	2	—	—	0.02	200	—
Textiles, dyeing	5	5–20	20	—	0.25	—	f
Textiles, general	5	20	20	—	0.50	—	—

* *Some hardness is desirable*
a Must conform to standards for potable water
b NaCl is no more than 275 mg/L
c SiO_2 no more than 10 mg/L; Ca and Mg bicarbonates, $Ca(HCO_3)_2$ and $Mg(HCO_3)_2$ are troublesome; sulfates and chlorides of Na, Ca, and Mg each no more than 300 mg/L
d No slime formation
e None corrosive
f Constant composition; residual alumina no more than 0.5 mg/L
Note: Stated values are general averages only; there is much local variance.
(American Society for Testing Materials, 1960; Fair at al., 1971, "Groundwater and Wells" Fletcher G. Driscoll, 2nd Ed., 1987)

Table 18-11. Recommended Maximum Concentrations of Trace Elements in Irrigation Waters

Elements	For waters used continuously on soils (mg/L)	For waters used up to 20 years on fine-textured soils pH 6.0–8.5 (mg/L)
Aluminum (Al)	5.0	20.0
Arsenic (As)	0.1	2.0
Beryllium (Be)	0.1	0.5
Boron (B)	*	2.0
Cadmium (Cd)	0.01	0.05
Chromium (Cr)	0.1	1.0
Cobalt (Co)	0.05	5.0
Copper (Cu)	0.2	5.0
Fluoride (F)	1.0	15.0
Iron (Fe)	5.0	20.0
Lead (Pb)	5.0	10.0
Lithium (Li)	2.5	2.5
Manganese (Mn)	0.2	10.0
Molybdenum (Mo)	0.01	0.05**
Nickel (Ni)	0.2	2.0
Selenium (Se)	0.02	0.02
Vanadium (V)	0.1	1.0
Zinc (Zn)	2.0	10.0

* *No problem when less than 0.75 mg/L. Increasing problem when between 0.75 and 2.0 mg/L. Severe problem when greater than 2.0 mg/L.*
** *For only acid fine-textured soils with relatively high iron oxide content.*
(National Academy of Sciences and National Academy of Engineering, 1972; "Groundwater and Wells" Fletcher G. Driscoll, 2nd Ed. 1987)

Table 18-12. Analytical Values Needed to Evaluate Irrigation Waters

Acidity — Alkalinity as $CaCO_3$
Ammonia-Nitrogen (NH_3-N)
Bicarbonate (HCO_3^-)
Boron (B)
Calcium (Ca)
Carbonate (CO_3^{2-})
Chloride (Cl)
Electrical conductivity
Iron (Fe)*
Lithium (Li)*
Magnesium (Mg)
Nitrate-Nitrogen (NO_3-N)
Phosphate-Phosphorus (PO_4-P)*
Potassium (K)*
Sodium (Na)
Sodium Absorption Ratio (SAR)
Sulphate (SO_4^{2-})

* *marked parameters are determined only in special situations*
Calculation of the Sodium Absorption Ration:

$$SAR = Na/\sqrt{(Ca + Mg)/2}$$

The concentration of Na, Ca, and Mg is in milliequivalent per liter (me/L).

Table 18-13. Maximum Concentration of Contaminants in Characterization of EP Toxicity

Contaminant	Maximum concentration (mg/L)
Arsenic	5.0
Barium	100.0
Cadmium	1.0
Chromium	5.0
Lead	5.0
Mercury	0.2
Selenium	1.0
Silver	5.0
Endrin	0.02
Lindane	0.4
Methoxychlor	10.0
Toxaphene	0.5
2,4-D	10.0
2,4,5-TP Silvex	1.0

2,4-D = 2,4-Dichlorophenoxyacetic acid; 2,4,5-TP = 2,4,5-Trichlorophenoxyacetic acid; EP Toxicity = Extraction Procedure Toxicity.
The Extraction Procedure Toxicity test was developed to characterize hazardous wastes based on the leaching ability of toxic substances in significant concentrations. A liquid extract (leachate) of the material is analyzed for the above 14 parameters. During the migration of the leachate, attenuation and dilution occur with a ratio factor of 100. This is used to establish the maximum concentration levels.
(U.S. Environmental Protection Agency "Extraction Procedure Toxicity Characteristics" C.F.R., Vol. 40, No. 261.24, May 19, 1980.)

The method by which states are to bring local ambient air quality into compliance with federal standards is the State Implementation Plan (SIP). The SIP is a document in which each state details to the federal government how, in what manner, and by which regulations the state will attain the air quality goals. The local air pollution control and air quality management districts operate under the authority of the different states.

The Clean Air Act also requires the EPA to establish National Emissions Standards for Hazardous Pollutants (NESHAP) to control the emission of substances so toxic

Table 18-14. Toxicity Characteristic Leachate Contaminants and Regulatory Levels

Contaminant	Regulatory level (mg/L)	Contaminant	Regulatory level (mg/L)
Acrylonitrile	5.0	Isobutanol	36.0
Arsenic	5.0	Lead	5.0
Barium	100.0	Lindane	0.06
Benzene	0.07	Mercury	0.2
Bis(2-chloroethyl)ether	0.05	Methoxychlor	1.4
Cadmium	1.0	Methylene chloride	6.6
Carbon disulfide	14.4	Methyl ethyl ketone	7.2
Carbon tetrachloride	0.07	Nitrobenzene	0.13
Chlordane	0.03	Pentachlorophenol	3.6
Chlorobenzene	1.4	Phenol	14.4
Chloroform	0.07	Pyridine	5.0
Chromium	5.0	Selenium	1.0
o-Cresol*	10.0	Silver	5.0
m-Cresol*	10.0	1,1,1,2-Tetrachloroethane	10.0
p-Cresol*	10.0	1,1,2,2-Tetrachloroethane	1.3
2,4-D	1.4	Tetrachloroethylene	0.1
1,2-Dichlorobenzene	4.3	2,3,4,6-Tetrachlorophenol	1.5
1,4-Dichlorobenzene	10.8	Toluene	14.4
1,2-Dichloroethane	0.40	Toxaphene	0.07
1,1-Dichloroethylene	0.1	1,1,1-Trichloroethane	30.0
2,4-Dinitrotoluene	0.13	1,1,2-Trichloroethane	1.2
Endrin	0.003	Trichloroethylene	0.07
Heptachlor (and its hydroxide)	0.001	2,4,5-Trichlorophenol	5.8
Hexachlorobenzene	0.13	2,4,6-Trichlorophenol	0.30
Hexachlorobutadiene	0.72	2,4,5-TP (Silvex)	0.14
Hexachloroethane	4.3	Vinyl chloride	0.05

** o (ortho), m (meta) and p (para)-Cresol concentrations are added together and compared to a threshold of 10.0 mg/L.*

In 1986, the EPA expanded the EP Toxicity characteristic substances (containing 8 metals, 4 insecticides, and 2 herbicides) with the additional 38 organic substances. The new procedure is called Toxicity Characteristic Leachate Procedure (TCLP) test. By the application of the TCLP test, the extract or leachate of the waste containing any of these 52 substances at or above the regulatory level qualified has hazardous, toxic waste. Also, the TCLP test used compound-specific dilution/attenuation factors instead of the 100 in EP Toxicity characterization. The listed parameters with the regulatory levels are in the U.S. Environmental Protection Agency, "Hazardous Waste Management System", C.F.R, Vol. 51, No. 114, June 13, 1986. The TCLP procedure is outlined in C.F.R. Vol.40, No 261.24, May 19, 1980.

that even small amounts may cause health problems. The EPA established NESHAPS for asbestos, beryllium, mercury, vinyl chloride, benzene, and arsenic.

Permit Systems

There are two types of permits to operate equipment:

- Permits for equipment that may *cause* air pollution
- Permits for equipment designed to *control* air pollution

Permits are required for new or modified facilities before construction begins. The permits are the primary tools to ensure that businesses follow air pollution control laws.

Federal Air Toxic Agencies and Their Rules

Environmental Protection Agency (EPA)

- Regulation of ambient air toxic
- Setting quality standards

Table 18-15. EPA Priority Toxic Pollutants

In 1976, EPA was blamed for failure to implement certain portions of the Federal Water Pollution Control Act. A "Consent Decree" demanded that the EPA study 65 compounds and classes of compounds for the development of regulations to control discharges into wastewater, using the best available technology (BAT) economically achievable. These compounds are referred to as Priority Pollutants. From the 129 Priority pollutants, 114 are organic and 15 are inorganic.

I. Alphabetical listing

1. *Acenaphtene
2. *Acrolein
3. *Acrylonitrile
4. *Benzene
5. *Benzidine
6. *Carbon tetrachloride (tetrachloromethane)

*Chlorinated benzenes (other than dichlorobenzenes)

7. Chlorobenzene
8. 1,2,4-trichlorobenzene
9. Hexachlorobenzene

*Chlorinated ethanes

10. 1,2-Dichloroethane
11. 1,1,1-Trichloroethane
12. Hexachloroethane
13. 1,1-Dichloroethane
14. 1,1,2-Trichloroethane
15. 1,1,2,2,-Tetrachloroethane
16. Chloroethane

*Chloroalkylethers (chloromethyl, chloroehtyl and mixed ethers)

17. bis(Chloromethyl) ether
18. bis(2-Chloroethyl) ether
19. 2-Chloroethyl-vinyl ether (mixed)

*Chlorinated naphthalene

20. 2-Chloronaphthalene

*Chlorinated phenols (other than those listed elsewhere; includes trichlorophenols and chlorinated cresols)

21. 2,4,6-Trichlorophenol
22. Parachloro meta-cresol
23. *Chloroform (Trichloromethane)
24. *2-Chlorophenol
25. 1,2-Dichlorobenzene
26. 1,3-Dichlorobenzene
27. 1,4-Dichlorobenzene

*Dichlorobenzidine

28. 3,3-Dichlorobenzidine

*Dichloroethylenes

29. 1,1-Dichloroethylene
30. 1,2-trans-Dichloroethylene
31. *2,4-Dichlorophenol

*Dichloropropane and Dichloropropene

32. 1,2-Dichloropropane
33. 1,3-Dichloropropylene (1,3-Dichloropropene)
34. *2,4-Dimethylphenol

*Dinitrotoluene

35. 2,4-Dinitrotoluene
36. *2,6-Dinitrotoluene
37. *1,2-Diphenylhydrazine
38. *Ethylbenzene
39. *Fluoranthene

*Haloethers (other than those listed elsewhere)

40. 4-Chlorophenyl phenyl ether
41. 4-Bromophenyl phenyl ether

42. bis(2-Chloroisopropyl) ether
43. bis(2-Chloroethoxy) methane
    *Halomethanes (other than those listed elsewhere)
44. Methylene chloride (Dichloromethane)
45. Methyl chloride (Chloromethane)
46. Methyl bromide (Bromomethane)
47. Bromoform (Tribromomethane)
48. Dichlorobromomethane
49. Trichlorofluoromethane
50. Dichlorodifluoromethane
51. Chlorodibromomethane
52. *Hexachlorobutadiene
53. *Hexachlorocyclopentadiene
54. *Isophorone
55. *Naphthalene
56. *Nitrobenzene
    *Nitrophenols
57. 2-Nitrophenol
58. 4-Nitrophenol
59. 2,4-Dinitrophenol
60. 4,6-Dinitro-o-cresol
    *Nitrosamines
61. N-Nitrosodimethylamine
62. N-Nitrosodiphenylamine
63. N-Nitrosodi-n-propylamine
64. *Pentachlorophenol
65. *Phenol
    *Phthalate esters
66. bis(2-ethylhexyle) phthalate
67. Butyl benzyl phthalate
68. di-n-Butyl phthalate
69. di-n-Octyl phthalate
70. Diethyl phthalate
71. Dimethyl phthalate
    *Polynuclear Aromatic Hydrocarbons
72. Benzo(a)anthracene (1,2-Benzanthracene)
73. Benzo(a)pyrene (3,4-Benzopyrene)
74. 3,4-Benzofluroanthene
75. Benzo(k)fluoranthene (11,12-Benzofluoranthene)
76. Chrysene
77. Acenaphtylene
78. Anthracene
79. Benzo(ghi)perylene (1,12-Benzoperylene)
80. Fluorene
81. Phenanthrene
82. 1,2,5,6-Dibenanthracene
83. Indeno(1,2,3-cd) pyrene (2,3-o-phenylenepyrene)
84. Pyrene
85. *Tetrachloroethylene
86. *Toluene
87. *Trichloroethylene
88. *Vinyl chloride (Chloroethylene)
    Pesticides and metabolites
89. *Aldrin
90. *Dieldrin
91. *Chlordane
    *DDT and metabolites
92. 4,4-DDT
93. 4,4-DDE (p,p-DDX)
94. 4,4-DDD (p,p-TDE)
    *Endosulfan and metabolites

95. Alpha endosulfan
96. Beta endosulfan
97. Endosulfan sulfate
    *Endrin and metabolites
98. Endrin
99. Endrin aldehyde
    *Heptachlor and metabolites
100. Heptachlor
101. Heptachlor epoxide
    *Hexachlorocyclohexane (all isomers)
102. Alpha BHC (Benzenehexachloride)
103. Beta BHC
104. Gamma BHC
105. Delta BHC
    *Polychlorinated biphenyls (PCB's)
106. PCB-1016 (Arochlor 1016)
107. PCB-1221 (Arochlor 1221)
108. PCB-1232 (Arochlor 1232)
109. PCB-1242 (Arochlor 1242)
110. PCB-1248 (Arochlor 1248)
111. PCB-1254 (Arochlor 1254)
112. PCB-1260 (Arochlor 1260)
113. *Toxaphene
114. *Antimony (total)
115. *Arsenic (total)
116. *Asbestos (fibrous)
117. *Beryllium (total)
118. *Cadmium (total)
119. *Chromium (total)
120. *Copper (total)
121. *Cyanide (total)
122. *Lead (total
123. *Mercury (total)
124. *Nickel (total)
125. *Selenium (total)
126. *Silver (total)
127. *Thallium (total)
128. *Zinc (total)
129 **2,3,7,8-Tetrachlorodibenzo-p-dioxin (TCDD)

** Specific compounds and chemical classes listed in the Natural Resources Defense Council (NRDC) consent decree and referenced in the Clean Water Act.*
***This compound was specifically listed in the consent decree; however, due to its extreme toxicity, the EPA recommends that laboratories not acquire an analytical standard for this compound.*
(Fletcher G. Driscoll, "Groundwater and Wells"
Second Edition, 1986, pp. 1067-68)

II. List by classification

Halogenated methanes

46. Methyl bromide
45. Methyl chloride
44. Methylene chloride (Dichloromethane)
47. Bromoform (Tribromomethane)
23. Chloroform (Trichloromethane)
48. Bromodichloromethane
51. Chlorodibromomethane
6. Carbon tetrachloride (Tetrachloromethane)

Chlorinated hydrocarbons

16. Chloroethane (Ethyl chloride)
88. Chloroethylene (Vinyl chloride)
10. 1,2-Dichloroethane (Ethylene dichloride)
13. 1,1-Dichloroethane

30. 1,2-trans-Dichloroethylene
29. 1,1-Dichloroethylene (Vinylidene chloride)
14. 1,1,2-Trichloroethane
11. 1,1,1-Trichloroethane
87. Trichloroethylene
85. Tetrachloroethylene
15. 1,1,2,2,-Tetrachloroethane
12. Hexachloroethane
32. 1,2-Dichloropropane
33. 1,3-Dichloropropylene
52. Hexachlorobutadiene
53. Hexachlorocyclopentadiene

Chloroalkyl ethers

18. bis(2-Chloroethyl) ether
42. bis(2-Chloroisopropyl) ether
19. 2-Chloroethylvinyl ether
43. bis(2-Chloroethoxy) methane

Haloaryl ethers

40. 4-Chlorophenyl phenyl ether
41. 4-bromophenyl phenyl ether

Nitrosamines

61. N-Nitrosodimethyl amine
62. N-Nitrosodiphenyl amine
63. N-Nitrosodi-n-propyl amine

Nitroaromatics

56. Nitrobenzene
35. 2,4-Dinitrotoluene
36. 2,6-Dinitrotoluene

Phenols

65. Phenols
34. 2,4-Dimethylphenol

Nitrophenols

57. 2-Nitrophenol
58. 4-Nitrophenol
59. 2,4-Dinitrophenol
60. 4,6-Dinitro-o-cresol

Chlorophenols

24. 2-Chlorophenol
22. 4-Chloro-m-cresol
31. 2,4-Dichlorophenol
21. 2,4,6-Trichlorophenol
64. Pentachlorophenol
114. 2,3,7,8-Tetrachlorodibenzo-p-dioxin (TCDD)

Benzidines, Hydrazine

5. Benzidine
28. 3,3-Dichlorobenzidine
37. 1,2-Diphenylhydrazine

Phthalate Esters

66. bis(2-Ethylhexyl) phthalate
67. Butylbenzyl phthalate
68. Di-n-butyl phthalate
69. Di-n-octyl phthalate
70. Diethyl phthalate
71. Dimethyl phthalate

Aromatics

4. Benzene
86. Toluene
38. Ethylbenzene

Polyaromatics

55. Naphthalene
1. Acenaphthene

77. Acenaphthylene
78. Anthracene
72. Benzo(a)anthracene (1,2-Benzanthracene)
73. Benzo(a)pyrene (3,4-Benzopyrene)
74. 3,4-Benzofluoranthene
75. Benzo(k)fluoranthene (11,12-Benzofluoranthene)
79. Benzo(ghi)perylene (1,12-Benzoperylene)
76. Chrysene
82. Dibenzo(a,h)anthracene (1,2,5,6-Dibenzanthracene)
80. Fluorene
39. Fluoranthene
83. Indeno(1,2,3-cd)pyrene (2,3-o-phenylene pyrene)
81. Phenanthrene
84. Pyrene

Chloroaromatics

7. Chlorobenzene
25. o-Dichlorobenzene
27. p-Dichlorobenzene
26. m-Dichlorobenzene
8. 1,2,4-Trichlorobenzene
9. Hexachlorobenzene
20. 2-chloronaphthalene

Polychlorinated Biphenyls (PCBs)

106–112. Seven listed

Pesticides

89. Aldrin
90. Dieldrin
91. Chlordane
95. alpha-Endosulfane
98. Endrin
99. Endrin aldehyde
100. Heptachlor
101. Heptachlor epoxide
102. alpha-BHC
103. beta-BHC
104. gamma-BHC
105. delta-BHC
92. 4,4-DDT
93. 4,4-DDE (p,p-DDX)
94. 4,4-DDD (p,p-TDE)
113. Toxaphene

Miscellaneous

2. Acrolein
3. Acrylonitrile
54. Isophorone
116. Asbestos
121. Cyanide

Metals

114. Antimony
115. Arsenic
117. Beryllium
118. Cadmium
119. Chromium
120. Copper
122. Lead
123. Mercury
124. Nickel
125. Selenium
126. Silver
127. Thallium
128. Zinc

Table 18-16. Ambient Air Quality Standards

Contaminant	Primary (federal) standard
Ozone	0.12 ppm
Carbon monoxide	35 ppm
	9 ppm (8 h)
Nitrogen dioxide	0.05 ppm (annual)
Sulfur dioxide	0.14 ppm (24 h)
	0.03 ppm (annual)
Particulate matter	150 ug/m^3 (24 h)
	50 ug/m^3 (annual)
Sulfates	25 ug/m^3 (24 h)*
Lead	1.5 ug/m^3 (3 months)
Hydrogen sulfide	0.03 ppm*
Vinyl chloride	0.010 ppm (24 h)*

Standard values are for one hour averages unless otherwise indicated.
** Standard values are California Ambient Air Quality Standards*
(Roger D. Griffin, "Principles of Hazardous Materials Management" Second printing, 1989, p. 44.)

- Establishing emission standards
- Enforcing Clean Air Act Section 112 (includes all air toxic studies and programs)
- Setting NESHAPs

Occupational Safety and Health Administration (OSHA)

- Deals with exposure of individuals to potential toxic air contaminants while on the job
- Operates under Department of Labor
- Regulates exposure in work places by using Permissible Exposure Levels (PELs) as its quality standards

Regulations for Air Toxicity

The EPA regulations specifically for air toxicity comprise the NESHAPS. These are required under authority of Section 112 of the Clean Air Act. Other Federal Regulations impacting air toxicity, such as Resource Conservation and Recovery Act (RCRA) and Toxic Substances Control Act (TSCA) are not too concerned with regulating air emissions. There are two exceptions: (1) Emissions from PCB incinerator. According to this regulation PCBs destruction standard is 99.9999% (all the PCBs going to incinerators, no more than 1 g in a million may be emitted). (2) Requirements for hazardous waste incinerators. These requirements correspond to a 99.99% destruction and removal efficiency for Principal Organic Hazardous Constituents (POHCs), 99% removal of hydrochloric acid (HCl) and the particulate emission not exceeding 180 mg/m^3.

18.9. Overview of Federal Regulations

Water

Clean Water Act (CWA)

The CWA is the current program of water pollution control. It is based upon the Federal Water Pollution Control Act Amendments of 1972. The major objective of the CWA is to restore and maintain the "chemical, physical, and biological integrity of the nation's waters".

- Controls the discharge of toxic materials into surface streams.
  The Act requires each state to set water quality standards for every surface water within its borders. To set these standards, states specify the uses of each surface water and restrict pollution to levels that permit those uses.
- Requirements for Wastewater Treatment Plant Operations
  The Act requires all municipal sewage systems to provide secondary treatment of the wastewater (biochemical process) before it is discharged. Financial programs provide grants under the CWA to the states.
- National Pollutant Discharge Elimination System (NPDES)
  To ensure that communities meet treatment requirements, sewage facilities must secure permits under the NPDES. Industries discharging waste into waterways or sewage treatment facilities require standards established by the EPA for their wastes. The NPDES permits are based on these specified effluent limitations. The EPA has 34 industrial categories covering about 130 toxic pollutants regulated for discharge into surface waters. Wastes from industries using public sewage systems must meet *pretreatment standards* designed for the industry.
- Water Quality Management Program
  Funding to the states for planning their control strategies on surface water quality. (See SWIM Program, 18.3.)
- Dredge and Fill Permit System, Wetland Protection
  Under section 404 of the Clean Water Act, the EPA and the U.S. Army Corps of Engineering are jointly responsible for protecting waters against degradation and destruction caused by disposal of dredged spoils or fills. Permits to carry out dredge and fill activities in wetlands areas are granted by the Corps of Engineers. Wetlands are vital elements of natural water systems because they control floods, provide natural filtration system for pollutants, and provide a habitat for fish and wildlife. The Office of Wetland Protection established by the EPA provides leadership in this project.
- Marine Protection, Research, and Sanctuary Act
  Protect the oceans from indiscriminate dumping of wastes.

Safe Drinking Water Act (SDWA)

The SDWA is designed to protect drinking water sources, groundwater, and surface waters. Under this act, the EPA established national standards for drinking water from both groundwater and surface water sources.

- Primary drinking water standards with MCLs and MCLGs
  The primary drinking water standards designed to protect human health. This law also required the EPA to establish MCLs and MCLGs at a level lower than MCLs for each contaminant that may have an adverse health effect on water consumers. (More details are in sections 18.2. and 18.3.)
- Protection of aquifers against contamination from the disposal of wastes by injection into deep wells
  These regulations assure that this disposal method does not damage the groundwater quality. States manage their own underground injection control programs with their regulatory systems.
- Groundwater protection strategy: Wellhead Protection Program and Sole Source Aquifer Demonstration Program
  The 1968 amendments to the Safe Drinking Water Act contain two programs

for groundwater assessment and protection. The Wellhead Protection Program is state developed and administered by the states. The EPA provides technical guidance and funding assistance. This program protects wells that supply public water systems. The goal of the Sole Source Aquifer Protection Program is to prevent contamination of aquifers designated as a sole source of water supply for an area or a community.

Hazardous Substances

Toxic Substance Control Act (TSCA)

It was designed as a catch-all regulation, to use laws that were related to environmental protection. It gives the EPA broad authority to regulate chemical substances that are dangerous to human health or the environment. Eight categories of chemical products are released from TSCA because they are regulated by other laws: pesticides, tobacco, nuclear materials, firearms and ammunition, food, food additives, drugs, and cosmetics.

- Premanufacture Notification Process
  Under TSCA, manufacturers are required to notify the EPA at least 90 days before producing or importing a new chemical substance. The new substance must be tested for different toxic effects and cancer causing potential. If the chemical is harmful, it must be regulated or the production of the chemical must be stopped.
- Chemical Inventory
  TSCA requires the EPA to develop and keep a current comprehensive inventory of commercially used chemicals. It contains all of the information, properties, health effects, and environmental risks of the substance. This inventory is called Material Safety Data Sheets (MSDS) and should be available in each workplace where chemicals are used in daily jobs.
- Good Laboratory Practices
  The EPA is required to establish approved standard methods for environmental testing laboratories, prepare guidelines for laboratories on how they can prove that their reported values are correct and defendable, and how to document all of the activities involved in the generation of analytical data and the final reports (laboratory QA/QC).
- Banned further production of toxic substances (Polychlorinated Biphenyls [PCBs], asbestos, dioxin)
  Some toxic substances require special attention because they are so widespread in the environment or because they pose serious health threats even in extremely low levels of contamination. Polychlorinated biphenyls (PCBs) were found to cause adverse reproductive effects, skin lesions, and tumors. When asbestos is inhaled, it causes lung cancer and mesothelioma (cancer of the membrane lining the chest and abdomen). Asbestos was frequently used in buildings as a fire retardant and insulator. The EPA's school asbestos rule, issued in May of 1982, required elementary and secondary school administrators to have their buildings inspected for asbestos.

Resource Conservation and Recovery Act (RCRA) and its Amendment Hazardous and Solid Waste Act (HSWA)

These acts deal with management of solid wastes with emphasis on hazardous wastes. The primary concern of RCRA is to protect groundwater supplies by creating regulations on treatment, storage, and disposal of hazardous wastes. The goal of the RCRA program is to regulate all aspects of the management hazardous wastes from production until treatment and disposal. These wastes include toxic substances, caustics, pesticides, and other flammable, corrosive, and explosive materials. Another major goal of the RCRA support states is to develop programs for non-hazardous solid waste management.

RCRA regulations include:

- Identifying and classifying hazardous waste
- Issuance of publications for generators about activities and developing standards for generators
- Developing standards for transporting hazardous wastes
- Developing standards for treatment, storage, and disposal
- Making enforcement standards available within the program

Chemical and physical properties characteristic to Hazardous Wastes are listed in Section 18.6.

Underground Storage Tanks

Leaking underground storage tanks containing petroleum products and other hazardous substances have been identified as a major source of groundwater contamination. (See Section 18.2.) Because of the national importance of this problem, Congress amended RCRA in 1984 and gave the responsibility to the EPA to develop programs and regulations for underground storage tanks. In 1986, further amendments of RCRA provided federal funds to clean up petroleum residues originated from leaking underground storage tanks. The fund was financed by tax on motor fuels.

Comprehensive Environmental Response, Compensation and Liability Act (CERCLA) (Superfund)

The Act is concerned primarily with funding cleanups of past activities and present spills. In 1983, CERCLA published an initial list of hazardous substances. In 1987, 191 substance were identified by the EPA as potential carcinogens.

There are four key concepts in CERCLA: (1) removal action, (2) remedial action, (3) funding, and (4) emergency planning.

1. *Removal action* is a short term action after one incident. The goal is to stabilize and clean up the site, and stop further release of hazardous substances.
2. *Remedial action* is a long term clean-up action at a site. The EPA established the National Priority List (NPL) for the sites that needed remedial actions based on site inspection and investigation. Remedial action is conducted under the guidance of the EPA, Army Corps of Engineering, or the state alone. Key changes to CERCLA under the Superfund Amendments and Reauthorization Act of 1986 (SARA) require that clean-up of Superfund sites to "assure protection of human health and the environment" should enrich the goal of the laws of other federal, state, and local agencies, as MCLs and MCLGs. Several sections of SARA pertain to carcinogen testing and regulations also.
3. *Funding.* Generators that contribute waste to the site are charged remedial actions with funding the activities. The funding process collects money also from taxes of petroleum and listed chemicals, from the environmental tax of corporations and from tax on hazardous waste production. The state has to contribute at least 10% of the total cost of the clean-up action.

4. Title III of SARA, which is the Emergency Planning and Community Right-to-Know Act deals with *emergency planning, emergency notification, toxic chemical release reporting, and the Community Right-to-Know demands.* It urges companies to report annually on the amounts of certain substances used and discharged into the environment. The Community Right-to-Know Act provides information to the public about the presence of hazardous chemicals in their communities.

One of the major information sources are the Material Safety Data Sheets (MSDSs) for every hazardous substance. It contains all the physical and chemical properties, the physical and health hazard and reactivity of the substance, and all of the precautions necessary for safe handling and use of the chemical.

Occupational and Safety Act (OSHA)
The reason for the act is personal protection. It was organized to give standards of allowable level to toxic chemicals for workers exposed to the hazard for 40 hours per week. OSHA issued Permissible Exposure Limits (PEL) based on exposure for a healthy 70 kg male. It also stipulates that on every work site, MSDSs should be available for complete information on every chemical substance associated with the work.

Federal Insecticide, Fungicide, and Rodenticide Act (FIFRA)
FIFRA was amended in 1972. Under this act, the EPA must control the manufacture and use of pesticides. According to FIFRA's regulation, manufacturers must register all new pesticides with the EPA, and those manufactured before 1972 must be reexamined to ensure that they meet the current safety standards. The EPA now requires carcinogenicity testing in two species for all food use pesticides (40 CFR 158). Canceled registration chemicals include DDT, Kepone, Ethylene Dibromide, and PCBs.

18.10. Overview of State Programs

Groundwater program development — Every state has received a groundwater strategy grant from the EPA for groundwater program development. Responsibilities of states include:

- Develop state groundwater protection strategy
- Coordinate the responsibilities of various state agencies involved in groundwater protection.
- Create a regulatory and legislative framework for groundwater protection programs.
- Develop technical activities designed to map and characterize groundwater resources, develop groundwater classification systems, and water quality standards.
- Develop data to compile and distribute groundwater information.
- Control non-point pollution sources.

Local land use control — Control the locations of industrial facilities and incorporate land use to groundwater protection.

State superfund program — Clean up hazardous waste sites.

Regulation on solid waste disposal sites — States are the primary regulators of solid waste disposal sites.

Regulations for commercial and industrial facilities — State controls discharge permits, storage and management of hazardous materials, recycling programs, and cleanup plans.

Waste disposal wells —Each state is responsible for an "in state" Underground Injection Control (UIC) program. The program includes waste characterization, monitoring wells, location of wells, sampling, and inspection. Each state program is unique.

Underground Storage Tank (UST) Program — Each state has its own UST program that is supplied by special funds to this purpose. The state program is responsible for permits, design of construction, installation, testing and monitoring of the quality, and for the selection of the out-of-service underground storage tanks.

Agricultural sources — State laws regulate pesticides, fertilizers, and irrigation waters. With cooperation from the Agricultural Department, states organize education and training programs for farmers. State agencies should inspect and keep records on chemigation equipments, and the lists and amount of used chemicals. (Chemigation = chemicals are added to the irrigation water). States also regulate pest and weed control programs and the proper disposal and treatment of animal wastes.

Mining regulations — Each state has its own mining regulations according to local requirements.

Septic systems — States are charged with the special design criteria, site-selection, operation and maintenance of residential septic systems, as well as septage disposal regulations. Regulations related to commercial and industrial septic systems are also state dependents. The list of industrial waste substances that septic systems are not designed to handle are as follows:

- Organic solvents and metal degreasers discarded by printers
- Organic and inorganic chemicals discarded by photoprocessing industry
- Soil and stain removers discarded by laundries and laundromats
- Solvents, such as trichloroethylene and perchloroethylene used by dry cleaners
- Waste oils, degreasers, and automotive fluids discarded by service stations
- The large volumes of grease and cleansers discarded by restaurants
- Dyes discarded by beauty salons

Transportation of hazardous materials — Transportation and special packaging of hazardous materials are regulated by the Department Of Transportation (DOT). According to DOT requirements, the state has to identify and document the following:

- What is the hazardous material and its quantity?
- What methods are used to transport the material?
- What are the potential hazards of the material being transported?
- How are local officials promoting the safety of the transportation of the hazardous material?

References

Sample Collection

FDER Wastewater Facilities Field Sampling and Support Procedures; Compiled and written by David Chasteen, September 2, 1991

DER Requirements for Quality Assurance; Presented by Florida Department of Environmental Regulation, Quality Assurance Section, May, June 1991

DER Manual for Preparing Quality Assurance Plans; DER - QA-001/90; Florida Department of Environmental Regulation, Quality Assurance Section, August 20, 1990

Handbook for Sampling and Sample Preservation of Water and Wastewater; EPA-600/4-82-029, September 1982

Methods for Chemical Analysis of Water and Wastes; EPA-600/4-79-020, Revised March 1983

Standard Methods for the Examination of Water and Wastewater; APHA-AWWA-WPCF 17th Edition, 1989

Test Methods for Evaluating Solid Waste (Vol.I-IV); EPA SW 846 Third Edition, November 1986

Groundwater Quality Protection; Larry W. Canter, Robert C. Knox, Deborah M. Fairchild; Lewis Publishers, 1989

Groundwater Contamination Sources, Control and Prevention; Chester D. Rail, Technomic Publishing Co., Inc. 1988

Principles of Hazardous Material Management; Roger D. Griffin, Lewis Publishers, 1989

Standard Operating Procedure Sampling Procedures for Groundwater; FDER Site Investigation Section Comprehensive Quality Assurance Plan; Compiled and written by Shekhar Melkote, July 22, 1991

Interim Methods for the Sampling and Analysis of Priority Pollutants in Sediments and Fish Tissues; EPA Environmental Monitoring and Support Laboratory, Cincinnati, OH; October 1980

FDER Soil Sampling SOP; Revision October 28, 1991

Methods of Air Sampling and Analysis; Third Edition, Lewis Publishers, 1989

Quality Assurance - Quality Control

DER Requirements for Quality Assurance; Presented by Florida Department of Environmental Regulation, Quality Assurance Section, May, June 1991

DER Manual for Preparing Quality Assurance Plans; DER-QA-001/90; Florida Department of Environmental Regulation, Quality Assurance Section, August 20, 1990

Standard Methods for the Examination of Water and Wastewater; APHA-AWWA-WPCF; 17th Edition, 1989

Test Methods for Evaluating Solid Waste; EPA SW 846 Third Edition, November 1986

Quality Assurance of Chemical Measurements; John Keenan Taylor; Lewis Publishers, 1988

Environmental Health, Toxicology

Introduction to Environmental Health; Daniel S. Blumenthal, M.D.; Springer Publishing Company, New York, 1985

Inorganics in Drinking Water and Cardiovascular Disease; Edward J. Calabrese, Robert W. Tuthill, Lyman Condie; Princeton Scientific Publishing, 1985

Toxicology and Biological Monitoring of Metals in Humans; Bonnie L. Carson, Harry V. Ellis III, Joy L. McCann; Lewis Publishers, 1987

Groundwater Pollution Microbiology; Gabriel Bitton and Charles P. Gerba; John Wiley & Sons, 1984

Chemistry and Ecotoxicology of Pollution; DES W. Connel, Gregory J. Miller; John Wiley & Sons, 1984

Environmental Ecology; Bill Freedman; Academic Press, Inc. 1989

Pollution Control and Conservation; Dr. M. Kovacs; Ellis Horwood Limited and Akademiai Kiado, 1985

Health Effects from Hazardous Waste Sites; Julian B. Andelman, Dwight W. Underhill; Lewis Publishers, 1988

Microbiology, An Introduction; Gerard J. Tortora, Berdell R. Funke, Christine L. Case; The Benjamin/Cummings Publishing Company, Third Edition, 1989

Chemistry

Chemistry, The Study of Matter and Its Changes; James E. Brady, John R. Holum; John Wiley & Sons, Inc. 1993

Introduction to General, Organic & Biochemistry; Frederick A. Bettelheim, Jarry March; Saunders College Publishing, Third Edition, 1991

Regulations, Standards

Identifying and Regulating Carcinogens; Office of Technology Assessment Task Force; Lewis Publishers, 1988

Drinking Water Quality Standards and Controls; John De Zuane; Van Nostrand Reinhold, 1990

Principles of Hazardous Materials Management; Roger D. Griffin; Lewis Publishers, 1989

You and Your Drinking Water; EPA Journal, Volume 12, Number 7, September 1986

Annual Report 1988; Northwest Florida Water Management District

Groundwater Contamination Sources, Control, and Preventive Measures; Chester D. Rail; Technomic Publishing Company, 1989

Hazardous and Toxic Materials Safe Handling and Disposal; Howard H. Fawcett, P.E.; John Wiley & Sons, Inc., Second Edition, 1988.

Glossary of Abbreviations Related to Regulations

AAQS	Ambient Air Quality Standards
APHA	American Public Health Agency
ASTM	American Society for Testing and Materials
CAA	Clean Air Act
CERCLA	Comprehensive Environmental Response Cooperation and Liability Act
CWA	Clean Water Act
DER	Department of Environmental Regulation
DEP	Department of Environmental Pollution (new name for DER)
DOT	Department of Transportation
EPA	Environmental Protection Agency
FIFRA	Federal Insecticide, Fungicide, and Rodenticide Act
HMTA	Hazardous Material Transportation Act
HRS	Health & Rehabilitation Services
HSWA	Hazardous and Solid Waste Act
MCLs	Maximum Contaminants Levels
MCLGs	Maximum Contaminants Levels Goals
NEPA	National Environmental Policy Act
NESHAP	National Emission Standards for Hazardous Air Pollutants
NIOSH	National Institute for Occupational Safety and Health
NPDES	National Pollutants Discharge Elimination System
NPL	National Priority List
NSPS	New Source Performance Standards
OSHA	Occupational Safety and Health Administration
OTA	Office of Technology Assessment (EPA)
OSWER	Office of Solid Waste Emergency Response
RCRA	Resource Conservation and Recovery Act
SARA	Superfund Amendments and Reauthorization Act
SDWA	Safe Drinking Water Act
SEPA	State Environmental Policy Act
SIP	State Implementation Plan
TSCA	Toxic Substance Control Act
VOCs	Volatile Organic Compounds
SOCs	Synthetic Organic Compounds
UIC	Underground Injection Control
UST	Underground Storage Tanks

Index

A

B

C

D

E

F

G

K

L

M

N

O

P

Q

R

S

T

U

V